Ecology of Desert Rivers

Desert or dryland regions cover about half the world's land surface and contain an extraordinarily diverse range of rivers. Despite their importance for people and wildlife, the ecology of these rivers is poorly known compared with that of mesic rivers of the world. Deserts are considerably more variable than mesic regions, with floods over vast floodplains followed by dry periods. This variability determines the behaviour and composition of organisms that can live in desert rivers. From algae to vertebrates, organisms have different strategies that equip them for the ecology of desert rivers: a 'boom and bust' ecology. This ecology changes when desert rivers are dammed to supply water to human communities. River regulation decreases hydrological variability, reducing habitat value for endemic species and favouring exotic species. Big challenges affect these unique rivers and their dependent ecosystems around the world; this book examines the threats and challenges.

PROFESSOR RICHARD KINGSFORD from the University of New South Wales, Australia, has wide experience in conservation biology. Born in East Africa in Kenya, his interest in wildlife began in childhood. His research over the past 20 years or so has focussed on the waterbirds, wetlands and rivers of arid Australia, which covers about 75% of the continent. These systems define the ecology of the Australian continent with their boom and bust periods, times of droughts and floods. Professor Kingsford's research has focussed on the wetlands of Cooper Creek, one of the world's most magnificent desert rivers, and the Paroo River, the last free-flowing river in the Murray–Darling Basin. His research has demonstrated the ecological value of many rivers in arid Australia, and the impacts of water resource development on desert rivers. In 2001 he was awarded a national science prize (Eureka) for environmental research for his work on Australian rivers.

Ecology of Desert Rivers

Edited by

Edited by

RICHARD KINGSFORD
University of New South Wales, Australia

CAMBRIDGE UNIVERSITY PRESS

Cambridge, New York, Melbourne, Madrid, Cape Town, Singapore, São Paulo

Cambridge University Press
The Edinburgh Building, Cambridge CB2 2RU, UK

Published in the United States of America by Cambridge University Press, New York

www.cambridge.org
Information on this title: www.cambridge.org/9780521818254

First published 2006

Printed in the United Kingdom at the University Press, Cambridge

A catalogue record for this publication is available from the British Library

ISBN-13 978-0-521-81825-4 hardback
ISBN-10 0-521-81825-7 hardback

Contents

Contributors

Paul C. E. Bailey
School of Biological Sciences and Australian Centre for
Biodiversity: Analysis Policy Management, Monash University,
Clayton, VIC 3800, Australia

Stephen R. Balcombe
Cooperative Research Centre for Freshwater Ecology, Centre for
Riverine Landscapes, Faculty of Environmental Sciences, Griffith
University, Nathan, Queensland 4111, Australia

P. J. Beyer
Department of Geography and Geosciences, Bloomsburg
University, Bloomsburg, PA 17815, USA

Paul I. Boon
Institute for Sustainability and Innovation, Victoria University,
P.O. Box 14428, Melbourne Central Mail Centre, Melbourne, VIC
8001, Australia

Dean W. Blinn
Department of Biological Sciences, Northern Arizona University,
Flagstaff, AZ 86011, USA

A. J. Boulton
Ecosystem Management, University of New England, Armidale,
NSW 2351, Australia

Margaret A. Brock
NSW Department of Natural Resources, P.O. Box U245, Armidale, NSW 2351, Australia; School of Environmental Sciences and Natural Resource Management, University of New England, Armidale, NSW 2351, Australia; and Co-operative Research Centre for Freshwater Ecology, Building 15, University of Canberra, ACT 2601, Australia

Stuart E. Bunn
Cooperative Research Centre for Freshwater Ecology, Centre for Riverine Landscapes, Faculty of Environmental Sciences, Griffith University, Nathan, Queensland 4111, Australia

Samantha J. Capon
Centre for Riverine Landscapes, Faculty of Environmental Sciences, Griffith University, Nathan, Queensland 4111, Australia; Co-operative Research Centre for Freshwater Ecology, Building 15, University of Canberra, ACT 2601, Australia; and School of Biological Sciences, Monash University, Clayton, Victoria 3800, Australia

Peter M. Davies
Centre of Excellence in Natural Resource Management, The University of Western Australia, Albany, Western Australia 6330, Australia

Christine S. Fellows
Cooperative Research Centre for Freshwater Ecology, Centre for Riverine Landscapes, Faculty of Environmental Sciences, Griffith University, Nathan, Queensland 4111, Australia

A. Georges
Applied Ecology Research Group and CRC Freshwater Ecology, University of Canberra, ACT 2601, Australia

K. M. Jenkins
Ecosystem Management, University of New England, Armidale, NSW 2351, Australia

R. T. Kingsford
School of Biological, Earth and Environmental Sciences,
University of New South Wales, Sydney, NSW 2052, Australia

A. D. Lemly
United States Forest Service, Southern Research Station Cold-
water Fisheries Research Unit, 1650 Ramble Road, Blacksburg, VA
24060, USA

Fiona J. McKenzie-Smith
Cooperative Research Centre for Freshwater Ecology, Centre for
Riverine Landscapes, Faculty of Environmental Sciences, Griffith
University, Nathan, Queensland 4111, Australia

John L. Porter
Science Branch, NSW Department of Environment and Conservation, PO Box 1967, Hurstville, NSW 2220, Australia; and School
of Environmental Sciences and Natural Resource Management,
University of New England, Armidale, NSW 2351, Australia

K. H. Rogers
Centre for Water for the Environment, Witswatersrand
University, Private Bag 3, 2052 Wits, South Africa

F. Sheldon
Centre for Riverine Landscapes, Faculty of Environmental
Sciences, Griffith University, Nathan, QLD 4111, Australia

J. R. Thompson
Wetland Research Unit, Department of Geography, University
College London, 26 Bedford Way, London WC1H 0AP, UK

M. C. Thoms
Cooperative Research Centre for Freshwater Ecology, University of
Canberra, ACT 2601, Australia

P. J. Unmack
School of Life Sciences, Arizona State University, PO Box 874501,
Tempe, AZ 85287-4501, USA

K. F. Walker
Cooperative Research Centre for Freshwater Ecology, School of Earth and Environmental Sciences DP312, The University of Adelaide, South Australia 5005, Australia

William D. Williams (Deceased)
Department of Environmental Biology, University of Adelaide, North Terrace, Adelaide, South Australia, Australia

M. J. Wishart
Freshwater Research Unit, University of Cape Town, Rondebosch 7701, South Africa

W. J. Young
CSIRO Land and Water, PO Box 1666, Canberra, ACT 2601, Australia. Current affiliation: IGBP Secretariat, Royal Swedish Academy of Sciences, Box 50005, S-104 05 Stockholm, Sweden

Preface

Rivers flow in and out of my childhood memories. There was the mountain river in Kenya where we caught butterflies and the contrasting desert river we passed on the way to the coast, a magnet for Africa's big game: elephants, buffalos and lions. In my teenage years, I would watch with fascination one of the world's most bizarre river creatures, the platypus, as it busily dived and bobbed to the surface in the nearby river, munching various invertebrate delicacies. Now my adulthood is immersed in the ecology of rivers.

Rivers weave their way through everyone's lives. We depend on them even if we don't realise it. They deliver most of our drinking water and, directly or indirectly, a considerable amount of our food, energy and clothing. People often first settled on the floodplain of a river, close to a permanent source of water. Relatively few of the world's cities and towns are far from rivers. But rivers serve more than such utilitarian functions. They give communities a sense of place. They occupy a central position in the cultural beliefs of many people, particularly indigenous communities. And they provide a home and resources for countless other species that coexist with us on this planet. Many organisms depend on rivers. Sometimes these bacteria, fungi, plants and animals provide 'ecosystem services' for us. They purify our water, give us food (fish and plants) and provide the goods (food and fibre) from the river that sustains us.

Rivers are places of contradiction: frenetic but peaceful. The inexorable power and direction of a flowing river is indescribably calming, but concentrate on one small part of the river and it is ever-changing. Currents churn, producing eddies, swirls and whirlpools where flotsam tosses and rolls unpredictably. From the bank of the river, it seems stable and predictable at one level but at another chaos reigns supreme. Even the perceived stability is only a reflection of relative awareness based on how far we look and how long we watch. Visit at another time and the whole river will be different, another path carved, a new rapid formed, a raging torrent, a trickle or perhaps even dry. The unpredictable nature of floods and dry periods and everything in between over time create an incredible diversity of different places for plants and animals to live. Rivers are always alive with life.

Most of us interact with only part of a river where we collect water, wash or swim. This is the enigma of a river: you seldom know more than a part of it. But to really understand the river or more importantly to look after it, we need to know the whole river, from its catchment to its end, from source to sea. Fiddling with parts of rivers has repercussions that extend well beyond 'our bit'. This begs the question: what is a river? The common perception is that a river comprises a channel filled with perennially flowing water. Many desert rivers would fail to meet this criterion. They don't flow all the time and it is sometimes hard to find a major channel. A river encompasses all of its dependent ecosystems: the catchment, tributaries, main channel, distributaries, lakes, swamps, floodplains, estuaries and the groundwater systems dependent on river flows. Groundwater systems are really underground rivers. Many decisions about rivers have ignored downstream effects on people and ecosystems. Sometimes we have not even known of the existence of this dependency until it is too late. This is the true vulnerability of the river.

Rainfall is the dominant force shaping the world's ecology, including people. Nearly half the world has low annual rainfall (less than 500 mm): the world's desert regions. Because of the scarcity of water and dependency of life on water, rivers in desert regions are the most dominant factor shaping the ecology of deserts. Despite the importance of desert rivers, our knowledge of their ecology is poor. Most people do not live in these regions; there is little dependable water. Affluent societies, which produce most of the world's science, live in well-watered regions. Even in Australia, with 75% of the land surface receiving less than 500 mm a year, most scientific effort on aquatic organisms and rivers has been done near our research institutions in well-watered parts of the country. There is still much to learn and understand about the basic ecology of desert rivers and their role in shaping the lives of dependent organisms.

Obviously, water is in short supply in desert regions. Here, the needs of human societies for water, predominantly abstracted from desert rivers, has had irreparable effects on the ecology of desert rivers. This is where the greatest environmental change has occurred, as our species appropriates this resource to meet our needs. Desert rivers and their dependent ecosystems rank high in a growing list of the world's ecological disasters caused by humans. They include the Aral Sea and its supply rivers, the Mesopotamian Marshlands, the wetlands of the Murray–Darling Basin, rivers in central China and the rivers of North America. The ecological effects are so severe that effects on local economies and quality of life are demonstrable. Understanding the full effects of such impositions is essential for current management and future decisions for rivers. Water does not just go into the desert to evaporate. Most rivers support some of the most biodiverse places on earth, places that if managed well can provide immeasurable ecosystem services but where current management of desert rivers is seldom sustainable.

This book attempts to provide some of the necessary knowledge to understand, appreciate and sustainably manage these magnificent ecosystems. There is one certainty about a desert river: it is always changing. This is part of its nature and reflects the variability of the climate in desert regions. Raging rivers in flood and dry river beds are an integral part of a river's life but often we have changed desert rivers forever, sometimes making some of them even disappear. There is no greater challenge for our generation and those to follow than the management of our rivers. Freshwater accounts for only about 2.5% of the world's water and most of this (69%) is locked up in snow and ice, although climate change is reducing this all the time. Only 0.26% of the total fresh water is found in lakes, reservoirs and rivers, and of this we can only access an even smaller amount. Access to fresh water will limit population increase as no other resource does. Currently more than a billion people do not have access to clean water, whereas more affluent nations are profligate with our water use. Populations growing in number and in quality of life will need water that can only come from our rivers. What will be the ultimate cost when we are long gone?

Richard Kingsford

Acknowledgements

I thank Paul Adam for encouraging me to take on this task. Alan Crowden showed great patience while Clare Georgy expertly guided me through the production process. Each chapter in this book was peer-reviewed and I am particularly indebted to Andrew Boulton who took on the onerous task of editing and coordinating reviews for all chapters that I authored or co-authored. Rachael Thomas helped compile and organise the chapters at the end of the process and Megan Head helped with the indexing. I am indebted to the managers in the New South Wales National Parks and Wildlife Service with the vision to realise that scientific research only reaps a good harvest when creativity is fostered. I especially thank the people of the rivers in inland Australia who have and continue to inspire me to pursue greater understanding and better management of our desert rivers.

I   Natural disturbance in desert river systems

I

Desert or dryland rivers of the world: an introduction

R. T. KINGSFORD AND J. R. THOMPSON

Rivers channel the world's rainfall into floodplains, lakes or groundwater basins, or out to sea. They provide habitats for diverse biota that often climax in floodplain wetlands, areas of incredible biodiversity. River flows are integral to many coastal and marine environments, processes and organisms (Gillanders & Kingsford, 2002). Their fresh water allows humans to penetrate and flourish in the most inhospitable parts of this planet. They are the arteries that define ecological landscapes and processes for many biota. Climate and the nature of land surfaces primarily govern the size, hydrology, geomorphology and ecology of rivers. For example, the Amazon River, which accounts for about 20% of the world's river flow, is a massive river system originating in areas of extremely high rainfall. Contrast this with rivers from the desert regions of the world (Fig. 1.1) where rainfall is less than 500 mm per year and is usually exceeded by evaporation. In these regions rivers often stop flowing for long periods, sometimes even years.

What makes desert rivers any different from other rivers or aquatic systems around the world? This book uses 'desert' and 'dryland' interchangeably to describe land areas and their rivers where there is less than 500 mm of annual rainfall: the arid and semi-arid

Ecology of Desert Rivers, ed. R. T. Kingsford. Published by Cambridge University Press.
© Cambridge University Press 2006.

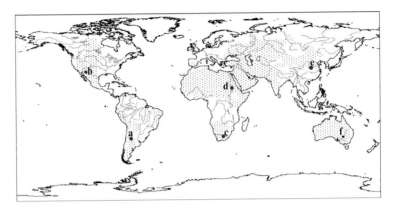

Figure 1.1. Desert or dryland regions of the world (dotted; annual rainfall < 500 mm), (after Middleton & Thomas, 1997) and some of their large rivers. Letters identify locations of six dryland rivers (dots are locations of gauges used), illustrating flow regimes (see Fig. 1.2): the San Juan River in South America (a), the Gila River in North America (b), the Orange River in South Africa (c), the Atbara River in northeast Africa (d), the Huanghe (Yellow River) in Asia (e) and the Darling River in Australia (f).

regions of the world (Fig. 1.1). This encompasses about 47% of the global land surface, including hyperarid, arid, semi-arid and dry humid regions (Table 1.1) (Middleton & Thomas, 1997). Why do we need a book about the ecology and management of these desert or dryland rivers? Intrinsic properties of scarcity and variability of these rivers and their associated floodplain habitats, combined with poor knowledge and increasing human pressures, demand attention. The story of desert or dryland rivers is one of changeable, changing and changed ecosystems, as humans progressively apply control.

Desert rivers do not have unique landforms (Nanson *et al.*, 2002) but their hydrology is much more variable than that of mesic rivers (McMahon *et al.*, 1992; Puckridge *et al.*, 1998; Peel *et al.*, 2001). We are only just beginning to understand the implications of such variability for the ecology of these rivers, the effects of river regulation and future management. Rivers in dry regions of the

Table 1.1. *Dry regions of the world, showing areas and percentages of each aridity zone in each global region of the world*

Region	Area[a]	Dry humid	Semi-arid	Arid	Hyper-arid	Total
Africa	2 965.6	9.1	17.3	16.9	22.7	66.0
Asia	4 256.0	8.3	16.3	14.7	6.5	45.8
Australasia	882.2	5.8	35.0	34.3	0	75.2
Europe	950.5	19.3	11.1	1.2	0	31.5
North America	2 190.9	10.6	19.1	3.7	<1	33.6
South America	1 767.5	11.7	15.0	2.5	1.5	30.6
World	13 012.7	10.0	17.7	12.1	7.5	47.2

[a] Millions of hectares.
Source: after Middleton and Thomas (1997).

world are the poor cousins in the knowledge base of river and wetland ecology. Their ecology is probably the least known of our freshwater resources (Williams, 1988; Nanson *et al.*, 2002), despite recent advances in understanding (see Bull & Kirkby, 2002), because relatively few people live in such inhospitable parts of the world. Scientific effort is often strongly biased towards humid regions, with most of our knowledge of aquatic ecology from temperate freshwater science (Ward *et al.*, 2001). Even in a relatively affluent country such as Australia, where arid regions dominate (75% of the land area), fresh-water scientific effort can be biased towards the mesic regions where most people live (Kingsford, 1995). Desert rivers and their ecology are often out of sight and out of mind, so it is important to consolidate our knowledge and provide a basic framework for ecological understanding of desert or dryland rivers.

DESERT OR DRYLAND RIVERS OF THE WORLD

Almost 50% of the world's land surface is either arid or semi-arid (Middleton & Thomas, 1997), occupying most continents (Comín & Williams, 1994). Many thousands of streams and large rivers flow wholly or partly through such areas (Fig. 1.1). Rivers and their

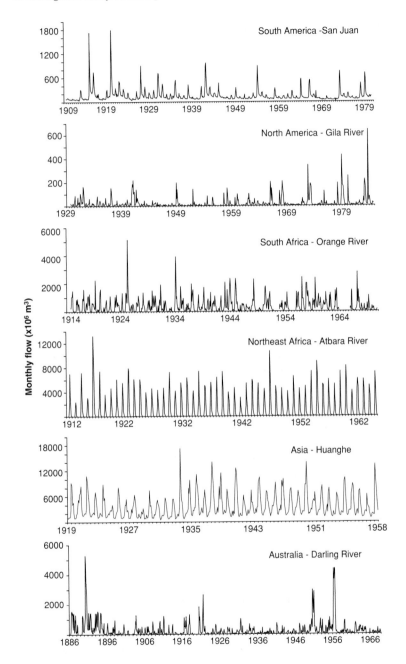

dependent ecosystems form a continuum of variability, seldom adequately captured by pigeonhole classifications. This variability is characteristically higher in dryland rivers. Rivers challenge us even more because their longitudinal dimensions seldom respect climatic regions; worse, for managers and policy makers, they do not respect jurisdictional or national borders (Postel, 1996; Kingsford *et al.*, 1998). Many large rivers that flow through desert regions (e.g. the Nile, Okavango and Murray) originate in mesic areas. This book adopts a broad definition of what constitutes a desert river because it is impossible to divorce a river from its catchment: desert rivers flow wholly or partly through desert or dryland regions of the world (annual rainfall < 500 mm).

Climate drives river flows and dependent ecological responses. Within desert regions rainfall is low and is often highly variable in both space and time (Peel *et al.*, 2001). Hydrology holds primacy in any treatment of rivers, their behaviour and their understanding. By way of introduction, we take the monthly flows of six unregulated desert rivers from different regions of the world: North America, South America, northeast Africa, South Africa, Asia and Australia (Fig. 1.2). Even a simple inspection of monthly flow regimes illustrates considerable differences among rivers from these different regions. Seasonal regularity, particularly in relation to wet and dry seasons in the tropics (Atbara River, northeast Africa) and snowmelt in temperate regions (Huanghe River, Asia) is translated into a clear seasonal signal in river flows (Fig. 1.2), which has considerable implications for ecology and management. Interannual variability is relatively small compared with that of rivers in other desert regions of the world such as South Africa, North and South America and Australia (Fig. 1.2). In these regions annual variability in the timing and volume of flows is also high. Some dryland rivers have periods of no flow or low flow. In some, such as the

Figure 1.2. (opposite) Monthly flow regimes for six desert or dryland rivers from around the world (see Fig. 1.1 for locations; dots indicate locations of gauges used for analysis). Rivers were chosen on the basis of the availability of at least 30 years of data for periods where there was relatively little river regulation (data provided courtesy M. Peel) (Peel *et al.*, 2001; McMahon *et al.*, 1992).

Atbara River in Northeast Africa (Fig. 1.2), these periods coincide with the marked dry season and their timing and duration are relatively uniform. Others, such as Australia's Darling River and the Gila River of North America (Fig. 1.2), exhibit less predictable periods of low or no flow. Such regions generally have highly stochastic rainfall that results in extremely variable river flows, a pattern particularly well known for Australian and South African rivers (McMahon *et al.*, 1992; Puckridge *et al.*, 1998; Peel *et al.*, 2001; Nanson *et al.*, 2002). Chapter 2 of this book extends this introduction into river ecology by examining in considerably more detail some of the differences in the hydrology of desert rivers and their implications for river ecology and water resource development.

It follows that hydrological disturbance patterns exert a dominant influence on the ecology of desert rivers, through the drying and flooding of river habitats: channels, waterholes, floodplains and estuaries. Hydrology affects geomorphological processes of rivers, which in turn drive the distribution of dependent vegetation (Chapter 3, this volume). The next section of the book has a series of chapters that examine ecological responses to variable flows in desert rivers. This begins with food webs and productivity (Chapter 4) and moves to higher levels of biota: plants (Chapter 5), invertebrates (Chapter 6) and vertebrates (Chapter 7). A new force, almost as important as climate, now governs the hydrology and ecology of many rivers: human control.

The human responses to water scarcity around the world are driving major changes to dryland rivers. Part II of this book concentrates on how humans have altered the behaviour of dryland rivers, affecting their ecology. Despite our lack of knowledge, we are busily exploiting dryland rivers, wreaking immeasurable ecological damage (Lemly *et al.*, 2000; Gillanders & Kingsford, 2002). Are we changing these unique systems forever? This part of the book begins with an examination of how we change desert river flows and the impact these changes have had on some of the more spectacular and biodiverse habitats in the world (Chapter 8). The next chapter shows the long-lasting and extensive hydrological and ecological effects of even relatively minor river regulatory structures, such as weirs, on the Lower River Murray (Chapter 9). Deserts are naturally salty places, but human land and river management is increasing the salinity of desert rivers with severe ecological consequences (Chapter 10). Expanding human populations represent the greatest pressure on the world's water resources (Postel, 2000), at their most extreme in desert regions (Chapter 11). Finally, a synthesis chapter (Chapter 12) examines the competing demands of the ecology of desert rivers and their changeable nature against our ever-increasing needs for water, imposing simplicity on incredibly complex ecosystems. Hopefully, this book will encourage an interest in the magnificent systems that are desert rivers, will raise awareness of the challenges that they face, and will in turn promote their future conservation.

REFERENCES

Bull, L. J. and Kirkby, M. J. (2002). *Dryland Rivers: Hydrology and Geomorphology of Semi-arid Channels.* Chichester: John Wiley and Sons.

Comín, F. A. and Williams, W. D. (1994). Parched continents: Our common future? In *Limnology Now: a Paradigm of Planetary Problems,* ed. R. Margalef, Amsterdam: Elsevier. pp. 473–527.

Gillanders, B. M. and Kingsford, M. J. (2002). Impact of changes in flow of freshwater on estuarine and open coastal habitats and associated organisms. *Oceanography and Marine Biology: Annual Review,* **40,** 233–309.

Kingsford, R. T. (1995). Occurrence of high concentrations of waterbirds in arid Australia. *Journal of Arid Environments,* **29,** 421–5.

Kingsford, R. T., Boulton, A. J. and Puckridge, J. T. (1998). Challenges in managing dryland rivers crossing political boundaries: lessons from Cooper Creek and the Paroo River, central Australia. *Aquatic Conservation: Marine and Freshwater Ecosystems,* **8,** 361–78.

Lemly, A. D., Kingsford, R. T. and Thompson, J. R. (2000). Irrigated agriculture and wildlife conservation: conflict on a global scale. *Environmental Management,* **25,** 485–512.

McMahon, T. A., Finlayson, B. L., Haines, T. A. and Srikanthan, R. (1992). *Global Runoff: Continental Comparisons of Annual and Peak Discharges.* Cremlingen-Destedt, Germany: Catena.

Middleton, N. J. and Thomas, D. S. G. (1997). *World Atlas of Desertification* (2nd edn). London: UNEP/ Edward Arnold.

Nanson, G. C., Tooth, S. and Knighton, A. D. (2002). A global perspective on dryland rivers: perceptions, misconceptions and distinctions. In *Hydrology and Geomorphology of Semi-arid Channels,* ed. L. J. Bull and M. J. Kirkby, Chichester: John Wiley and Sons Ltd. pp. 17–54.

Peel, M. C., McMahon, T. A., Finlayson, B. L. and Watson, F. G. R. (2001). Identification and explanation of continental differences in the variability of annual runoff. *Journal of Hydrology,* **250,** 224–40.

Postel, S. (1996). *Dividing the Waters: Food Security, Ecosystem Health, and the New Politics of Scarcity.* (Worldwatch paper 132). Washington, DC: Worldwatch Institute.

Postel, S. L. (2000). Entering an era of water scarcity: the challenges ahead. *Ecological Applications,* **10,** 941–8.

Puckridge, J. T., Sheldon, F., Walker, K. F. and Boulton, A. J. (1998). Flow variability and the ecology of arid zone rivers. *Marine and Freshwater Research,* **49,** 55–72.

Ward, J. V., Tockner, K., Uehlinger, U. and Malard, F. (2001). Understanding natural patterns and processes in river corridors as the basis for effective river restoration. *Regulated Rivers: Research and Management*, **17**, 311–23.

Williams, W. D. (1988). Limnological imbalances: an antipodean viewpoint. *Freshwater Biology*, **20**, 407–20.

2

Flow variability in large unregulated dryland rivers

W. J. YOUNG AND R. T. KINGSFORD

INTRODUCTION

The presence or absence of water defines much of the ecology of semi-arid and arid landscapes (Stafford Smith & Morton, 1990; Chapter 7, this volume), and rivers and their flows mediate much of the ecological variability (Walker *et al.*, 1995). Dryland rivers drain or traverse the hyperarid, arid and semi-arid regions of the world where mean annual rainfall is less than 500 mm and where mean annual evaporation is equivalent to at least 95% of the rainfall and is often even higher (Meigs, 1953; Chapter 1, this volume); these are the 'B' category climates in the Köppen climate classification (Köppen & Geiger, 1930). Dryland regions represent over half of the world's land area (Thomas, 1989) and are drained by many of the world's major rivers (Kingsford, 2000a; Chapter 1, this volume).

Although dryland regions are widespread in distribution, their harsh climates and limited water resources have often meant that they support comparatively small human populations, and thus dryland rivers are generally less developed than their humid counterparts. For the same reasons, far fewer scientific studies have occurred in dryland regions than in humid regions (Williams, 1988; Kingsford, 1995). As the global population and the associated food and energy demands grow

Ecology of Desert Rivers, ed. R. T. Kingsford. Published by Cambridge University Press.
© Cambridge University Press 2006.

(Postel, 2000), dryland rivers are increasingly the focus of water resource development, sometimes with catastrophic human and ecological consequences (Lemly et al., 2000; Chapter 8, this volume). Unfortunately, scientific knowledge from humid river studies has been inadequate for guiding the sustainable development of dryland rivers (Davies et al., 1992; Wishart et al., 2000), resulting in increased recent attention focused on dryland rivers and their catchments (see, for example Graf, 1988; Agnew & Anderson, 1992; Davies et al., 1992; Walker et al., 1995; Kingsford, 1999; Bull & Kirkby, 2002).

Our aim in this chapter is to describe the nature of flow variability in large (drainage area greater than 20 000 km^2), unregulated rivers in arid and semi-arid (dryland) regions, and contrast this with that of large, unregulated rivers in humid regions. First, we broadly identify the hydrological and geomorphic characteristics of dryland rivers and the ecological implications of flow variability. We then illustrate these with case study analyses of the flow regimes of eight rivers: six from dryland regions and two from humid regions. We compare and contrast the dryland and humid flow regimes, and identify key characteristics and differences among the dryland flow regimes. Finally, we discuss the implications of these flow regimes for the ecology of dryland rivers, and for the development of their water resources.

HYDROLOGIC CHARACTER

Evaporation and transpiration typically account for at least 95% of the rainfall in dryland regions, making the hydrologic balance a delicate one (Pilgram et al., 1988). Dryland regional rainfall varies strongly in time and space at several scales, and a significant fraction often occurs as localised intense storms (Stafford Smith & Morton, 1990) followed by long periods of little or no rain. Although rainfall in some dryland regions – particularly in monsoonal climates–exhibits strong seasonal patterns, more commonly the seasonal variations in dryland region rainfall are irregular among years. In many dryland regions of the world, the interannual variability of rainfall is high. These variations are at least partly a result of large-scale quasi-regular atmospheric circulation patterns such as the El Niño–Southern Oscillation (ENSO) (see, for example, Puckridge et al., 2000).

Evapotranspiration from evergreen vegetation (characteristic of dryland regions) is higher than from deciduous vegetation, contributing to the high variability of runoff in dryland regions (Peel et al., 2001).

The sparse and variable plant cover of arid regions means that interception is often minimal, and evaporation from bare soil assumes greater importance in the water balance than it does in humid regions. Furthermore, hydrophobia and surface sealing of soils can lead to rapid runoff following rainfall. The delicate balance between rainfall and evaporation means that the spatial patterns in plant cover and soil properties lead to high spatial variability in surface runoff. High temporal and spatial variability of rainfall and runoff translates into highly variable flow regimes (Walker et al., 1995; Puckridge et al., 1998; Peel et al., 2001; Boulton et al., 2000). Interannual and multidecadal variations dominate the temporal flow variations (Boulton et al., 2000), with smaller and irregular seasonal variations.

Dryland rivers can be hydrologically classified as either allogenic or endogenic; the former pass through dryland regions but source most of their flow from outside dryland regions, whereas the latter source most of their flow from dryland regions (Nanson et al., 2002). Allogenic rivers commonly sustain perennial flow, whereas endogenic rivers are usually intermittent with little or no baseflow and high channel transmission losses. With this intermittency in endogenic rivers, the interplay between the spatial and temporal variations in rainfall, high evaporation and channel transmission losses can lead to extreme variations in flow. For example, monthly flows in Cooper Creek, central Australia, ranged from zero to nearly 60 times the median monthly flow between 1940 and 1987 (Puckridge et al., 2000).

In endogenic dryland rivers, even moderate floods are largely captured by waterholes, floodplain wetlands or riverine lakes in the lower basin, with little or no streamflow leaving the basin. For example, the Bulloo River, Cooper Creek and the Diamantina River in central Australia terminate in large shallow lakes and wetlands. Even the Paroo River in central Australia (see Fig. 2.2), although geomorphically connected to downstream drainage networks, exhibits little hydrologic connection to these areas except during large floods, predominantly dissipating in large freshwater lakes (Kingsford et al., 2001). In large dryland catchments, the high variability of runoff generation and transmission losses translates into high spatial variability in streamflow, with streamflow in the lower basin sometimes poorly correlated with streamflow in the upper basin (W. J. Young, K. Brandis & R. T. Kingsford, unpublished data). Small and moderate floods may be generated by runoff in only a subset of the tributary catchments (Kotwicki, 1986), or more locally in the mid- or lower catchment (Kingsford et al.,

1999). Typically, only the largest floods result from widespread rainfall across the entire upper catchment or a sequence of floods (Puckridge *et al.*, 2000). The spatial variations in topography, soil properties and vegetation types across the floodplains of the lower regions of many dryland basins contribute to this behaviour (Kingsford *et al.*, 2001; Roshier *et al.*, 2001b). Spatial differences in streamflow and the inundation of floodplains and wetlands creates a mosaic of hydrological conditions and intermittently connected habitats. This drives the ecology of dryland rivers and their dependent ecosystems (Sheldon *et al.*, 2002).

The nature of floods in dryland rivers is also strongly dependent on catchment relief and channel morphology. Small, steep, upland or piedmont channels are characterised by flash floods, which have in the past been considered characteristic of dryland rivers (Grimm & Fisher, 1989). However, floods in lowland floodplain dryland rivers are usually slow in responding to rainfall and may last for many months (Puckridge *et al.*, 2000; Roshier *et al.*, 2001b; Nanson *et al.*, 2002). Downstream reduction in peak discharge per unit area is far more pronounced in dryland rivers than in humid rivers (McMahon, 1979). Furthermore, even without normalising by catchment area, floods in dryland rivers, even the large infrequent floods, initially increase in the downstream direction as tributaries join, and then decrease owing to large channel transmission losses and a reduction in tributary additions (Knighton & Nanson, 1994a). Relative flood magnitudes are far more variable in dryland rivers than in humid rivers (Nanson *et al.*, 2002). For example, the 50-year flood is typically two or three times the size of the annual flood in many humid rivers, but it can exceed ten times the annual flood in dryland rivers. Finally, floods in at least some dryland rivers can occur in clusters associated with La Niña episodes (Puckridge *et al.*, 2000).

GEOMORPHIC CHARACTER

Globally, channels of dryland rivers show similar diversity of forms to humid rivers, and average channel geometries are not distinctly different (Nanson *et al.*, 2002). There are two important exceptions: waterholes and flood-outs. Waterholes are self-scouring sections of channel or floodplain that perennially hold water in dryland riverscapes (Knighton & Nanson, 1994b; Nanson *et al.*, 2002) and represent important refuge habitats for aquatic vertebrates (Chapter 7, this volume). They have a globally limited distribution, but are found in a range of physiographic settings in Australia (see, for example, Argue & Salter, 1977; Tooth, 1999), including the low-gradient anastomosing rivers of the Channel Country in central Australia (Knighton & Nanson, 1994). Flood-outs are unchannelled features at the terminus of rivers where flow dissipates (Tooth, 1999). They occur at a wide range of scales (1–1000 km^2) and are common in many dryland rivers of Australia and Southern Africa (Nanson *et al.*, 2002). Flood-outs form when there is flow reduction, often with a barrier to the flow, such as aeolian dunes, bedrock outcrops or prior alluvial deposits (Nanson *et al.*, 2002). Although flood-outs are characteristic of many dryland rivers, the alluvial fans of some humid rivers also dissipate flow.

In spite of the lack of a strong distinction between dryland and humid river-channel forms, some channel forms are more typical of dryland channels than of humid channels. For example, a downstream reduction of channel dimensions, especially width, is more characteristic of dryland rivers, sometimes exacerbated by avulsions and flood-outs (see, for example, Dunkerly, 1992; Tooth, 1999). Early studies (see, for example, Wolman & Gerson, 1978) suggested that lack of riparian vegetation caused channels in some dryland rivers to widen during large floods and remain in this state between floods. Although some dryland rivers (see, for example, Cooper Creek), are wide and shallow this shape is common in unconsolidated sand or gravel bed channels, especially braided channels (Rosgen, 1994), and is no more typical of dryland rivers than of humid rivers.

IMPLICATIONS OF FLOW VARIABILITY

Prolonged periods without flow and extreme floods constitute natural disturbances in dryland river ecosystems. In dryland basins of high or moderate relief, intense rainstorms produce rapid flow rises, high flow velocities and high shear stresses. This results in extreme bedload (sand, gravel and cobble) transport, and downstream deposition in the main channel, in distributaries or as overbank deposits (Laronne & Reid, 1993; Chapter 3, this volume). These high-energy floods disrupt and physically restructure many riverine habitats (Pettit *et al.*, 2001; Chapter 3, this volume). In dryland rivers of low relief, flood events are less flashy and in the lower catchment may last for many months, depending on basin topography, channel–floodplain morphology and vegetation cover. These prolonged flood events greatly increase the connectivity of the riverine system, allowing waterborne dispersal of seeds, spores and biota (Jenkins & Boulton, 2003) and the migration of large aquatic fauna (Roshier *et al.*, 2001a; Chapter 7, this volume). This reconnection also homogenises many aspects of water chemistry, particularly salinity, across the system (Sheldon *et al.*, 2002). The abundance of water and the increase in nutrient (including organic carbon) availability initiate a rapid 'boom' cycle in some dryland rivers and associated floodplain wetlands (Kingsford *et al.*, 1999; Chapter 4, this volume). Aquatic macrophytes and other flood-dependent vegetation tend to grow and reproduce (Chapter 5, this volume). These changes stimulate rapid reproduction of many riverine fauna, including invertebrates (Chapter 6, this volume), fish (Puckridge *et al.*, 2000; Chapter 7, this volume), waterbirds (Kingsford & Porter, 1993) and other vertebrates (Chapter 7, this volume). The 'bust' occurs through the drying phase as wetted habitats shrink, increasing population densities and consequently predation pressure (Boulton *et al.*, 2000; Chapter 7, this volume), and inducing physiological stresses on many organisms as water temperature and solute concentrations rise and oxygen concentrations decrease. Some plant and animal species survive in dry habitats until the next flood, but many die or rely on passive or active dispersal from more permanent refuges, when the next flood arrives (see Chapters 5, 6 and 7, this volume).

FLOW VARIABILITY OF EIGHT LARGE UNREGULATED RIVERS

For this analysis, we defined large rivers as those draining more than 20 000 km^2 but, to ensure the rivers' comparability in scale, we limited our selection to rivers draining less than 200 000 km^2. Water resource development has affected most large rivers in the world for some time (Dynesius & Nilsson, 1994; Vörösmarty *et al.*, 1997); long flow records for large unregulated rivers are therefore rare. Dryland rivers are typically less developed than their humid counterparts, but flow records for dryland rivers are generally fewer and shorter, often because they are remote and undeveloped (see, for example, Cooper Creek). For this reason, we imposed a minimum record length of only 20 years.

We accessed the datasets of McMahon *et al.* (1992), Vörösmarty *et al.* (1998), Slack and Landwehr (1992) and the online database of the Queensland Department of Natural Resources, Mines and Energy (QDNRM, 2004). Very few records met our selection criteria. For example, only the Lundi River (Fig. 2.1) (Tokwe Confluence, record length 20.4 years) of the 557 river records used by McMahon *et al.* (1992) met the criteria. We restricted our analyses of dryland rivers to just six flow records: two rivers from each of the Australian, African and North American continents. All are from steppe environments of the Köppen climate classification (Köppen & Geiger, 1930); four rivers (the Paroo, Warrego, Lundi and Limpopo) (Figs. 2.1–2.4) have dry hot climates (mean annual temperature greater than 18 °C) and two (the Little Colorado and Green; Fig. 2.5) have dry cool climates (mean annual temperature less than 18 °C) (Table 2.1). Flow records from two humid European rivers were selected for comparison: the Vuoksi and the Vistula (Fig. 2.6), which have snow climates and sufficient precipitation in all months of the year (Table 2.1). These were selected instead of records from temperate climate rivers, because there are few large unregulated temperate

Figure 2.1. The Lundi River, South Africa, has moderate monthly flow variability and moderately strong, albeit unpredictable seasonal variations. It is dry for about 26% of months (Image courtesy of Earth Sciences and Image Laboratory, NASA Johnson Space Center).

Figure 2.2. The Paroo River, arid Australia, is a river that dries to a series of waterholes but during floods inundates up to almost one million hectares. It also floods a series of large terminal overflow lakes that retain water for up to three years. It has a highly variable flow regime (Photo R. T. Kingsford).

Figure 2.3. The Warrego River, arid Australia, has extremely variable monthly flows, higher than all other flows analysed. Peak flows exceed 3000 times the median flow (Photo R. T. Kingsford).

Figure 2.4. The Limpopo River, South Africa, has a reasonably strong seasonal flow signal, punctuated by periods of intermittency (14% of months) with median monthly flows significantly higher during *El Niño* months than in *Other* months, (see text). (Photo K. Hall).

Figure 2.5. The Little Colorado River, North America, is a dryland endogenic river which has a weak seasonal signal due to moderate, but unpredictable seasonal variations. Flows are somewhat intermittent (14% of months have zero flow), and median flows are significantly higher in *El Niño* months than in *Other* months (Photo M. Habersack).

rivers (Dynesius & Nilsson, 1994) and none were found with records of sufficient length and quality.

Table 2.1. *Names, locations, drainage areas, Köppen climate types and gauge station record details for the six dryland and two humid rivers used in analyses*

Dryland rivers were identified as either endogenic (runoff primarily sourced from dryland regions) or allogenic (runoff primarily sourced from outside dryland regions). Köppen climate types: BSh, semi-arid, steppe (hot); BSk, semi-arid, steppe (cool); Dfb, humid continental, mild summer, year-round rainfall; Dfc, subarctic, cool summer, year-round rainfall.

Region	Country	River	Type	Station	Latitude	Longitude	Area (km²)	Köppen climate	Period	Years
Dryland	Australia	Paroo	Endogenic	Caiwarro[a]	−28.69	144.79	23 600	BSh	1967–2003	36.0
	Australia	Warrego	Endogenic	Wyandra[a]	−27.25	145.97	42 900	BSh	1967–2003	36.1
	USA (Arizona)	Little Colorado	Endogenic	Cameron[b]	35.93	111.57	68 600	BSk	1948–1988	41.0
	USA (Utah)	Green	Allogenic	Green River[b]	38.97	110.15	116 100	BSk	1906–1962	57.0
	Zimbabwe	Lundi	Endogenic	Tokwe Confluence[c]	−21.13	31.27	23 000	BSh	1959–1980	20.4
	Zimbabwe/South Africa	Limpopo	Endogenic	Beitbridge Pumpstation[c]	−22.22	29.98	196 000	BSh	1961–1980	19.0[d]
Humid	Finland	Vuoksi		Imatra[c]	61.15	28.77	61 300	Dfc	1847–1984	138.0
	Poland	Vistula		Warsaw[c]	52.25	21.03	84 700	Dfb	1921–1984	56.0

[a]Data from QDNRM (2004). [b]Data from Slack & Landwehr (1992). [c]Data from Vörösmarty et al. (1998). [d]Record length criterion relaxed slightly.

Figure 2.6. The Vistula River, Poland, is a perennial humid region river with highly predictable, albeit low-magnitude, seasonal flow variations driven by seasonal pattern in precipitation and snowmelt (Photo A. Rosner).

The selected rivers have an average drainage basin area of approximately 77 000 km^2, and an average record length of slightly over 50 years (Table 2.1). The discrepancy between record length and period for the Vistula (Fig. 2.6) record is because of seven missing years around the Second World War (1938–45 inclusive).

Characterising flow variability

Variability within a dataset is the magnitude and frequency of deviations from a central tendency (Townsend & Hildrew, 1994), but there are different measures of central tendency and of deviation. Flow data – especially from dryland rivers – are non-normal or skewed, and so the median is a better measure of central tendency than is the mean. Similarly, dispersion measures about the median are preferable to those around the mean, as the latter (see, for example, coefficient of variation) are more sensitive to the size of the dataset (Richards, 1989, 1990).

There are many different datasets that can be used for characterising flow variability, with the temporal resolution of the data determining which aspects of flow variability can be described. For example, daily data can be used to characterise the shape of a flow event hydrograph, whereas monthly or daily data allow characterisation of the seasonal patterns of flow. Given the wide range of possibilities, it is unsurprising that there is no standard approach for characterising flow variability. Rather, the measures used vary with the questions, with assumptions about what measures of variability are 'ecologically relevant'. For example, a 'habitat template' of streams in the USA was created by using 12–15 flow measures of flood frequency, flood predictability and overall flow variability, presumed to be ecologically relevant (Poff & Ward, 1989). Typically, experimental studies with well-defined hypotheses focus on one or a few specific flow measures, for example measures of depth variations in studies of littoral biofilms (Sheldon, 1994), or measures of daily flow variability and base flow stability in studies of the functional traits of fish (Poff & Allan, 1995). Conversely, regional or global analyses of flow regimes use a range of measures at different timescales and samples of these datasets (see, for example, variability of annual flow volumes and annual peak flows among continents (McMahon *et al.*, 1992; Peel *et al.*, 2001). Some studies have used a large number of measures of flow variability, covering temporal (daily to annual) and ecological aspects on many rivers (Richter *et al.* 1996, 1997; Puckridge *et al.*, 1998). For example, in fish, hydrological variability in the temporal flood pulse may be linked to reproduction, duration to longevity, and flood magnitude to mobility and colonisation (Puckridge *et al.*, 1998).

As well as analyses of the statistical distributions of data, analyses of the time series of data are also informative. Because most ecosystems are affected by the change of seasons, seasonal flow patterns are an obvious and well-studied aspect of flow sequences. But many longer flow periodicities may also be important, invoking the concept of persistence, or the 'significant dependence of observations a long time span apart' (Hosking, 1984). For example, flood clusters in Cooper Creek in central Australia are associated with occurrences of La Niña, persistent positive values of the Southern Oscillation Index (SOI) (Puckridge *et al.*, 2000). Similarly, ENSO affects flow regimes of dryland rivers around the world (see, for example, McMahon *et al.*, 1992; Nicholls *et al.*, 1996; Walker *et al.*, 1995), establishing the importance of persistence in characterising flow variability.

Because of the paucity of long daily flow records, we restricted our analyses to monthly flow data. To indicate the gross water balance for the case study rivers, we calculated the mean monthly flow. The typical flow magnitude in the case study rivers is indicated by the median monthly flow; the variance of the median is indicated by its relative standard error (RSE, standard error divided by the median). Because the datasets are extremely non-normal, the standard error of the median cannot be calculated by using a standard formula. We used a delete-one jackknife replication (Wolter, 1985) to estimate the standard error, with calculations performed by using the WesVar 4.2 software (Westat, 2002). WesVar 4.2 uses the Woodruff method (Särndal *et al.*, 1992) to indirectly calculate the standard error of the median by first calculating the confidence intervals of the median from the estimated cumulative distribution function.

We used measures of variability based on the median as the measure of central tendency, and measures of spread about the median as measures of the deviation. We consider monthly and seasonal variability, focussing on characterising intermittency because of the importance of flow cessation and drying as a disturbance regime in dryland rivers. We also consider flow persistence and its connection to aspects of the global climate system. We assess flow variability by using three graphical presentations of the data and six quantitative measures derived from these presentations: monthly intermittency (MI), duration of intermittency (DI), monthly variation (MV), seasonal variation (SV), seasonal predictability (SP) and seasonal signal (SS). Five of these measures are non-dimensional ratios; the sixth has dimensions of time. This study therefore focusses on flow variability and not on flow magnitude.

The time series of monthly values are plotted together with the 12-month moving median, the latter series indicating persistence (Fig. 2.7). The cumulative frequency distributions or monthly flow duration curves (Fig. 2.8) clearly show the monthly intermittency (MI), with the duration of intermittency (DI) calculated from the time series. MI is the proportion of months with zero flow, with high values indicating greater intermittency of monthly flow. DI is the average duration (in months) of the periods for which the monthly flow is equal to zero. Monthly variability (MV) is also based upon the flow duration curves, and is the overall variability of monthly flows measured by the spread of cumulative distribution function of monthly flows. It is the ratio of the difference between the tenth and ninetieth exceedance percentiles (EP) to the fiftieth exceedance percentile (the median) (Equation 2.1). High values indicate greater overall variability of monthly flows. MV has been used in previous studies of flow variability (see, for example, Puckridge *et al.*, 1998).

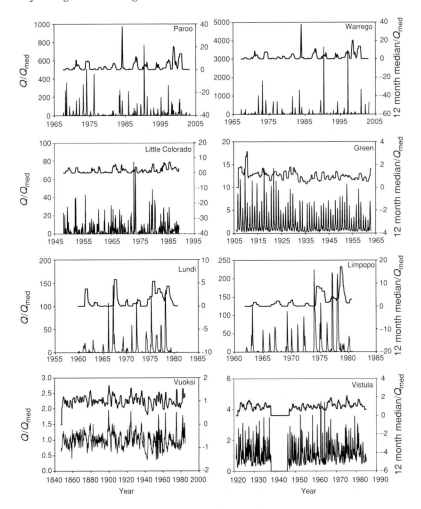

Figure 2.7. Time series of monthly flow values standardised by the full-record median (lower lines, left-hand ordinate) and 12-month moving median flow values standardised by the full-record median (upper lines, right-hand ordinate). Data are for the eight case study rivers, including six dryland rivers (Paroo, Warrego, Little Colorado, Green, Lundi and Limpopo) and two humid rivers (Vuoksi and Vistula).

$$MV = \frac{EP_{10th} - EP_{90th}}{EP_{50th}}. \qquad (2.1)$$

The box plots of month-by-month flow percentiles, among years, highlight the seasonal patterns (Fig. 2.9) and allow calculation of

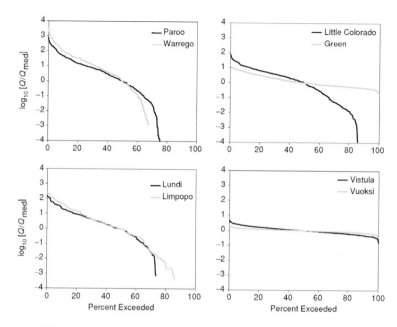

Figure 2.8. Flow duration curves of monthly flow values standardised by the full-record median (base 10 logarithm) for the eight case study rivers.

seasonal variation (SV), seasonal predictability (SP) and the overall seasonal signal (SS). SV and SS are new measures of flow variability developed for this study. SV measures the relative magnitude of the seasonal variation (among calendar months) and is the ratio of the difference between the maximum of the calendar month medians and minimum of the calendar month medians to the median of the calendar month medians (Equation 2.2). High values of SV indicate strong seasonal variation.

$$SV = \frac{\text{Max}_{i=1}^{12}(\text{Median } Q_i) - \text{Min}_{i=1}^{12}(\text{Median } Q_i)}{\text{Median}_{i=1}^{12}(\text{Median } Q_i)}. \qquad (2.2)$$

where i are the 12 calendar months (January–December, and median Q_i is the median flow for month i).

SP measures the relative predictability of the seasonal pattern among years and is the median among the calendar months of the variability among years, within a calendar month. The variability within a calendar month is again measured as the ratio of the difference between the maximum and the minimum to the median

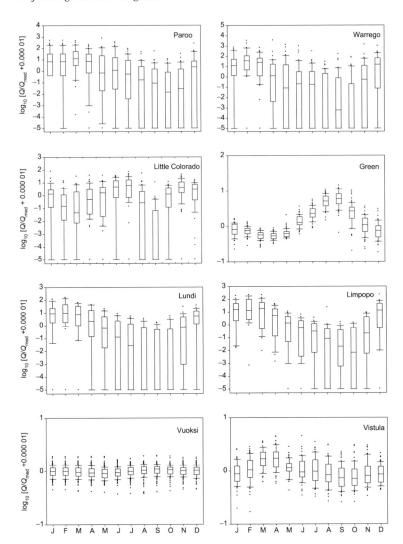

Figure 2.9. Box plots (median, quartiles, fifth and ninety-fifth percentiles) by calendar month of monthly flow values standardised by the full-record median (base 10 logarithm) for the eight case study rivers. A small constant value (0.000 01) has been added to the standardised values before logarithmic transformation, so that zero values are plotted as −5.

(Equation 2.3). High values indicate a low predictability of the seasonal variation among years. SP has been used in previous studies of flow variability (see, for example, Puckridge *et al.*, 1998).

$$SP = Median_{i=1}^{12} \left(\frac{Max_{j=1}^{m}(Q_j) - Min_{j=1}^{m}(Q_j)}{Median_{j=1}^{m}(Q_j)} \right)_i \qquad (2.3)$$

where i are the 12 calendar months (January–December) and m = number of years of record.

SS measures the overall seasonal pattern or signal: a combination of the magnitude of the seasonal variation (SV) and the predictability of this variation (SP). SS is defined as the ratio of SV to SP (Equation 2.4). High values of SS indicate a more predictable seasonal pattern. SP provides an alternative to Colwell's (1974) index of predictability and has been used before for flow variability (see, for example, Poff & Ward, 1989). Problems have been shown with the use of Colwell's indices in hydrological analyses (see, for example, Poff, 1996; Gan et al., 1991). For example, the index of predictability is sensitive to the time class-interval underlying the frequency distribution (Walker et al., 1995) and the length of record.

$$SS = \frac{SV}{SP}. \qquad (2.4)$$

The cumulative deviation plots show the deviations from the full record median, standardised by the full record median (Fig. 2.10). These illustrate the size and sequencing of discursions away from the record median.

To assess ENSO-related persistence in the flow records, the median monthly flow, the tenth percentile flow and the proportion of months with zero flow were compared among *El Niño*, *La Niña* and *Other* months. Significant differences in the median values were determined by using a Mann–Whitney rank sum test ($p < 0.05$). *El Niño* months have large positive values of the Southern Oscillation Index (SOI), whereas *La Niña* months have large negative SOI values. The monthly SOI is the difference between the mean sea-level atmospheric pressures measured at Tahiti and Darwin. We used the SOI values of CPC (2002), and defined *El Niño* months as those that had a moving average SOI for the previous 12 months greater than the median of positive 12-month moving average SOI values (January 1900–December 2002). Similarly, we defined *La Niña* months as those that had a moving average SOI for the previous 12 months less than the median of the negative 12-month moving average SOI values across the record (Fig. 2.11). Months that were neither *El Niño* nor *La Niña* were termed *Other* for this analysis.

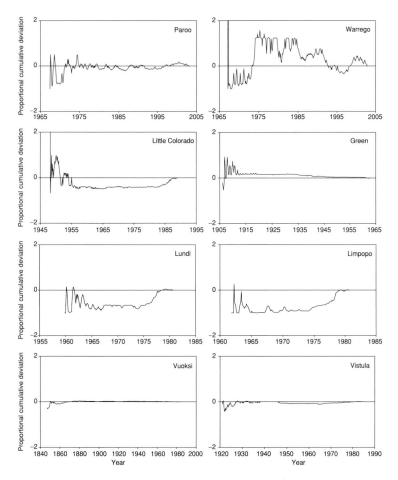

Figure 2.10. Cumulative monthly flow deviations for the eight case study rivers. Values are ratios of partial-record medians to the full-record median.

Flow variability: descriptions and analyses

The median monthly flows show the range in typical flow magnitude among the eight rivers (Table 2.1) and highlight the comparatively low yield from the catchments in Bsh climate zones, owing to higher evaporative losses. The relative standard errors show that the medians are only reasonably well defined for the two humid rivers and the Green River (Table 2.1; Fig. 2.5). For the remaining dryland rivers the variances

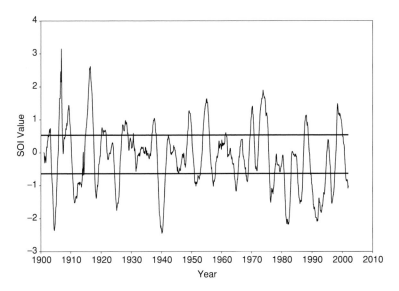

Figure 2.11. Twelve-month moving average Southern Oscillation Index
(SOI) values from December 1900 to December 2002. *El Niño* months are
defined in this chapter as those that plot above the upper threshold line,
La Niña months are those that plot below the lower threshold line, and
Other months are those that plot between the two threshold lines.

in the estimates of the median monthly flow are extremely high, particularly for the Warrego River (Fig. 2.3). As the measures of variability
used herein are all based on the median as the measure of central
tendency, it should be noted that with the available datasets for the
endogenic dryland rivers the standard errors of the estimates for these
measures are very high. A further discussion of this issue is presented at
the end of this section.

The time series of monthly flows, standardised by the median
monthly flow, reveals key characteristics of the flow regimes. This
includes frequent periods of zero flow and irregular floods in the
endogenic dryland rivers, in contrast to the more regular flood periodicity of the dryland allogenic Green River, and the highly stable flows of
the two humid rivers (Fig. 2.7). These plots also clearly demonstrate the
differences in the relative magnitudes of peak flows among the rivers.
In the humid rivers, peaks are 2–3 times the median, whereas in the
allogenic Green River peaks are 10–15 times the median. Peaks in
the endogenic dryland rivers range from around 70 times the median

in the Little Colorado, through 100–200 times the median in the two South African rivers, up to over 600 and over 3,000 times the median in the Paroo (Fig. 2.2) and Warrego Rivers (Fig. 2.3), respectively. The 12-month moving median series (standardised by the median monthly flow) reveal a periodicity of higher median flows in the vicinity of three years in all rivers (Fig. 2.7). Once again, the magnitude of these variations differs greatly from less than twice the median in the humid rivers to over 20 times in the two Australian rivers.

The monthly flow duration curves, standardised by the median monthly flow, show that monthly flows are far more variable in the endogenic dryland rivers than in either the allogenic Green River or the humid Vistula (Fig. 2.6) and Vuoksi rivers (Fig. 2.8, Table 2.1). The intermittency of endogenic dryland rivers is revealed in the significant proportions of time for which there is zero flow (Fig. 2.8), with MI values showing the Warrego (Fig. 2.3) is the most intermittent of the eight rivers (Table 2.1). The DI values follow a pattern similar to that of MI values: non-zero only for the endogenic dryland rivers (Table 2.1). However, they are higher for the African case study rivers than the Australian rivers, because of their stronger seasonal signal, reflecting better-defined wet and dry seasons.

Monthly flow percentiles by calendar month highlight the differences in the seasonal flow patterns among the rivers (Fig. 2.9). Flows in these plots are standardised by the median monthly flow, with a small constant value (0.000 01) added so that zero values plot on the logarithmic scale as minus five. The Green River (Fig. 2.5) has the strongest seasonal signal (SS) owing to a strong (SV) and predictable (SP) seasonal variation (Table 2.1, Fig. 2.9). The Limpopo River (Fig. 2.4) ranks second in seasonal signal (SS), with a strong but less predictable seasonal variation; the Vuoksi River has a highly predictable seasonal variation, but little variation (Table 2.1; Fig. 2.9). The Paroo and Warrego rivers have large seasonal variations (as also reflected in MV), with differences among calendar months (Fig. 2.9), but the predictability is so low that the overall seasonal signal is considerably weaker than in any of the other rivers (Table 2.1). The much greater SV value for the Warrego River compared with the adjacent Paroo River (Table 2.1) is because its more northerly headwaters are more influenced by tropical weather patterns.

The proportional cumulative deviation from the median reveals the long-term behaviour of the flow regimes for the rivers (Fig. 2.10). Clearly, the two humid rivers have stable flow regimes, with no significant deviation from the median (Fig. 2.10). The two African rivers, the

Limpopo (Fig. 2.4) and the Lundi (Fig. 2.1), with similar periods of record (Table 2.1), show a reduction in the median early in the record (negative gradient), with a sustained increase in the median (positive gradient) only occurring after the floods of the late 1970s (Fig. 2.10). The Little Colorado River is similar but the flow regime of the Green River (Fig. 2.5) is more stable, with slightly higher flow than the long-term median early in the record, followed by a gradual decline to the long-term median (Fig. 2.10). The strong and predictable seasonal variations of the Green River are apparent early in the record, before the length of record damps variation. The adjacent Paroo and Warrego rivers, with similar periods of record (Table 2.1), show quite different behaviour (Fig. 2.10). The Paroo River regime is reasonably stable, but the Warrego River regime displays large excursions above and below the full-record median throughout the record.

The ENSO effects are variable: median monthly flows are significantly higher in *El Niño* months than in *Other* months in the Little Colorado, Limpopo and Vistula rivers, and significantly higher in *La Niña* months than in *Other* months in the Warrego, Green and Lundi rivers (Table 2.1). Although differences in the tenth percentile flows and the percentage of months with no flow cannot be tested easily for statistical significance, the tenth percentile flows in the Lundi and the Limpopo rivers are substantially greater in *El Niño* months and *La Niña* months compared with *Other* months. Similarly, there are substantially fewer months with no flow during *La Niña* months compared with *Other* months in the Warrego, Lundi and Limpopo rivers, and during *El Niño* months compared with *Other* months in the Little Colorado River. ENSO has significant effects on these dryland and humid river regimes, with larger proportional effects in the dryland rivers chosen.

In terms of overall flow regime variability, the endogenic dryland rivers ranked from least to most variable as: (i) the Little Colorado River, (ii) the Lundi, Limpopo and Paroo Rivers (Figs. 2.1, 2.2 and 2.4), and (iii) the Warrego River (Tables 2.2 and 2.3; Figs. 2.3, 2.8–2.10). The Little Colorado River is the least intermittent of the endogenic dryland rivers, with the lowest monthly and seasonal variability, and little apparent ENSO influence on high flows. There are strong contrasts between the flow regimes of different river types (humid, endogenic and allogenic dryland, but there is also a significant range in flow regime character among the endogenic dryland rivers.

The high variance in the estimates of the median and other quantile-derived statistics for the endogenic dryland rivers is worthy

Table 2.2. *Descriptive statistics and flow variability measures for the eight case study rivers*

Descriptive statistics are: Mean, mean monthly flow; MED, median monthly flow; and RSE (%), relative standard error, i.e. the standard error of the median divided by the median (x 100). Flow variability measures include monthly intermittency (MI), duration of intermittency (DI), monthly variation (MV), seasonal variation (SV), seasonal predictability (SP) and seasonal signal (SS). See text for detailed descriptions.

Region	River	Mean (m^3 × 10^{-6})	MED (m^3 × 10^{-6})	RSE (%)	MI	DI (months)	MV	SV	SP	SS
Dryland	Paroo	44.26	2.06	484.4	0.25	2.00	45.3	13.2	216.2	0.06
	Warrego	44.10	0.66	12 218.4	0.32	2.50	96.3	96.6	2652.8	0.04
	Little Colorado	554.13	122.20	81.03	0.14	1.00	12.6	4.69	40.5	0.12
	Green	14 768.5	7329.2	5.88	0.00	0.00	4.9	5.74	2.15	2.67
	Lundi	186.1	21.41	351.7	0.26	3.00	20.8	13.7	36.1	0.38
	Limpopo	174.9	11.53	85.15	0.14	3.00	43.9	19.1	14.1	1.36
Humid	Vuoksi	1552.5	1541.1	1.93	0.00	0.00	0.58	0.19	1.42	0.13
	Vistula	1488.6	1270.2	4.21	0.00	0.00	1.64	1.12	2.72	0.41

Table 2.3. *Results of the ENSO influence analysis for El Niño, La Niña and Other months*

Results for median monthly flow, tenth percentile monthly flow and the percentage of months with no flow are all expressed as ratios of the respective values for the *Other* years. Asterisks indicate medians that are significantly different ($p < 0.05$) from values in *Other* months.

Region	River	Percentage of months			Median monthly flow		Tenth percentile monthly flow		Percentage of months with no flow	
		El Niño	La Niña	Other	El Niño	La Niña	El Niño	La Niña	El Niño	La Niña
Dryland	Paroo	38.2	21.5	40.3	1.21	2.54	0.98	1.44	1.21	0.78
	Warrego	38.1	21.3	40.6	0.94	7.60*	0.89	2.18	1.10	0.58
	Little Colorado	26.4	21.4	52.2	4.37*	0.43	1.02	0.71	0.49	1.22
	Green	21.6	26.2	52.2	1.14	1.26*	1.07	1.45	0	0
	Lundi	17.9	21.8	60.3	3.01	3.68*	4.68	2.99	0.88	0.48
	Limpopo	19.7	24.1	56.2	3.95*	2.48	5.29	3.36	0.81	0.44
Humid	Vuoksi	23.6	23.5	52.9	0.97	0.97	0.97	1.00	0	0
	Vistula	20.8	21.0	58.2	1.16*	1.00	1.02	1.05	0	0

of further comment. Clearly, to determine the median (or any other statistic) to a given level of accuracy requires a longer record for endogenic dryland rivers than for less variable regimes. The available flow record is only a sample of the entire population of flows. If we assume that the data spread in these samples is equal to the data spread in the respective populations, then the record length required to obtain a given relative standard error (RSE, standard error divided by the median) can be calculated, because the standard error is the square root of the ratio of the variance to the sample size. For example, to obtain an RSE of 10% in the monthly median flow would require less than ten years of record for the two humid rivers and only about 20 years for the Green River. However, for the endogenic dryland rivers the required record length to achieve this relative standard error goes into at least the thousands or even tens of thousands of years, and so goes beyond the timescale within which any river regime could be considered stationary.

Rather than focussing simply on record length, it is instructive to consider the degree of intermittency. In particular, we see a monotonic increasing trend in RSE with increasing MI for the endogenic dryland rivers (Table 2.1). The higher the proportion of zero flows, the less useful the median flow becomes as a measure of central tendency, and clearly for MI > 0.5 the median flow is zero. As the proportion of zero flows in the record increases, the amount of actual information contained in the record decreases for a given length of record. Hence as *MI* increases it becomes increasingly appropriate to determine the degree of intermittency (MI) (both overall and seasonally) but to calculate flow variability statistics across only the non-zero flows. For the Warrego River, RSE for the non-zero proportion of the record is 21.2%, compared to over 12 000% for the full record, whereas for the Paroo River RSE for the non-zero proportion of the record is 22.1%, compared with 484% for the full record. Clearly, more reliable assessments of flow variability can be obtained by using only the non-zero portion of an intermittent flow record.

This approach treats the flow record as a series of discrete flow episodes or clusters of hydrographs separated by periods of zero flow, a perspective that has underpinned some modelling of these types of stream flow record (see, for example, Cigizoglu *et al.*, 2001). The degree of intermittency at which this approach becomes preferred will depend on the context: the questions being asked, or the comparisons being made. There have been several attempts to define or classify degrees of flow intermittency (see, for example, Hedman & Osterkamp, 1982; Boulton & Lake, 1988; Comín & Williams, 1994). Hedman & Osterkamp (1982) use perhaps the simplest, defining 'perennial' as flowing for more than 80% of the time, 'intermittent' as flowing for 10%–80% of the time, and 'ephemeral' as flowing for less than 10% of the time. Even from our monthly data, our endogenic dryland rivers would be considered intermittent in this classification. Even the Warrego, for which daily data are available, flows for 18% of the time on a daily basis and would be intermittent rather than ephemeral. Basing analyses only on the non-zero flows is clearly appropriate for ephemeral regimes, but is also appropriate for some intermittent regimes. The RSE (%) and ME values (Table 2.1) suggest that once the proportion of zero flows exceeds 0.15, statistical analyses of flow variability could be sensibly based only on the non-zero flows.

Ecological implications for the eight rivers

Aquatic biota survive in riverine ecosystems where suitable habitat exists for survival and for reproduction. The hydrological regime of a river determines the spatial and temporal availability of habitat and its predictability. For the rivers we analysed, the different flow regimes (Fig. 2.7) will provide different opportunities and constraints for riverine biota.

Dryland rivers had higher measures of intermittency and monthly and seasonal variation than did humid rivers (Table 2.1). Water is always available in the humid rivers and in the Green River (Figs. 2.5 and 2.7), thus providing habitat for some species that need continuous access to water. Contrast this with the flow regimes of the African and Australian rivers (Fig. 2.7). Intermittency in the dryland rivers ranged from the Limpopo (Fig. 2.4) and the Little Colorado with zero flow for 14% of the months on record, to the Warrego (Fig. 2.3) where nearly one third of all months had zero flow (Fig. 2.7; Table 2.1). The dry spells are critical phases for aquatic biota to survive, not just in terms of the proportion of dry time, but also in terms of the seasonal predictability. For example, the Warrego (Fig. 2.3) is the most intermittent (highest MI), but it does not have the longest median dry spells (highest DI) because of a very low seasonal predictability (Table 2.1). In the Lundi and Limpopo Rivers (Figs. 2.1 and 2.4), the combination of the degree of intermittency (MI) and the seasonal predictability (SP) leads to a longer duration of intermittency (DI) with a median dry-spell duration of three months, typically in the dry season. We expect that rivers with long dry spells will favour particular aquatic organisms capable of surviving these periods.

Dry periods represent a challenge for aquatic organisms. Sedentary organisms have no choice and need to survive periods with little to no water. Algae, bacteria and fungi commonly regenerate from desiccated spores. Similarly, many aquatic plants on dryland rivers can survive for years on the floodplain or in the seed bank (Chapter 5, this volume), just as eggs of some invertebrates have desiccation-resistant stages (Chapter 6, this volume). More mobile animals such as aquatically dependent insects or vertebrates may retreat to remnant waterholes that persist in some dryland rivers (Chapter 7, this volume). Desert frogs are an exception, burying themselves in cocoons through the dry periods (Chapter 7, this volume). Drying periods bring ecological stresses other than desiccation, such as increased salinity (Kingsford & Porter, 1993; Kingsford et al., 1999) that affect biota (Chapter 10, this volume).

Waterbirds can fly to refuge habitats during dry periods (chapter 7, this volume). Lakes on dryland rivers can retain water for long periods. For example, Lake Hope on Cooper Creek, central Australia, may only fill once every four years, but retains water for about the same time, making it the last of the wetlands to dry (Kingsford et al., 1999). Similarly, as floodplains along the Paroo River in central Australia dry out, waterbird density on the remaining floodplain wetlands markedly increases (Kingsford, 1996). The hydrology of dryland rivers drives an ecological response in waterbirds that completely contrasts with that of similar species in more humid regions. Australian and South African waterbirds are often referred to as nomadic because their ecology reflects the variability of the dryland rivers on which they depend (Siegfried, 1970; Kingsford & Norman, 2002). Northern hemisphere waterbirds have predictable breeding seasons and migratory behaviour, governed by the high seasonal predictability of their aquatic habitats (Fig. 2.7; Table 2.1). In contrast, waterbirds in desert regions of Australia and South Africa breed when there is available habitat, delivered by unpredictable flow regimes of desert rivers. Similarly, their movements are governed by the extent of the available habitat, concentrating on the remaining aquatic habitats during dry periods (Kingsford, 1996; Roshier et al., 2002; Chapter 7, this volume). There is low predictability in seasonal flow in endogenic dryland rivers and so recruitment, regeneration and movement of biota must cope with the absence of seasonal cues (e.g. photoperiod or rainfall). The absence of an annual wet season in many dryland rivers means that many species may not be able to breed every year see, for example colonial waterbirds (Kingsford & Johnson, 1998). This may have significant long-term effects on survival of species.

Although dryland rivers may have times of low to no flows, they have contrastingly significant flood periods where extensive floodplains are inundated (Roshier et al., 2001b; Kingsford et al., 2001; Bunn et al., 2003; Chapter 4, this volume). These floods determine community composition by defining the temporal sequences of habitat availability for different species, and successional processes (Jenkins & Boulton, 2003; Kingsford et al., 2004). Sometimes multiple floods can occur in a single year, or several years may pass without a single flood; the flood history affects the responses of biota and the ecosystem (Puckridge et al., 2000). Extreme flows in dryland rivers are large disturbances, maximal in the Warrego River (Fig. 2.3) across our case study rivers (Table 2.1). The larger the extreme flows, the longer their influence will persist in the 'memory' of the system, driving geomorphological processes such as erosion and sedimentation (Chapter 3, this volume). Structural aspects of the habitat mosaic, and hence entire aquatic communities, reflect the history of major flow events. The larger the flow event, the longer its imprint is likely to be reflected in the ecosystem.

During flood periods, many aquatic biota living on dryland rivers rapidly exploit large areas of flooded habitat. Floodplains are highly productive (Chapter 4, this volume), providing opportunities for fish populations to expand rapidly (Puckridge et al., 2000; Chapter 4, this volume). Such areas are also rapidly colonised by aquatic invertebrates during floods (Jenkins & Boulton, 2003), and often support significant breeding events for biota (Kingsford & Johnson, 1998; Kingsford et al., 1999; Puckridge et al., 2000). It is not simply the temporal variability of floods, but also the complex spatial patterns of inundation depth and frequency, that create a mosaic of different floodplain habitats that supports a high diversity of species (Ward, 1998; Ward et al., 1999). This spatial variability is enhanced by the temporal variability of flow, creating a gradient in inundation frequency laterally across the floodplain and longitudinally downstream. Typically, only the largest infrequent floods reach the floodplains furthest from the channel system and fill the terminal lakes and wetlands of the floodplains at the end of large endoreic rivers (Kingsford et al., 1999; Roshier et al., 2001b). For example, Lake Eyre in central Australia is the terminal basin for several rivers (Diamantina River, Cooper Creek, Neales River and some minor creeks) and it receives some water about once every two years on average, although it seldom fills completely (Kotwicki, 1986).

Although desert river systems may seem inhospitable, given their unpredictable aquatic habitat, such an interpretation may be simplistic. Australian waterbirds are known to retreat to more mesic areas during dry times (Frith, 1982). However, when desert regions flood again, these waterbirds flock back, deserting the mesic regions, whatever the time of the year (Kingsford & Norman, 2002). Part of the reason for this may be the extraordinary diversity and abundance of the food web sustained by dryland rivers (Chapters 4–7, this volume). Shallow water and high temperatures provide ideal conditions for life to flourish. At these times such areas are considerably more productive than many aquatic habitats that are constantly flooded. For example, floodplain lakes in arid Australia, with constant water levels, had foraging waterbird densities that were much lower than similar lakes that dried and flooded unpredictably (Kingsford et al., 2004).

Whereas dryland rivers may differ significantly from humid rivers in terms of unpredictability, they also differ significantly amongst themselves in terms of seasonality and intermittency. There are differences of orders of magnitude in these aspects among our case study rivers (Table 2.1). Most are geographically separated, but the adjacent Paroo and Warrego Rivers (Figs. 2.2 and 2.3) with considerably different flow regimes (order-of-magnitude difference in seasonal predictability) may have intriguing ecological differences that reflect such differences. How different are the sedentary biota of these systems? One might predict that significant ecological differences exist, given their hydrological regimes. Unfortunately relatively little is known of the invertebrates or flood-dependent vegetation on the Warrego River, in contrast to the Paroo River.

Alternatively, organisms may respond differently to a changing environment. For example, waterbirds in humid and desert regions of the world are physiologically and taxonomically similar, and yet their behaviour is considerably different (Kingsford & Norman, 2002). Some Australian waterbird species (for example, Australian Wood Duck *Chenonetta jubata*) held in captivity in the northern hemisphere adapt to the seasonal predictability of their environment, breeding during the spring each year. So, where biota are similar in humid and dryland rivers, they may simply alter their response to variability.

The ways of describing a river's hydrology are many (McMahon et al., 1992; Puckridge et al. 1998; Peel et al., 2001) and the ecological relevance of these approaches is even more difficult to define beyond broad statements of connection (Puckridge *et al.*, 1998). Part of the problem is that our knowledge of the ecology of dryland river biota is in its infancy, particularly in terms of interpreting long-term (decadal) variations that may be essential for understanding species' survival, particularly for long-lived species. Understanding the relation between ecology and the flow regime for different species will be essential for future management. Whereas such patterns may be easier to study and more predictable in humid rivers, in dryland rivers they will require longer periods of investigation and probably remain as unpredictable as the rivers on which they depend.

Implications for water resource development

People need reliable supplies of water: the antithesis of the variability of dryland rivers (Walker *et al.*, 1997). The variable disturbance patterns of dry periods and flood periods and everything in between determine the ecology of dryland rivers (Chapters 5–7, this volume). Water resource development usually requires the damming of rivers to store water that can then be reliably supplied each year to off-stream uses.

To provide an annual diversion volume (or draft) with a given reliability, storage volume is a function of the variability of annual flows. For normally distributed annual flows, storage and draft are related by the square of the coefficient of variation of annual flows and the square of the probability of failure (complement of reliability) (McMahon, 1976). The maximum draft for a given probability of failure increases monotonically with storage volume and is asymptotic to a draft equal to the mean annual flow. The actual storage required to achieve this maximum is a function of the variability of annual flows. This relation ignores evaporation from storages, which is substantial in dryland regions. Including evaporation substantially changes the relation between storage and draft (for typical storage morphometries) in the dryland regions of Australia and North America, reducing the maximum draft, with this maximum attained at comparatively low storage (McMahon, 1978). For a 95% reliability for the Paroo and Warrego basins, the maximum draft is zero (McMahon, 1978). Streamflow regulation of dryland rivers is therefore problematic, with lower reliability implicit. In rivers, such as the Colorado, deep, narrow gorges reduce evaporative loss and thus increase the potential for regulation. Conversely, in floodplain dryland rivers, there is little terrain for river impoundment, and so storage volume to surface area ratios are low and evaporation losses high. Furthermore, a significant fraction of flood waters pass across the floodplain rather than down the channel, and so storages built on the floodplain away from the channel can direct flood waters into storage. Storage volumes may be high, but supply is not reliable, so water use is opportunistic (Kingsford, 2000a).

River regulation can affect dryland river flow regimes in multiple ways. Where the main river channel is used to convey water to downstream users, zero flows will be less common and the seasonality of downstream water demand may alter the seasonality of the flow regime (see, for example, Maheshwari et al., 1995). In this situation, a dryland river's flow regime will tend to become more and more similar to that of an equivalent-sized humid river as the degree of regulation increases (Fig. 2.7, Gehrke et al., 1995). McMahon & Finlayson (2003) coined the term 'anti-drought' to describe the removal of natural zero and low flows that often follows regulation; they contend that, in Australian river management, both induced droughts and anti-droughts are serious concerns. Changes to the seasonality of flow can affect breeding and recruitment patterns of biota. More regular flow may encourage alien species, as has happened with European carp *Cyprinus carpio* in the rivers of the Murray–Darling Basin, Australia (Gehrke et al., 1995). Where the main channel is not used to deliver water to downstream users, for example where water is diverted to large off-channel storages, and downstream of water users (irrigation or drinking water) where the main channel is used for delivery, total flow decreases and flow intermittency increases, with ecological consequences for the river and its floodplains (Kingsford, 2000b; Lemly et al., 2000; Chapter 8, this volume). Only large floods tend to reach such areas and even these may be considerably reduced, depending on the nature of upstream storage and/or water demands. Regulation and diversion reduces flow variability at all temporal and spatial scales for downstream ecosystems, and has geomorphologic consequences (Ligon et al., 1995; Thoms 2003). Boom and bust cycles are less likely to occur at the same scale in regulated dryland rivers. This will affect the food web and can reduce the diversity and density of waterbirds (Kingsford et al., 2004). Small and medium floods are typically the most affected, as these often provide much of the supply for human needs in dryland regions. Small and medium floods can be essential for replenishing the waterholes and wetlands that are the refuges for biota during dry periods. Elimination of these habitats may prevent the long-term survival of many vertebrates that depend on these dry-period refuges (Chapter 8, this volume).

CONCLUSIONS

Our analyses of eight rivers demonstrate the considerable temporal variability of flow regimes in dryland rivers compared with humid rivers. Flow regimes determine the ecology of dryland rivers (Walker et al., 1995). Humid rivers are perennial with lower monthly variability and a relatively predictable seasonal pattern (Fig. 2.7; Tables 2.2 and 2.3). Even within dryland rivers, we see considerable variability in flow regimes (Fig. 2.7; Puckridge et al., 1998). The allogenic dryland Green River is intermediate between endogenic dryland rivers and humid rivers. It is fully perennial, with intermediate monthly and seasonal variability, but a predictable seasonal pattern (Figs. 2.7–2.10; Tables 2.2 and 2.3). In contrast, the five endogenic dryland rivers are intermittent with high monthly and seasonal variability, have low seasonal predictability, and most are significantly influenced by ENSO (Figs. 2.7–2.10; Tables 2.2 and 2.3). The ecological implications of these flow characteristics will be profound and are becoming better known with increasing research. High hydrological variability, reflected in a considerable range of spatial and temporal variability, distinguishes dryland rivers from their humid counterparts, but flow regulation and diversions will inevitably mean loss of ecological complexity and biota (Chapter 8, this volume). Hydrologically, flow regulation and diversion will in some situations reduce dryland river flow variability, making flow regimes more similar to those of humid rivers, whereas in other situations they will increase aspects of flow variability: particularly intermittency. In the first situation, many biota may lose their competitive edge over biota from less harsh environments; in the second, the environment may become too harsh even for many endemic biota.

ACKNOWLEDGEMENTS

We thank Murray Peel for access to the dataset of Peel et al. (2001) that builds on that of McMahon et al. (1992). We also thank Andrew Boulton and two anonymous referees whose advice improved this chapter.

REFERENCES

Agnew, C. and Anderson, E. (1992). *Water Resources in the Arid Realm*. London: Routledge.

Argue, J. R. and Salter, L. E. M. (1977) Waterhole development: a viable water resource option in the arid zone? In *Hydrology Symposium: The Hydrology of Northern Australia* (National Conference Publication No.77/5), pp. 35–9. Australia: Institution of Engineers.

Boulton, A. J. and P. S. Lake. (1988) Australian temporary streams - some ecological characteristics. *Verhandlungen der Internationale Vereinigung für theoretische und angewandte Limnologie*, **23**, 1380–3.

Boulton, A. J., Sheldon, F., Thoms, M. C. and Stanley, E. H. (2000). Problems and constraints in managing rivers with variable flow regimes. In *Global Perspectives on River Conservation: Science, Policy and Practice*, ed. P. J. Boon, B. R. Davies and G. E. Petts, pp. 415–30. Chichester: John Wiley and Sons.

Bull, L. J. and Kirkby, M. J. (2002). *Dryland Rivers: Hydrology and Geomorphology of Semi-arid Channels*. Chichester: John Wiley and Sons.

Bunn, S. E., Davies, P. M. and Winning, M. (2003). Sources of organic carbon supporting the food web of an arid zone floodplain river. *Freshwater Biology*, **48**, 619–35.

Cigizoglu, H. K., Adamson, P. T. and Metcalfe, A. V. (2001). Bivariate stochastic modelling of ephemeral streamflow. *Hydrologic Processes*, **16**, 1451–65.

Colwell, R. K. (1974). Predictability, constancy and contingency of periodic phenomena. *Ecology*, **55**, 1148–53.

Comín, F. A. and Williams, W. D. (1994). Parched continents: Our common future? In *Limnology Now: a Paradigm of Planetary Problems*, ed. R. Margalef, pp. 473–527. Amsterdam: Elsevier Science.

CPC (2002). Climate Prediction Center of the National Weather Service of the US National Oceanic and Atmospheric Administration. http://www.cpc.ncep. noaa.gov.

Davies, B. R., Thoms, M. C., Walker, K. F., O'Keeffe, J. H. and Gore, J. A. (1992). Dryland rivers: their ecology, conservation and management. In *The Rivers Handbook Vol.2: Hydrological and Ecological Principles*, ed. P. Calow and G. E. Petts, pp. 484–511. Oxford: Blackwell Scientific.

Dunkerly, D. L. (1992). Channel geometry, bed material and inferred flow conditions in ephemeral stream systems, Barrier Range, western NSW, Australia. *Hydrological Processes*, **6**, 417–33.

Dynesius, M. and Nilsson, C. (1994). Fragmentation and flow regulation of river systems in the northern third of the world. *Science*, **266**, 753–62.

Frith, H. J. (1982). *Waterfowl in Australia*. Sydney: Angus and Robertson.

Gan, K. C., McMahon, T. A. and Finlayson, B. L. (1991). Analysis of periodicity in streamflow and rainfall data by Colwell's indices. *Journal of Hydrology*, **123**, 105–18.

Gehrke, P. C., Brown, P., Schiller, C. B., Moffatt, D. B. and Bruce, A. M. (1995). River regulation and fish communities in the Murray-Darling River system, Australia. *Regulated Rivers: Research and Management*, **11**, 363–75.

Graf, W. L. (1988) *Fluvial Processes in Dryland Rivers*. Berlin: Springer-Verlag.

Grimm, N. B. and Fisher, S. G. (1989). Stability of periphyton and macroinvertebrates to disturbance by flash floods in a desert stream. *Journal of the North American Benthological Society*, **8**, 293–307.

Hedman, E. R. and Osterkamp, W. R. (1982). *Streamflow characteristics related to channel geometry of streams in Western United States*. US Geological Survey Water-Supply Paper 2193: pp. 1–17.

Hosking, J. R. M. (1984). Modelling persistence in hydrologic time series using fractional differencing. *Water Resources Research*, **20**, 1898–908.

Jenkins, K. M. and Boulton, A. J. (2003). Ecological connectivity in a dryland river: short-term aquatic microinvertebrate recruitment following floodplain inundation. *Ecology*, **84**, 2708–23.

Kingsford, R. T. (1994). Waterholes and their significance in the anastomosing channel system of Cooper Creek, Australia. *Geomorphology*, **9**, 311–24.

Kingsford, R. T. (1995). Occurrence of high concentrations of waterbirds in arid Australia. *Journal of Arid Environments*, **29**, 421–5.

Kingsford, R. T. (1996). Wildfowl (Anatidae) movements in arid Australia. In *Proceedings of the Anatidae 2000 Conference, Strasbourg, France, 5–9 December 1994*. ed. M. Birkan, J. van Vessem, P. Havet, J. Madsen, B. Trolliet and M. Moser. *Gibier Faune Sauvage (Game Wildlife)*, **13**, 141–55.

Kingsford, R. T. (1999). Managing the water of the Border Rivers in Australia: irrigation, government and the wetland environment. *Wetlands Ecology and Management*, **7**, 25–35.

Kingsford, R. T. (2000a). Protecting or pumping rivers in arid regions of the world? *Hydrobiologia*, **427**, 1–11.

Kingsford, R. T. (2000b). Ecological impacts of dams, water diversions and river management on floodplain wetlands in Australia. *Austral Ecology*, **25**, 109–27.

Kingsford, R. T. (2001). An event based approach to the hydrology of arid zone rivers in the Channel Country of Australia. *Journal of Hydrology*, **254**(1–4), 102–23.

Kingsford, R. T. and Johnson, W. J. (1998). The impact of water diversions on colonially nesting waterbirds in the Macquarie Marshes in arid Australia. *Colonial Waterbirds*, **21**, 159–70.

Kingsford, R. T. and Norman, F. I. (2002). Australian waterbirds - products of the continent's ecology. *Emu*, **102**, 47–69.

Kingsford, R. T. and Porter, J. L. (1993). Waterbirds of Lake Eyre. *Biological Conservation*, **65**, 141–51.

Kingsford, R. T., Curtin, A. L. and Porter, J. (1999). Water flows on Cooper Creek in arid Australia determine 'boom' and 'bust' periods for waterbirds. *Biological Conservation*, **88**, 231–48.

Kingsford, R. T., Jenkins, K. M. and Porter, J. L. (2004). Imposed hydrological stability on lakes in arid Australia and effect on waterbirds. *Ecology*, **85**, 2478–92.

Kingsford, R. T., Thomas, R. F. and Curtin, A. L. (2001). Conservation of wetlands in the Paroo and Warrego catchments in arid Australia. *Pacific Conservation Biology*, **7**, 21–33.

Knighton, A. D. and Nanson, G. C (1994a). Flow transmission along an arid zone anastomosing river, Cooper Creek, Australia. *Hydrological Processes*, **8**, 137–54.

Knighton, A. D. and Nanson, G. C (1994b). Waterholes and their significance in the anastomosing channel system of Cooper Creek, Australia. *Geomorphology*, **9**, 311–24.

Kotwicki, V. (1986). *Floods of Lake Eyre*. Adelaide: Engineering and Water Supply Department.

Köppen, W. and Geiger, R. (1930). *Handbuch der Klimatologie*. Berlin: Gebrüder Bornträger.

Laronne, J. B. and Reid, I. (1993). Very high rates of bedload transport by ephemeral desert rivers. *Nature*, **366**, 148–50.

Lemly, A. D., Kingsford, R. T. and Thompson, J. R. (2000). Irrigated agriculture and wildlife conservation: conflict on a global scale. *Environmental Management*, **25**, 485–512.

Ligon, F. K, Dietrich, W. E. and Trush, W. J. (1995). Downstream ecological effects of dams. *Bioscience*, **45**, 183–92.

Maheshwari, B. L., Walker, K. F. and McMahon, T. A. (1995). Effects of regulation on the flow regime of the River Murray, Australia. *Regulated Rivers: Research and Management*, **10**, 15–38.

McMahon, T. A. (1976). Preliminary estimation of reservoir storage for Australian streams. *Civil Engineering Transactions, The Institution of Engineers, Australia*, **CE18**, 55–9.

McMahon, T. A. (1978). Australia's surface water resources: potential development based on hydrologic factors. *Civil Engineering Transactions, The Institution of Engineers, Australia*, **CE20**, 155–164.

McMahon, T. A. (1979). Hydrological characteristics of arid zones. In *The Hydrology of Areas of Low Precipitation*. Proceedings of the Canberra Symposium. IAHS Publication No. 128, pp. 105–23.

McMahon, T. A. and Finlayson, B. L. (2003). Droughts and anti-droughts: the low-flow hydrology of Australian Rivers. *Freshwater Biology*, **48**, 1147–60.

McMahon, T. A., Finlayson, B. L., Haines, T. A. and Srikanthan, R. (1992). *Global Runoff: Continental Comparisons of Annual and Peak Discharges*. Germany: Catena. Cremlingen-Destedt.

Meigs, P. (1953). World distribution of arid and semi-arid homo-climates. Arid Zone Hydrology (UNESCO Arid Zone Research Series) **1**, 203–9.

Nanson, G. C., Tooth, S. and Knighton, A. D. (2002). A global perspective on dryland rivers: perceptions, misconceptions and distinctions. In *Hydrology and Geomorphology of Semi-arid Channels*, ed. L. J. Bull and M. J. Kirkby, pp. 17–54. Chichester: John Wiley and Sons Ltd.

Nicholls, N., Lavery, B., Frederiksen, C., Drosdowsky, W. and Torok, S. (1996). Recent apparent changes in relationships between the El Niño-Southern Oscillation and Australian rainfall and temperature. *Geophysical Research Letters*, **23**, 3357–60.

Peel, M. C., McMahon, T. A., Finlayson, B. L. and Watson, F. G. R. (2001). Identification and explanation of continental differences in the variability of annual runoff. *Journal of Hydrology*, **250**, 224–40.

Pettit, N. E., Froend, R. H. and Davies, P. M. (2001). Identifying the natural flow regime and the relationship with riparian vegetation for two contrasting Western Australian rivers. *Regulated Rivers: Research and Management*, **17**, 201–15.

Pilgram, D. H., Chapman, T. G. and Doran, D. G. (1988). Problems of rainfall-runoff modelling in arid and semi-arid regions. *Hydrological Sciences*, **33**(4), 379–400.

Poff, N. L. (1996) A hydrogeography of unregulated streams in the United States and an examination of scale-dependence in some hydrological descriptors. *Freshwater Biology*, **36**, 71–91.

Poff, N. L. and Allan, J. D. (1995). Functional organisation of stream fish assemblages in relation to hydrological variability. *Ecology*, **76**, 606–27.

Poff, N. L. and Ward, J. V. (1989). Implications of streamflow variability and predictability for lotic community structure: a regional analysis of streamflow patterns. *Canadian Journal of Fisheries and Aquatic Sciences*, **46**, 1805–18.

Postel, S. L. (2000). Entering an era of water scarcity: the challenges ahead. *Ecological Applications*, **10**, 941-8.

Puckridge, J. T., Sheldon, F., Walker, K. F. and Boulton, A. J. (1998). Flow variability and the ecology of arid zone rivers. *Marine and Freshwater Research*, **49**, 55-72.

Puckridge, J. T., Walker, K. F. and Costelloe, J. F. (2000). Hydrological persistence and the ecology of dryland rivers. *Regulated Rivers: Research and Management*, **16**, 385-402.

QDNRM (2004). http://www.nrm.qld.gov.au/watershed/ Online hydrologic datasets of Queensland Department of Natural Resources, Mines and Energy, Australia.

Richards, R. P. (1989). Measures of flow variability for Great Lakes tributaries. *Environmental Monitoring and Assessment*, **12**, 361-77.

Richards, R. P. (1990). Measures of flow variability and a new flow-based classification of Great Lakes tributaries. *Journal of Great Lakes Research*, **16**, 53-70.

Richter, B. D., Baumgarter, J. V., Powell, J. and Braun, D. P. (1996). A method for assessing hydrologic alteration within ecosystems. *Conservation Biology*, **10**, 1163-74.

Richter, B. D., Baumgarter, J. V., Wigington, R. and Braun, D. P. (1997). How much water does a river need? *Freshwater Biology*, **37**, 231-49.

Rosgen, D. L. (1994) A classification of natural rivers. *Catena*, **22**, 169-99.

Roshier, D. A., Robertson, A. I. and Kingsford, R. T. (2002). Responses of waterbirds to flooding in an arid region of Australia and implications for conservation. *Biological Conservation*, **106**, 399-411.

Roshier, D. A., Robertson, A. I., Kingsford, R. T. and Green, D. G. (2001a). Continental-scale interactions with temporary resources may explain the paradox of large populations of desert waterbirds in Australia. *Landscape Ecology*, **16**, 547-56.

Roshier, D. A., Whetton, P. H., Allan, R. J. and Robertson, A. I. (2001b). Distribution and persistence of temporary wetland habitats in arid Australia in relation to climate. *Austral Ecology*, **26**, 371-84.

Särndal, C.-E., Swensson, B. and Wretman, J. (1992). *Model Assisted Survey Sampling*. New York: Springer-Verlag.

Sheldon, F. (1994). Littoral ecology of a regulated dryland river (River Murray, South Australia) with reference to Gastropoda. Ph.D. Thesis, University of Adelaide.

Sheldon, F., Boulton, A. J. and Puckridge, J. T. (2002). Conservation value of variable connectivity: aquatic invertebrate assemblages of channel and floodplain habitats of a central Australian arid-zone river, Cooper Creek. *Biological Conservation*, **103**, 13-31.

Siegfried, W. R. (1970). Wildfowl distribution, conservation and research in southern Africa. *Wildfowl*, **21**, 89-98.

Slack, J. R. and Landwehr, J. M. (1992) Hydro-Climatic Data Network (HCDN): A U. S. Geological Survey streamflow data set for the United States for the study of climate fluctuations, 1874-1988. (U. S. Geological Survey Open-File Report 92-129, 19 pp.) http://water.usgs.gov/pubs/wri/wri934076/1st_page.html.

Stafford Smith, D. M. and Morton, S. R. (1990). A framework for the ecology of arid Australia. *Journal of Arid Environments*, **18**, 255-78.

Thomas, D. S. G. (1989) The nature of arid environments. In *Arid Zone Geomorphology*, ed. D. S. G. Thomas, pp. 1-10. London: Belhaven Press.

Thoms, M. C. (2003). Floodplain-river ecosystems: lateral connections and the implications of human interference. *Geomorphology*, **56**, 335-49.

Tooth, S. (1999). Floodouts in central Australia. In *Varieties of Fluvial Form*, ed. A. J. Miller and A. Gupta, pp. 219–47. Chichester: John Wiley and Sons.

Townsend, C. R. and Hildrew, A. G. (1994). Species traits in relation to a habitat template for river systems. *Freshwater Biology*, **31**, 265–75.

Vörösmarty, C. J., Fekete, B. M. and Tucker, B. A. (1998). *Global River Discharge, 1807–1991*, Version 1.1 (RivDIS). Data set available online: http://www.daac. ornl.gov from Oak Ridge National Laboratory Distributed Active Archive Center, Oak Ridge, Tennessee, USA.

Vörösmarty, C. J., Sharma, K. P., Fekete, B. M., *et al.* (1997). The storage and aging of continental runoff in large reservoir systems of the world. *Ambio*, **26**, 210–19.

Walker, K. F., Puckridge, J. T. and Blanch, S. J. (1997). Irrigation development on Cooper Creek, central Australia – prospects for a regulated economy in a boom-and-bust ecology. *Aquatic Conservation: Marine and Freshwater Ecosystems*, **7**, 63–73.

Walker, K. F., Sheldon, F. and Puckridge, J. T. (1995). A perspective on dryland river ecosystems. *Regulated Rivers: Research and Management*, **11**, 85–104.

Ward, J. V. (1998). Riverine landscapes: biodiversity patterns, disturbance regimes and aquatic conservation. *Biological Conservation*, **83**, 267–78.

Ward, J. V., Tockner, K. and Schiemer, F. (1999). Biodiversity of floodplain river ecosystems: ecotones and connectivity. *Regulated Rivers: Research and Management*, **15**, 125–39.

Westat (2002). *WesVar® 4.2 Users Guide*. USA: Westat. 344 pp.

Williams, W. D. (1988). Limnological imbalances: an antipodean viewpoint. *Freshwater Biology*, **20**, 407–20.

Wishart, M. J., Davies, B. R., Boon, P. J. and Pringle, C. M. (2000). Global disparities in river conservation: 'First World' values and 'Third World' realities. In *Global Perspectives on River Conservation: Science, Policy and Practice*, ed. P. J. Boon, B. R. Davies and G. E. Petts, pp. 335–69. Chichester: John Wiley and Sons.

Wolman, M. G. and Gerson, R. (1978). Relative scales of time and effectiveness of climate in watershed geomorphology. *Earth Surface Processes and Landforms*, **3**, 189–208.

Wolter, K. M. (1985). *Introduction to Variance Estimation*. New York: Springer-Verlag.

3

Variability, complexity and diversity: the geomorphology of river ecosystems in dryland regions

M. C. THOMS, P. J. BEYER AND K. H. ROGERS

INTRODUCTION

Dryland regions cover approximately 47% of the world's land surface (Middleton & Thomas, 1997). The rivers that drain these regions show great diversity in their physical character. Our understanding of the processes and morphologies of 'dryland rivers' have increased with the recent reviews of these systems (Graf, 1987; Thornes, 1994; Reid & Frostick, 1997; Tooth & Nanson, 2000; Tooth, 2000). An evident trend in river research over the past 10 years has been the increased emphasis on studying large river systems, particularly ecosystems, stressing the interconnections between physical and biological components. Research on dryland rivers has followed a similar trend (Thomas, 1997): between 1982–1992 and 1993–2003 there was a 40% increase in the number of published scientific articles on dryland river geomorphology in 20 international and national journals concerned. Despite this, many of our beliefs about dryland river processes come from rivers in humid and temperate regions, transferred to dryland settings with often amusing consequences (see Nanson et al., 2002).

The geomorphology of dryland rivers reflects their environment: climate, topography and vegetation. Climatically, dryland rivers are characterised by low but highly variable precipitation and sparse,

Ecology of Desert Rivers, ed. R. T. Kingsford. Published by Cambridge University Press.
© Cambridge University Press 2006.

unevenly distributed vegetation cover. Although these are important attributes that make dryland regions distinct, categorising dryland rivers into different types may be futile because of flow and sediment transport variability and the resulting physical complexity (Nanson *et al.*, 2002). Moreover, there is limited knowledge about these rivers; many 'distinct dryland river morphologies' are also apparent in rivers from other climatic regions. In a recent perspective on dryland rivers, Nanson *et al.* (2002) suggest that there may only be two landforms unique to dryland rivers: flood-outs, and ephemeral channels with relatively permanent waterholes (see details in Tooth (1999) and Knighton & Nanson (2000)). To improve our understanding, and to develop conceptual models for dryland rivers, we require data from a range of dryland settings. This chapter focusses on large, alluvial, low-gradient dryland rivers, particularly their in-channel environment (Fig. 3.1). It adds to the dryland river studies on high-energy, upland systems (Schick, 1993; Thornes, 1994; Reid & Frostick, 1997), large, anastomosing river-floodplain systems (Nanson & Huang, 1999; Tooth, 1999; Knighton & Nanson, 2000;) (Fig. 3.2), and highly confined, meandering systems (Graf, 1987).

Figure 3.1. The main channel of the Barwon–Darling River, Australia, is highly regular with low complexity, resembling a large canal. This reach of the Barwon–Darling River is heavily affected by water resources development (Photo M. C. Thoms).

Figure 3.2. An inset floodplain or in-channel bench on the Culgoa River, Australia (one of the rivers of the Lower Balonne floodplain). These features are bank-attached temporary sediment stores that develop at intermediate elevations between the main channel and floodplain.

Alluvial rivers are process–response systems, freely adjusting their channel dimensions (form) in response to changes in flow and sediment regimes (processes). Changes in discharge (Q_w) and sediment load (Q_s) induce changes in channel width (w), depth (d), slope (S), meander wavelength (λ), sinuosity (ρ) and width:depth ratio (F) (Schumm, 1977). Bankfull width and depth and channel shape (F) are the primary channel adjustment variables, whereas changes in planform (λ, ρ) and channel slope (S) are second- and third-order variables, respectively. Empirical process–form relations have been established through flume experiments or field observations, with most field data on alluvial river channel morphology coming from rivers in temperate or humid regions (see, for example, Church, 1992). However, much early work from dry regions has contributed strongly to the development of fluvial geomorphology. Recent work on rivers in dryland regions emphasises the variability of fluvial form and process in time and space. Dryland rivers often have extended periods of no flow interrupted by brief periods of high-magnitude flows. Unlike the closely linked feedback and relatively frequent adjustment of perennial systems, the process–form relations of dryland rivers may be discontinuous in time and space.

Both equilibrium and non-equilibrium conditions exist in dryland rivers, depending on the scale of observation (Tooth & Nanson, 2000).

This chapter also further highlights the importance of understanding the ecogeomorphology of dryland river systems (Thoms & Parsons, 2002); integrating hydrology, geomorphology and ecology, it makes four sequential points. The first point is that alluvial rivers are process–response systems (as above), where flow and sediment transport regimes are process drivers producing responses in river morphologies (dimensions of the cross-section, planform and slope as well as flood-plain structure). Second, there is a direct relationship between process variability and the diversity or complexity of physical form in alluvial dryland rivers. High flow variability is a characteristic feature of dry-land rivers (McMahon, 1979; Puckridge *et al.*, 1998; Chapter 2, this volume) producing more diverse or complex river morphologies than exist in less variable, humid and temperate rivers. Third, decreases in process variability simplify river morphology, thereby reducing physical diversity. Many dryland rivers are affected by large-scale water re-sources development (Walker *et al.*, 1997; Chapter 8, this volume), which changes flow regimes, especially reducing flow variability (Thoms & Sheldon, 2000a). Fourth, the loss of physical diversity has implications for the functioning of aquatic ecosystems. Aquatic ecology has focussed on the premise that the physical structure of river systems — habitat — provides the templet within which evolution develops characteristic species traits (Southwood, 1977, 1988). Conceptually, there is a direct link between temporal heterogeneity (process), spatial heterogeneity (physical structure or habitat), and biodiversity in riverine habitats (Townsend & Hildrew, 1994). Dryland rivers are biologically highly productive (Bunn *et al.*, 2003; Chapter 4, this volume); empirical evidence suggests direct associations between biological diversity and environmental variability (Sheldon *et al.*, 2002; Kingsford *et al.*, 2004). We illustrate these interactions with two case studies from different continents: the Sabie River in South Africa and the Barwon–Darling River in Australia (Fig. 3.1).

PROCESS VARIABILITY

Hydrological regimes

Dryland river systems differ in hydrological character from rivers in other climatic zones, in terms of water availability, flow variability, and the frequency of floods (Alexander, 1985). Dryland environments have relatively little precipitation (<500 mm) with minimal conversion of this precipitation to actual runoff because of high rates of evaporation. Ratios of mean annual precipitation to mean annual runoff for Australia and southern Africa for example, are 9.8% and 8.6% respectively, compared with a world average of 48%.

Dryland river systems are also more variable than those in humid regions. The mean coefficient of variation (CV) of annual flows for dryland rivers is: Australia, 1.27; southern Africa, 1.14; the Mediterranean, 1.25 (McMahon, 1979; Finlayson & McMahon, 1988). Areas with little precipitation (hence little runoff) generally exhibit high flow variability, with CVs twice those of humid regions, irrespective of drainage basin size (Finlayson & McMahon, 1988). Moreover, an extensive analysis of flow variability in 20-year hydrographs of 52 rivers was able to separate tropical rivers from dryland rivers (Puckridge et al., 1998). Indeed, large dryland rivers such Cooper Creek and the Diamantina River in central Australia have distinctively variable flow regimes, markedly different from flow regimes in humid and temperate regions. Dryland rivers tend to have long-term hydrographs characterised by periods of extreme flooding followed by extensive periods of low or no flow.

Flood flows in dryland river systems are also more highly variable than flood flows in other climatic regions (Chapter 2, this volume). For example, the ratio of the 100-year recurrence interval flood to the mean annual flood (Q_{100}/Q) is 1.5 times higher in dryland regions than in other climatic regions (Finlayson & McMahon, 1988). Australian and southern African dryland rivers exhibit the greatest degree of variability, with values greater than 12.

Sediment regimes

Sediments are an active component of fluvial ecosystems. Sediment regimes reflect catchment conditions (e.g. climate, geology and land use) and influence river morphology and habitat assemblages. There is no simple relation between climate and the amount of sediment generated within a catchment, because of the influences of vegetation, basin lithology and soils. Effective annual precipitation is the annual precipitation required to generate the observed annual runoff at a standardised annual mean temperature (Langbein & Schumm, 1958). The relation between sediment yield and effective precipitation shows that the greatest sediment yields occur with an annual precipitation of 400–500 mm (i.e. characteristic of dryland regions). Langbein & Schumm (1958) argue that lower sediment yields occur in low-precipitation areas (<200 mm) because there is not enough runoff (energy) to initiate erosion. With high precipitation (>800 mm), vegetation retards erosion, reducing sediment yields. Data from 1246 drainage basins demonstrate maximum sediment yield for dryland regions (Walling & Webb, 1983).

Rivers draining dryland catchments generally have substantial loads of sediment. A global review of sediment regimes of rivers (Holeman, 1968) found that the Yellow River, which drains the dryland loess terrain of China, generates the greatest sediment load (1890 t yr^{-1} or 2640 t km^{-2}). Moreover, dryland rivers transport much of their sediment load in relatively few, large discharges (Baker, 1977; Graf, 1987). Up to 60% of the total load is transported by flow events with return intervals longer than 10 years, whereas those flow events carry only 10% of the total load in the rivers of more humid regions (Neff, 1967). Therefore, dryland rivers have higher 'dominant discharges' (Andrews, 1980) where the largest fraction of the annual sediment load, over a period of time, occurs in very large flood events. Indeed, Graf (1985) suggests that the concepts of magnitude and frequency are climatically dependent. The more variable the flow regime, the greater the fraction of the total load that will be transported by infrequent, large-magnitude flow events.

PHYSICAL HETEROGENEITY

'Equilibrium' channel forms are maintained by the transfer of energy (water) and mass (sediment). However, variability of discharge and sediment movements in dryland river systems can inhibit morphological equilibrium (Stevens *et al.*, 1975; Tooth & Nanson, 2000). Characteristics of complex dryland river channels are well illustrated by the morphology of rivers in Kruger Park, South Africa (Rogers & O'Keefe, 2003) (Fig. 3.3). These rivers are generally contained within a relatively stable 'macro-channel' but the within-macro-channel environment is dynamic, with five main river zones or channel types (alluvial single channel, alluvial braided, mixed pool–rapid, mixed anastomosing and bedrock anastomosing) (Moon *et al.*, 1997). The complexity of the

Figure 3.3. The main channel of the Bokhara River, Queensland, Australia (one of the rivers of the Lower Balonne floodplain). For most of the time the Bokhara and many other dryland rivers in Central Austraila exist as a series of Semi-permanent waterholes (photo M. C. Thoms).

within-channel morphology arises from interactions between a highly variable sediment supply, hydrology, and a complex long profile. Indeed, in the Sabie River, there is an intricate downstream pattern of areas of high and low sediment storage (Heritage *et al.*, 1997). The three zones of relatively high sediment storage are associated with a braided section, an alluvial single-channel section, and a mixed anastomosing section. Variations in discharge and sediment supply often result in large physical adjustments to the within-channel environment, but this differs between river zones. A major flood event in 1996 on the Sabie River produced non-linear and spatially variable patterns of change (Rountree *et al.*, 2000). Mixed anastomosing channels were the most stable; pool–rapid channels were the most dynamic; and there were frequent changes between alluvial- and bedrock-dominated states. The other river zones generally became more alluvial, although at different rates, but seldom linear. There may be multiple channel

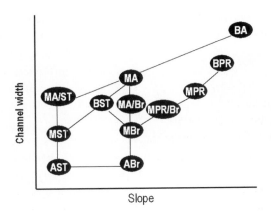

Figure 3.4. The Olifants River, South Africa, experienced an array of changes to different states in these reaches of the Kruger National Park, depending on the slope and width of the macro-channel, between 1947 and 2000 (after Rountree *et al.*, 2000). Abr, alluvial braided; AST, alluvial single channel; BA, bedrock anastomosing; BPR, bedrock pool–rapid; Br, braided; BST, bedrock single channel; MA, mixed anastomosing; MBr, mixed braided; MPR, mixed pool–rapid; MST, mixed single-channel; ST, single channel.

change pathways for rivers in the Kruger Park (Fig. 3.4) (Rogers & O'Keefe, 2003). Such a model highlights the complex morphological response to variations in the discharge and sediment regime along a river and provides a mechanism for the understanding of river-channel heterogeneity in dryland river systems.

Dryland rivers often have 'compound channels' (Graf, 1987; Thoms & Walker, 1992). There can be a single low-flow channel and a wider high-flow channel, representing adjustments to different dominant flow regimes. There are different forms of compound channel. The Gila River, Arizona, has a well defined inner low-flow channel that meanders within a much larger outer flood channel, which is often braided in planform (Graf, 1987). By comparison, the rivers of the Murray–Darling system in Australia have 'nested compound channels' where the low-flow channel is contained or nested within a series of higher-flow channels (Thoms & Walker, 1992). Nested compound channels reflect channel adjustments to the highly variable flow and sediment regimes of Australian dryland rivers (Thoms & Olley, 2004). In-channel benches are prominent morphological features of the rivers of the Barwon–Darling River (Fig. 3.1), Australia (Riley & Taylor, 1978). They are depositional, flat, elongated and often crescent-shaped in planform and the lower surfaces are formed by suspended-load deposition, forming as point, concave, convex and lateral benches with laminations (0.1–14 cm) of fine inorganic sediments and enriched organic mud (Thoms & Olley, 2004). The upper surfaces are relict surfaces, part of the present floodplain, and are inundated about once in every 15 years (Woodyer, 1968; Riley, 1975).

Frequency histograms of stage heights for two gauging stations in the Barwon–Darling River produced a banding of flows at distinct stage heights (Fig. 3.5) (Thoms & Sheldon, 1997). Water levels and discharges associated with these flow bands correspond closely to in-channel morphological features. At Wilcannia, for example, the majority of flows, between 0.5 and 1.5 m, corresponds to the low-water mark on the channel banks and the lowest in-channel feature. Other groupings of stage levels occur between 3.25 and 3.75 m, 5.5–6 m, 7.5 m and 9.75 m, and all of them are closely associated with in-channel morphological benches. Similar associations between flow banding and channel morphology occur at Walgett, albeit at different flow levels. Similar interaction between flow and channel morphology and these in-channel benches is found in many of the rivers in inland Australia. These 'inset floodplains' (Warner, 1994) probably reflect channel adjustments to long-term climatic fluctuations.

Dryland channel morphology and its response to hydrological and sediment regimes illustrate the complexity of these systems. Half

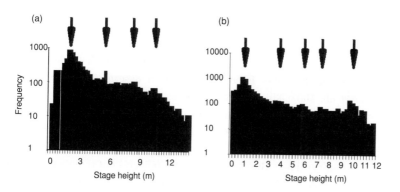

Figure 3.5. Frequency histogram of daily stage heights (water levels) for gauging stations ($n = 80$ years for both stations), at (a) Walgett and (b) Wilcannia; the levels of the in-channel features observed at each station are indicated.

of the channel cross-sections studied (seven of 14) in the Verde River, Arizona, exhibited distinct low-flow channels within a large braided channel, indicating that not even this aspect of dryland river morphology is ubiquitous. The low-flow channel of the Verde River had a capacity less than the two-year return flow. The return intervals of capacity flows were highly variable, representing discharges from less than $1 \text{ m}^3 \text{ s}^{-1}$ to over $300 \text{ m}^3 \text{ s}^{-1}$ (Beyer, 1998). Clearly, no single volume or recurrence interval of flow dominated the compound channel morphology, emphasising how a range of flows shapes the whole system. Stability ratios (a given recurrence interval of flow over the critical shear stress determined from particle size) highlight the role of moderate flows in dryland systems. In all 14 Verde River cross-sections, all geomorphic surfaces within the broader channel had stability ratios greater than 1 (i.e. over 50% of the sediment was moving) for the five-year recurrence interval discharge. Under more frequent flows, such as the two-year return flow, the surfaces of the side benches in half of the cross-sections were stable, but there was no clear pattern of occurrence of this stability (Beyer, 1998).

Much of the early research on the geomorphology of dryland river floodplains comes from the American south-west and has been added to by the recent work of Nanson & Huang (1999), Tooth (1999, 2000), Tooth & Nanson (1999, 2000) and Thoms (2003). Tooth & Nanson (2000) show the extent and diverse array of different river-floodplain types that exist in dryland environments. Indeed, Nanson *et al.* (2002) comment on the difficulty of generalising about dryland river patterns and their associated floodplains.

The diversity of floodplain features, at a scale smaller than that used by Nanson and Tooth, is well illustrated by the Lower Balonne Floodplain in south-west Queensland, Australia (Figs. 3.6 & 3.7). The geomorphology of this floodplain has been described in detail (see Foster *et al.*, 2002; Sims & Thoms, 2002; Thoms, 2003). These studies note an array of different morphological units, and at the scale of the entire floodplain there appears to be a greater diversity of morphological units in the upstream reaches of the floodplain than in the downstream reaches. The distribution is a reflection of the geomorphological history of the Lower Balonne Floodplain and contemporary hydrological variables. Using the Shannon diversity index, the diversity of morphological units between St George and 30 km downstream of the bifurcation of the Condamine–Balonne River was four times that of the remaining floodplain. Flat floodplain, distributary channels, and

Figure 3.6. The anastomosing river channels of Cooper Creek, Australia, allow inundation of extensive floodplains (Photo S. Bunn).

flood runners dominated the lower reaches of the Lower Balonne Flood-plain (Fig. 3.6). In addition, classification of the current flow regime of the region reveals two groups of stations with similar hydrological character; the difference between the groups was reflected by different flooding character (Thoms, 2003). In summary, the upper regions of the Lower Balonne floodplain have a more variable flooding regime than does the lower region, suggesting that the flooding process shapes the surface of the floodplains.

RESPONSE TO HUMAN DISTURBANCE

Transfers of energy and mass are governing influences in rivers. A river's physical templet is proscribed by these two transfers (Southwood, 1977, 1988), but their effects on physical and ecological functioning are accentuated in dryland rivers because of their variability. Floods maintain the rhythm of an ecosystem, albeit erratically in dryland rivers.

The water resources of dryland regions are heavily exploited, and many major rivers in these regions are regulated via dams and other structures such as weirs, canals and levees (Thoms, 2003; Chapter 8, this

Figure 3.7. This section of the Olifants River in Kruger National Park, South Africa, marks the beginning of a bedrock anastomosing river reach. Plant distribution reflects availability of substratum and its interaction with variable flow regimes (Photo M. C. Thoms).

volume). Control over this resource will only increase because of the rising demand for water by people inhabiting arid and semi-arid regions (Davies *et al.*, 1994). Over half the world's human population inhabits dryland environments. The consequences of water resources development on flow and sediment regimes and the subsequent down-stream morphological adjustment are well known (see, for example, Petts, 1984). Most examples are from humid and temperate regions, but responses to water resources development may differ in dryland rivers (Walker & Thoms, 1993). Dryland rivers have high dominant discharges in comparison with humid and temperate rivers with high–magnitude, low-frequency events (Davies *et al.*, 1994). Water resources developments reduce the dominant discharge of dryland rivers, changing the hydro-logical, sediment transport and disturbance regimes (Fig. 3.8), thus affecting the physical and ecological functioning of these systems. Rivers in the Murray–Darling Basin, Australia, subjected to large scale water resources development have experienced reductions in 'domin-ant discharge' between 20 and 81 percent and this has been primarily through changes in the frequency of floods. Overall, the distribution of

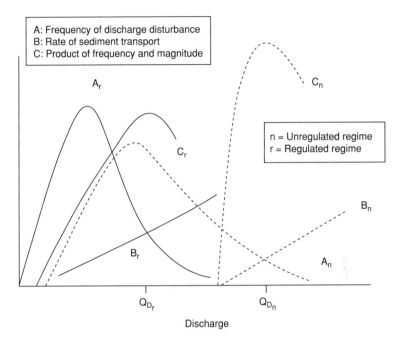

Figure 3.8. A conceptual model of the impact of water resources development on the dominant discharge and sedimentation of dryland rivers (after Thoms & Walker, 1992).

flows have become skewed towards small floods (Fig. 3.8), resulting in a reduction in the dominant discharge.

Regulated flow regimes can be so different from natural flow regimes that the environmental impact of flow regulation may be greater on dryland systems than elsewhere. In the Murray–Darling Basin, for example, there are over 20 large headwater dams, in excess of 6000 smaller regulating weirs, an interbasin water transfer scheme, and abstraction of large volumes of water by the irrigation industry. Irrigation extraction has affected the entire hydrological regime (flow regime, flow history and flood pulse), not just the low and 'average' flows of the rivers in the Barwon–Darling Basin (Thoms & Sheldon, 2000b). Water extraction licences increased from 20 in 1960 to 267 in 1994. By 1994, diversions were equivalent to over 60% of the natural flow at Menindee (Table 3.1).

Comparison of historical (surveyed in 1886) and contemporary cross-sections along two river reaches of the Barwon–Darling River (Fig. 3.1), Australia, illustrates the influence of flow changes on the

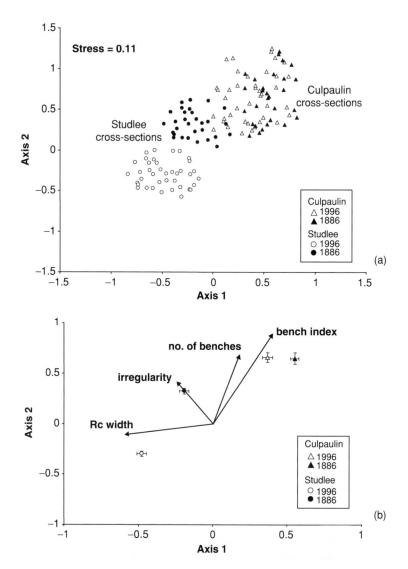

Figure 3.9. (a) Semi-strong hybrid ordination in two dimensions of the morphological variables collected from cross-sections of sites from two reaches (Culpaulin and Studlee Stations) at two time periods (1886 and 1996) of the Barwon–Darling River. Sites have been plotted and labelled according to their time–reach combination. Stress = 0.06. (b) Centroids (mean ± SE of the *x*, *y* coordinates for each site–reach combination) have been plotted along with the significant Principal Axis Correlation associations, shown as arrows.

Table 3.1. *Hydrological changes in the Barwon–Darling River, New South Wales, Australia, associated with water resources development*

Flow period	Scale	Documented change
Flow regime	> 100 years	48% reduction in long-term median annual flows
		68% reduction in annual flows with $AEP^a < 1.01$
Flow history	1–100 years	58% reduction in some monthly flows
		An increase in the predictability of monthly flows
		91% reduction in the magnitude of the annual flood event
Flood pulse	< 1 year	Doubling of the rate of fall of some flood events
		Extractions equivalent to 65% of daily flow during some flood events

[a]AEP, Annual Exceedance Probability of flood events.
Source: (summarised from Thoms & Sheldon (2000b).

in-channel complexity of a large dryland river (see Sheldon & Thoms, 2006). The cross-sectional character of the river channel was assessed by using 12 variables and compared by using multivariate statistical analyses (Sheldon & Thoms, 2006). The reach not influenced by water resources development (Culpaulin) contained five distinct morphological features — in-channel benches — at different elevations within the channel. By comparison, cross-sections from the regulated reach (Studlee) had a shallow channel with few horizontal in-channel features. Cross-sections from the 1886 and 1996 surveys of the Culpaulin reach clustered together, whereas the cross-sections from the 1886 survey of the Studlee reach clustered separately from the 1996 survey (Fig. 3.9). Morphological differences between the reaches and survey times were considerable (Fig. 3.9). The Culpaulin reach for the 1886 and the 1996 surveys was characterised by great channel complexity, particularly in number of benches and bench index (Fig. 3.9). The 1886 survey of the Studlee reach found irregular channel complexity, whereas the 1996 survey discovered increased bankfull widths (Fig. 3.9). The historic and contemporary Culpaulin cross-sections are more complex than those of the Studlee reach, suggesting that water resources developments have reduced or simplified in-channel complexity by destroying in-channel benches. The greater channel complexity in the unregulated Culpaulin reach is probably associated with higher flow variability at this site compared with the Studlee reach.

Much of the published research on the influence of water resources development has focussed on the within-channel environment (cf. Petts, 1984), with limited studies on floodplains. The Lower Balonne floodplain in Australia (Figs. 3.6 & 3.7) is affected by water resources development, in the form of extractions upstream, some damming, and floodplain development in the form of levees and off-river storages (Thoms, 2003). Modelled data suggest that water resources development reduced quantities of dissolved organic carbon released from the Lower Balonne Floodplain by nearly 5000 t during 1986–1996. Approximately 33 300 t of dissolved organic carbon would have been made available under 'natural' conditions, compared with 25 629 t under 'current' conditions, during that period. Annual reductions in dissolved organic carbon released range from 21 t to 1293 t for those years, or 8%–79% (Fig. 3.10). Moreover, water resources development and floodplain development had different impacts on the potential supply of dissolved organic carbon from the Lower Balonne Floodplain. Large floods potentially make more dissolved organic carbon available than small floods because they inundate more of the floodplain surface. For example, under the natural-flow scenario, a flood with an average recurrence interval (ARI) of 10 years would initiate the potential release of 520 t of dissolved organic carbon on average, compared with 90 t from a small flood (ARI of 1.5 years) (Fig. 3.10). Small flood events, especially those with an ARI of less than 2 years, have been reduced by 22%–48% compared with reductions of 3.98%–5.78% for floods with an ARI of more than 5 years. Hence, the estimated potential release of dissolved organic carbon from the floodplain surface varies with size of flood.

Estimated quantities of organic carbon released in floods of different magnitudes change under 'natural' and 'current' flow scenarios, plus under the influence of floodplain development (Fig. 3.10). A change in the hydrological regime alone reduces the potential supply of dissolved organic carbon from the floodplain surface by 7.6%–50%; floodplain development may reduce this potential supply by a further 23%–50%. The combined impact of water resources and floodplain developments eliminates the role of floods with an ARI < 2 years in

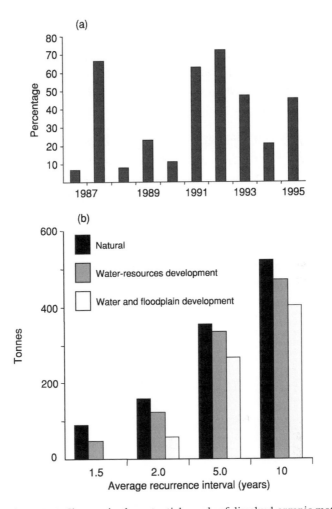

Figure 3.10. Changes in the potential supply of dissolved organic matter from the Lower Balonne Floodplain. (a) Annual reductions for the period 1986–96. (b) Difference among natural, water-resources (diversions) and floodplain development (levee banks) for four floods of different magnitudes.

dissolved organic carbon release from the original floodplain surface. So, under 'current' hydrological and floodplain-development conditions the potential supply of dissolved organic carbon from the Lower Balonne Floodplain appears to be reduced by 23%–100%, depending on the size of flood.

ECOLOGICAL IMPLICATIONS: CASE STUDIES

Two examples, one from South Africa and the other from Australia, underline the importance of understanding flow–habitat–ecology interactions in dryland rivers.

Morphology–plant associations in the Sabie River, Kruger Park, South Africa

In the rivers of the Kruger Park, South Africa, strong environmental gradients in water availability and substratum type interact to provide a patchy physical (geomorphological), diverse and dynamic environment for plant colonisation (Figs. 3.3 and 3.11). Increases in physical heterogeneity are associated with increased vegetation heterogeneity. Along the Sabie River, abundance of riparian tree species varies with elevation above the riverbed (Fig. 3.12), ranging from *Breonadia salicina* at the lowest elevation to *Spirostachys africana* at higher elevations

Figure 3.11. Interactions between the common reed *Phragmites mauritianus* and the mobile sands of the Sand River, Kruger National Park, South Africa, produce an intricate mosaic of series of in-channel vegetated islands (Photo M. C. Thoms).

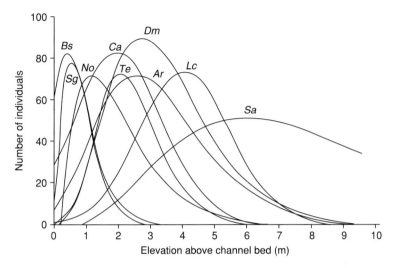

Figure 3.12. Distribution curves of the abundance of tree species (stem diameter >3 cm) above the bed of the Sabie River, Kruger Park, South Africa. Ar, *Acacia robusta*; Bs, *Breonadia salicina*; Ce, *Combretum erythrophyllum*; Dm, *Diospyros mespiliformis*; Lc, *Lonchocarpus capassa*; No, *Nuxia oppositifolia*; Sa, *Spirostachys africana*; Sg, *Syzygium guineense*; Te, *Trichelia emetica*. After van Coller (1993).

within the macro-channel (van Coller, 1993). Species are lost at three different elevations, corresponding to the limits of active, seasonally flooded and ephemeral features. Two main species grow on the banks of the macro-channels. *Spirostachys africana* is terrestrial, occurring primarily in the riparian zone at highest elevations; it forms the boundary with terrestrial vegetation (Fig. 3.12). *Diospyros mespiliformis* occurs lower within the macro-channel and forms the transition to macro-channel floor communities (Fig. 3.12).

There are four vegetation assemblages within the active channels of the Kruger Park (Fig. 3.13). The active channel experiences frequent flooding, sedimentation and erosion, which generate an irregular topography. The amount of bedrock outcropping influences the vegetation of the various river zones (Rogers & O'Keefe, 2003). *Combretum erythrophyllum* occurs at the highest elevations of the active channel on fine-textured substrata of alluvial islands with other main woody species: *Ficus sycomorus*, the invasive shrub *Lantana camara* and *Pavetta lanceolata*. *Phylanthus reticulatus* is at lower elevations associated with coarser sediment islands. The reed *Phragmites mauritianus* dominates

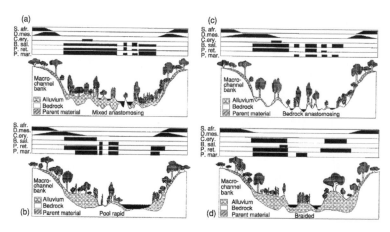

Figure 3.13. Cross-sectional diagrams of the proportional thickness of six vegetation communities in four different channel zones on the Sabie river, Kruger Park, South Africa. B.sal, *Breonadia salicina*; C.ery., *Combretum erythrophyllum*; D.mes., *Diospyros mespiliformis*; P.mar., *Phragmites mauritianus*; P.ret., *Phylanthus reticulatus*; S.afr., *Spirostachys africana*. After Rogers & O'Keefe (2003).

low-level, loose coarse alluvium and exposed bedrock close to the active channel, where *Ficus caprifolia, Syzygium guineense* and *Securinega virosa* are also common. The evergreen *Breonadia salicina* community also occurs on bedrock sections of the active channel with *Nuxia oppositifolia, S. guineense, S. cordatum* and *P. mauritianus*. The differential species distribution and differences in the complex vegetation assemblages between the different channel types are directly related to the range of hydrogeomorphic conditions present in the Kruger Park rivers (Figs. 3.13, 3.3 and 3.11) (Rogers & O'Keefe, 2003).

Temporal patterns of vegetation change in the Kruger Park rivers are also highly complex (Carter & Rogers, 1989) and are directly related to river channel adjustments (Rountree *et al.*, 2000). Analyses of aerial photographs (1940 and 2000) reveal three long-term changes in vegetation response to major flood events in the Sabie and Olifants Rivers (Carter & Rogers, 1989; Rountree *et al.*, 2000) (Fig. 3.3). The probabilities of a change in river landscape state have been assessed by recording changes in six river states (rock, reed, sand, herbaceous vegetation, terrestrial shrubs, and trees) in 20 m grid squares on the photos. Changes are complex with no simple linear patterns (Fig. 3.14). In 1940 the river had a wide sandy and rocky bed and was probably 'recovering' from the large floods of 1925. The period 1940–1965 was

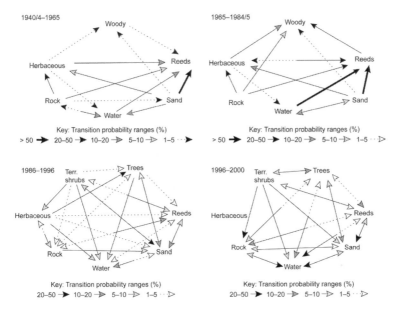

Figure 3.14. River landscape state changes observed in the Olifants River, South Africa, over four time periods. After Rogers & O'Keefe, (2003).

dynamic with a small overall shift towards more herbaceous and reed cover and wider channels. Between 1965 and 1985 river channel change was more frequent and directional with major reductions in the area of physical substratum (rock, sand, water), as it was colonised by reeds (Fig. 3.14). Reed beds were then invaded by trees to form large patches of closed-canopy forest. Between 1986 and 1996, changes were even more complex and less directional than in either of the preceding periods. There was still a trend towards vegetation establishment despite a severe drought in 1991–1992 which killed trees and a 1 : 50 year flood in 1996 which removed some vegetation and caused small-scale erosion and deposition patches along the macro-channel floor. Changes from the drought and flood were most evident in bedrock anastomosing channels (Rountree *et al.*, 2000). Changes were also evident between 1996 and early 2000 as a result of a 1 : 100 year flood. This flood reversed the vegetation establishment trend of the preceding 60 years to a state similar to that in 1940 (Fig. 3.14). Large areas of riparian forest, shrubs and reed stands (Fig. 3.11) were removed and replaced by a complex mosaic of rock, sand and water.

Although state change was different in all four periods (Fig. 3.14) an episodic stripping model of geomorphic change emerges. Progressive sediment storage between the two big floods of 1925 and 2000 reduced the surface area of water and provided an additional substratum for vegetation establishment. The general colonisation sequence was reeds establishing on rocky and sandy substrata, followed later by trees, which formed gallery forest. There was also a stochastic pattern of change in the years soon after the 1925 flood, followed by directional change and then stabilisation in the past twenty years. The potential for resetting by floods seems to become less likely until a very large and infrequent disturbance resets the system.

Channel complexity and organic matter retention in the Barwon–Darling River, Australia

In-channel benches act as small floodplains, increasing the habitat complexity of the main channel, providing surfaces for organic matter accumulation, retention and transformation, and promoting invertebrate colonisation and productivity (Sheldon & Thoms, 2006). Organic matter provides complex habitat and is a vital food source for aquatic invertebrates (Prochazka *et al.*, 1991). In humid and temperate streams experiencing regular overbank flows, a large proportion of the allochthonous organic biomass enters from the surrounding floodplain. However, floodplain inundation in some dryland rivers does not occur as frequently (Walker *et al.*, 1997). Thus in-channel features are relatively more important for ecological processes within the channel (Thoms & Sheldon, 1997). In-channel morphological features accumulate and trap significant quantities of organic material. Amounts of coarse particulate organic matter (CPOM) among and between in-channel benches and nearby steeper surfaces were different at two sites on the Barwon-Darling River (Table 3.2; Fig. 3.1). Horizontal surfaces retained more CPOM and FPOM (fine particulate organic matter) than vertical surfaces on all in-channel features except Level 3 (Table 3.2). Interestingly, the sediment composition of both surfaces on Level 3 of the Culpaulin site was predominantly sand, whereas the sediment composition of all other features was dominated by silt and clay. Differences in the capacity of features to trap organic matter may relate to their formation (Thoms & Olley, 2004), surface roughness and or the relation between input of organic matter and the time of deposition. Differences in the amount of CPOM retained on the horizontal surfaces of the range of features may reflect differing proximities to overhanging

Table 3.2. *Mean (SE) mass (g) of the coarse particulate organic matter (CPOM) fraction and the percentage organic content of the fine particulate organic matter (FPOM) fraction for in-channel benches sampled at two sites on the Barwon–Darling River*

All comparisons between vertical and horizontal surfaces were significant (p < 0.01) (Mann–Whitney U test), apart from Calpaulin Level 3, organic content (FPOM).

Site	Feature	Surface	CPOM >2 cm (g)		Organic content (%)	
Culpaulin	Level 1	horizontal	62.47	$U = 100, p <$	11.18	$U = 100, p <$
			(10.49)	0.001	(0.67)	0.001
		vertical	4.23		3.53	
			(1.17)		(0.19)	
	Level 2	horizontal	21.29	$U = 98, p <$	8.38	$U=100, p <$
			(5.69)	0.001	(0.58)	0.001
		vertical	0.57		4.99	
			(0.31)		(0.16)	
	Level 3	horizontal	61.87	$U = 90, p <$	3.68	$U = 58, p >$
			(15.14)	0.001	(0.57)	0.05
		vertical	4.11		2.59	
			(1.26)		(0.28)	
	Level 4	horizontal	11.09		7.28	
			(2.55)		(0.15)	
Studlee	Level 2	horizontal	63.62	$U = 86, p <$	9.79	$U = 90, p <$
			(11.19)	0.001	(0.32)	0.001
		vertical	11.47		6.25	
			(3.95)		(0.15)	

Source: [Data from Sheldon and Thoms (2006)].

riparian vegetation. The greater the complexity of dryland river chan-nels, the greater the surface area available to trap organic matter to provide a food source and a habitat for low-order aquatic organisms such as macroinvertebrates. Conversely, a reduction in the degree of complexity, such as when these systems are subjected to water resources development, may affect the overall integrity of these ecosystems.

CONCLUSIONS

Large rivers have been considered less complex than small streams with fewer retentive structures (Naiman *et al.*, 1987; Webster *et al.*,

1994). This ignores the role of large-scale in-channel complexity in large dryland rivers. The varying geomorphic surfaces along river valleys create physical patterns, reflected in the development of riparian plant communities, the distributions of aquatic biota and the retention of organic matter (Gregory *et al.*, 1991). Dryland streams with high channel irregularity contain habitats just as complex as upland gravel-bed streams. Vegetation patterns vary with habitat, as shown in the Sabie and Olifants Rivers (Fig. 3.3), South Africa, and these irregular channels have the capacity to retain more organic material than low-complexity streams (Prochazka *et al.*, 1991) as demonstrated for the Barwon–Darling River (Fig. 3.3), Australia.

Physical heterogeneity in dryland rivers is important for ecosystem health. Variable channel morphologies play an important role in vegetation diversity and in energy transfers within river systems. In terms of aquatic processes, litter accumulates on the flat surfaces of the in-channel features during low-flow periods. As flow levels rise, the accumulated organic matter is inundated, becoming available to aquatic organisms for processing and forming part of the food chain. The in-channel features are functionally similar to wetlands, although inundated more frequently. The greater the complexity of the channel, the greater the surface area available for organic matter, which provides a food source and a habitat for the low-order aquatic organisms such as macroinvertebrates. At a larger scale, in-channel features are important surfaces for the establishment of vegetation, and the spatial arrangement of vegetation can, in part reflect the physical heterogeneity of the physical channel environment. Hence, reductions in the degree of physical complexity could have repercussions for overall ecosystem health.

Water resources development is one factor that changes the character of riverine ecosystems. To assess the environmental flow requirements for a developed dryland river, it is essential to understand how the physical environment affects aquatic ecosystems. Dryland river ecosystems are the result of complex interacting physical, chemical and biological factors. The integrity of river ecosystems relies on this complexity: a factor often overlooked for a focus on 'flows'. Given the uniqueness of dryland river ecosystems, there is a need for specialist policies for the management of these systems (Thoms & Cullen, 1998).

The formulation of policies based on our present knowledge of these ecosystems may seem premature, especially because we cannot easily extrapolate from experiences gained in other river systems. The management of dryland river ecosystems should focus on managing variability. Historically, dryland rivers have been managed with a single priority: human demand for water (Biswas, 1996) with disastrous ecological consequences (see, for example, Davies *et al.*, 1994). Four main problems face scientists and managers concerned with management of river heterogeneity (Rogers & O'Keefe, 2003): (1) an interdisciplinary understanding, so that there can be sensible interactions among scientists and managers; (2) a hierarchical vision and set of objectives for river management, tested against public opinion; (3) a proposed flow regime that satisfies the objectives; and (4) the capacity to predict and monitor the response of river biodiversity to the proposed flow regime, allowing auditing to ensure management objectives are met.

Most river science once followed a unidisciplinary approach, often focussing on managing single species. Interdisciplinary studies are essential for effective biodiversity management at the ecosystem level (see, for example, Poff *et al.*, 1997; Richter *et al.*, 1997). For example, the new discipline ecogeomorphology (Thoms & Parsons, 2002) provides an interdisciplinary framework that can bring about fresh solutions to environmental problems in dryland rivers. It recognises the interactions between hydrological, geomorphological and ecological components of rivers at multiple spatial and temporal scales. For dryland rivers, high process variability (flow and sediment regime) promotes physical heterogeneity (morphological complexity), considerably influencing biological factors. Identification of the key spatial and temporal scales of this interaction is essential for effective management of dryland rivers. Few environmental management strategies adequately identify the part(s) of the river system that can be or need to be managed, and many fewer invest in the required scientific knowledge at the appropriate scale. Management — the interface between science and policy — is turbulent but potentially exciting.

REFERENCES

Alexander, W. J. R. (1985). Hydrology of low latitude Southern Hemisphere landmasses. *Hydrobiologia*, **125**, 75–84.
Andrews, E. D. (1980). Effective and bankfull discharge of streams in the Yampa River Basin, Colorado and Wyoming. *Journal of Hydrology*, **46**, 311–30.
Baker, V. R. (1977). Stream-channel response to floods, with examples from central Texas. *Geological Society of America Bulletin*, **88**, 1057–77.
Beyer, P. J. (1998). Spatial variability in stability thresholds, Verde River, Arizona. *Middle States Geographer*, **31**, 1–14.

Biswas, S. P. (1996). Global water scarcity: issues and implications with special reference to India. *Verhandlungen der Internationalen Vereinigung für theoretische und angewandte Limnologie*, **26**, 115–21.

Bunn, S. E., Davies, P. M. and Winning, M. (2003). Sources of carbon supporting the food web of an arid zone floodplain river. *Freshwater Biology*, **48**, 619–35.

Carter, A. J. and Rogers, K. H. (1989). *Phragmites* reedbeds in the Kruger National Park: The complexity of change in riverbed state. *Proceedings of the Fourth South African National Hydrological Symposium*, pp. 339–40.

Church, M. (1992). Channel morphology and typology. In *Rivers Handbook*, ed. P. Calow and G. E. Petts, pp. 126–43. London: Blackwell Scientific.

Collier, A. L. van (1993). *Riparian vegetation of the Sabie River: relating spatial distribution patterns to characteristics of the physical Environment*. M.Sc. thesis, University of the Witswatersrand, Johannesburg, South Africa.

Davies, B. R., Thoms, M. C., Walker, K. F., O'Keefe, J. H. and Gore, J. A. (1994). Dryland rivers: their ecology, conservation and management. In *Rivers Handbook*, ed. P. Calow and G. E. Petts, pp. 484–511. London: Blackwell Scientific.

Finlayson, B. L. and McMahon, T. A. (1988). Australia vs the World: A comparative analysis of stream flow characteristics. In *Fluvial Geomorphology of Australia*, ed. R. F. Warner, pp. 17–40. Sydney: Academic Press.

Foster, J. M., Thoms, M. C. and Parsons, M. E. (2002). Multivariate statistical techniques and the interpretation of floodplain sediments. *International Association of Hydrological Sciences*, **276**, 451–62.

Graf, W. L. (1985). Magnitude and frequency effects. In *The Encyclopaedic Dictionary of Physical Geography*, ed. A. Goudie, pp. 272–4. London: Blackwell.

(1987). *Fluvial Processes in Dryland Rivers*. Berlin: Springer-Verlag.

Gregory, S. V., Swanson, F. J., McKee, W. A. and Cummins, K. W. (1991). An ecosystem perspective of riparian zones. *BioScience*, **41**, 540–51.

Heritage, G. L., van Niekerk, A. W. and Moon, B. P. (1997). The geomorphological response to changing flow regimes of the Sabie and Letaba River systems. *Water Research Commission Report* No. 376/1/97. Pretoria.

Holeman, J. N. (1968). The sediment yield of major rivers of the world. *Water Resources Research*, **4**, 737–47.

Kingsford, R. T., Jenkins, K. M. and Porter, J. L. (2004). Imposed hydrological stability imposed on lakes in arid Australia and effect on waterbirds. *Ecology*, **85**, 2478–92.

Knighton, A. D. and Nanson, G. C. (2000). Waterhole form and process in the anastomosing channel system of Cooper Creek, Australia. *Geomorphology*, **35**, 101–17.

Langbein, W. B. and Schumm, S. A. (1958). Yield of sediment in relation to mean annual precipitation. *Transactions of the American Geophysical Union*, **39**, 1076–84.

McMahon, T. A. (1979). Hydrological characteristics of arid zones. The hydrology of areas of low precipitation. *International Association of Hydrological Sciences*, **128**, 105–23.

Middleton, N. J. and Thomas, D. S. G. (1997). *World Atlas of Desertification* (2nd edn). London: United Nations Environment Programme, and Edward Arnold.

Moon, B. P., van Niekerk, A. W., Heritage, G. L., Rogers, K. H. and James, C. S. (1997). A geomorphological approach to the ecological management of rivers in the Kruger National Park. *Transactions of the Institute of British Geographes*, **22**, 31–48.

Naiman, R. J., Melillo, J. M., Lock, M. A., Ford, T. E. and Reice, S. R. (1987). Longitudinal patterns of ecosystem processes and community structure in a subarctic river continuum. *Ecology*, **68**, 1139–56.

Nanson, G. C. and Huang, H. Q. (1999). Anabranching rivers: divided efficiency leading to fluvial diversity. In *Varieties of Fluvial Form*, ed. A. J. Miller and A. Gupta, pp. 477–94. Chichester: John Wiley and Sons.

Nanson, G. C., Tooth, S. and Knighton, A. D. (2002). A global perspective on dryland rivers: perceptions, misconceptions and distinctions. In *Dryland Rivers: Hydrology and Geomorphology of Semi-arid Channels*, ed. L. J. Bull and M. J. Kirkby, pp. 17–54. Chichester: John Wiley and Sons.

Neff, E. L. (1967). Discharge frequency compared to long term sediment yields. *International Association of Hydrological Sciences*, **75**, 236–42.

Petts, G. E. (1984). *Impounded Rivers: Perspectives for Ecological Management*. Chichester: John Wiley and Sons.

Poff, L. J., Allan, J. D. and Bain, M. B. (1997). The natural flow regime: a paradigm for river conservation and restoration. *BioScience*, **47**, 769–84.

Prochazka, K., Stewart, B. A. and Davies, B. R. (1991). Leaf litter retention and its implications for shredder distribution in two headwater streams. *Archive für Hydrobiologie*, **120**, 315–25.

Puckridge, J. T., Sheldon, F., Walker, K. F. and Boulton, A. (1998). Flow variability and the ecology of large rivers. *Marine and Freshwater Research*, **49**, 55–72.

Reid, I. and Frostick, L. E. (1997). Channel form, flows and sediment in deserts. In *Arid Zone Geomorphology: Process, Form and Change in Drylands*, ed. D. S. G. Thomas, pp. 205–99. Chichester: John Wiley and Sons.

Richter, B. D., Baumgartner, J. V., Wigington, R. and Braum, D. P. (1997). How much water does a river need? *Freshwater Biology*, **37**, 231–49.

Riley, S. J. (1975). *The development of distributary channel systems with special reference to channel morphology*. Ph.D. Thesis, University of Sydney.

Riley, S. J. and Taylor, G. (1978). The geomorphology of the Upper Darling River system with special reference to the present fluvial system. *Proceedings of the Royal Society of Victoria*, **90**, 89–102.

Rogers, K. H. and O'Keefe, J. (2003). River heterogeneity: ecosystem structure, function and management. In *The Rivers of Kruger National Park*, ed. R. Du Toit, K. H. Rogers and B. Biggs, pp. 189–218. Washington, DC: Island Press.

Rountree, M. W., Rogers, K. H. and Heritage, G. L. (2000). Landscape state change in the semi-arid Sabie River, Kruger National Park, in response to flood and drought. *South African Geographical Journal*, **82**, 173–81.

Schick, A. P. (1993). Geomorphology in Israel. In *The Evolution of Geomorphology*, ed. H. J. Walker and W. E. Grabau, pp. 231–7. Chichester: John Wiley and Sons.

Schumm, S. A. (1977). *The Fluvial System*. New York: Wiley.

Sheldon, F. and Thoms, M. C. (2006). Geomorphic in-channel complexity: the key to organic matter dynamics in large dryland rivers? *Geomorphology*, (in press).

Sheldon, F., Boulton, A. J. and Puckridge, J. T. (2002). Conservation value of variable connectivity: aquatic invertebrate assemblages of channel and floodplain habitats of a central Australian arid zone river, Cooper Creek. *Biological Conservation*, **103**, 13–31.

Sims, N. C. and Thoms, M. C. (2002). What happens when floodplains wet themselves: Vegetation response to inundation on the Lower Balonne Floodplain. *International Association of Hydrological Sciences*, **276**, 195–202.

74 M. C. Thoms *et al.*

Southwood, T. R. E. (1977). Habitat, the templet for ecological strategies? *Journal of Animal Ecology*, **46**, 337–65.

(1988). Tactics, strategies and templets. *Oikos*, **52**, 3–18.

Stevens, M. A., Simons, D. B. and Richardson, E. V. (1975). Non-equilibrium river form. *Journal of the Hydraulics Division, American Society of Civil Engineers*, **101**, 557–66.

Thomas, D. S. G. (ed.) (1997). *Arid Zone Geomorphology: Process, Form and Change in Drylands*. London: Belhaven Press.

Thoms, M. C. (2003). Floodplain-river ecosystems: lateral connections and the implications of human interference. *Geomorphology*, **56**, 335–50.

Thoms, M. C. and Cullen, P. (1998). The impact of irrigation withdrawals on inland river systems. *Rangelands Journal*, **20**, 226–36.

Thoms, M. C. and Olley, J. M. (2004). The stratigraphy, mode of deposition and age of inset floodplains on the Barwon Darling River, Australia. *International Association of Hydrological Sciences*, **228**, 316–25.

Thoms, M. C. and Parsons, M. E. (2002). Ecogeomorphology: an interdisciplinary approach to river science. *International Association of Hydrological Sciences*, **276**, 113–20.

Thoms, M. C. and Sheldon, F. (1997). River channel complexity and ecosystem processes: the Barwon–Darling River, Australia. In *Frontiers in Ecology: Building the Links*, ed. N. Klomp and I. Lunt, pp. 193–206. Oxford: Elsevier.

(2000a). Lowland rivers: an Australian introduction. *Regulated Rivers: Research and Management*, **16**, 375–83.

(2000b). Water resource development and hydrological change in a large dryland river: the Barwon-Darling River, Australia. *Journal of Hydrology*, **228**, 10–21.

Thoms, M. C. and Walker, K. F. (1992). Channel changes related to low-level weirs on the River Murray, South Australia. In *Lowland Floodplain Rivers*, ed. P. A. Carling and G. E. Petts, pp. 235–49. Chichester: John Wiley and Sons.

Thornes, J. B. (1994). Channel processes, evolution and history. In *Geomorphology of Desert Environments*, ed. A. D. Abrahams and A. J. Parsons, pp. 288–317. London: Chapman and Hall.

Tooth, S. (1999). Floodouts in central Australia. In *Varieties of Fluvial Form*, ed. A. J. Miller and A. Gupta, pp. 219–48. Chichester: John Wiley and Sons.

(2000). Process, form and change in dryland rivers: a review of recent research. *Earth-Science Reviews*, **51**, 67–107.

Tooth, S. and Nanson, G. C. (1999). Anabranching rivers on the northern plains of arid central Australia. *Geomorphology*, **29**, 211–33.

(2000). Equilibrium and non-equilibrium conditions in dryland rivers. *Physical Geography*, **21**, 2–29.

Townsend, C. R. and Hildrew, A. G. (1994). Species traits in relation to a habitat templet for river systems. *Freshwater Biology*, **31**, 265–75.

Walker, K. F. and Thoms, M. C. (1993). Environmental effects of flow regulation on the lower River Murray, Australia. *Regulated Rivers: Research and Management*, **8**, 103–19.

Walker, K. F., Puckridge, J. T. and Blanch, S. J. (1997). Irrigation developments on Cooper Creek, central Australia — prospects for a regulated economy in a boom-and-bust ecology. *Aquatic Conservation: Marine and Freshwater Ecosystems*, **7**, 63–73.

Walling, D. E. and Webb, B. W. (1983). Patterns of sediment yield. In *Background to Palaeohydrology: A Global Perspective*, ed. K. J. Gregory, pp. 69–100. Chichester: John Wiley and Sons.

Warner, R. F. (1994). Instability in channels and floodplains in south-east Australia: Natural processes and human activity impacts. *Revue de geographie de Lyon*, **69**, 17–24.

Webster, J. R., Covich, A. P., Tank, J. L. and Crockett, T. V. (1994). Retention of coarse organic particles in streams in the southern Appalachian Mountains. *Journal of the North American Benthological Society*, **13**, 140–50.

Woodyer, K. D. (1968). Bankfull frequency in rivers. *Journal of Hydrology*, **6**, 114–42.

Woodyer, K. D., Taylor, G. and Crook, K. A. W. (1979). Depositional processes along a very low-gradient, suspended-load stream: the Barwon River, New South Wales. *Sedimentary Geology*, **22**, 97–120.

4

Aquatic productivity and food webs of desert river ecosystems

S. E. BUNN, S. R. BALCOMBE, P. M. DAVIES, C. S. FELLOWS

AND F. J. MCKENZIE-SMITH

INTRODUCTION

A fundamental consideration in the study of stream and river ecosystems is the identification of the sources of organic matter that enter the food web and ultimately sustain populations of fish, waterbirds and other aquatic or semi-aquatic vertebrates. Much of our knowledge in this regard has been derived from small temperate forest streams, particularly those in the northern hemisphere. These studies have identified the importance of terrestrial sources of organic carbon and, in particular, highlighted the strong linkages between streams and their riparian zones (Cummins, 1974; Gregory et al., 1991). Terrestrial sources of organic carbon, derived either from upstream processes or in the case of floodplain rivers from lateral exchange during floods, have also been considered to be a major contributor to the food webs of large rivers (Vannote et al., 1980; Junk et al., 1989). However, there is a growing view that these models of ecosystem function have understated the role of autochthonous (i.e. produced within the system) sources in large rivers (Lewis et al., 2001; Thorp & Delong, 2002; Bunn et al., 2003; Winemiller, 2006).

Very little information is, however, available for dryland river systems. This is unfortunate, given that over 40% of the world's land

Ecology of Desert Rivers, ed. R. T. Kingsford. Published by Cambridge University Press.
© Cambridge University Press 2006.

mass is semi-arid and another 25% is arid or hyper-arid (Davies *et al.*, 1994; Middleton & Thomas, 1997), with many dryland rivers (Chapter 1, this volume). In Australia, over 90% of the 3.5 million kilometres of river channels (measured at the 1 : 250 000 scale) are lowland rivers and most of these are characterised as dryland systems (Thoms & Sheldon, 2000). The sparse vegetation of dryland catchments and riparian zones un-doubtedly influences the quantity and quality of terrestrial inputs to rivers, as will the unpredictable and highly variable nature of their flow regimes (Puckridge *et al.*, 1998; Chapter 2, this volume). The character-istic flow extremes of desert rivers are also considered to be the major drivers of "boom and bust" cycles of productivity, especially in systems with extensive floodplains and associated wetland systems (Walker *et al.*, 1995; Kingsford *et al.*, 1999). The high turbidity of some dryland river systems also has a marked influence on the distribution and productivity of algae and other aquatic plants (Bunn *et al.*, 2003).

In this chapter, we review available information on the sources and fate of organic carbon in arid and semi-arid zone streams and rivers, from Australia and overseas. Much of the overseas information comes from the cool and warm deserts of the western USA, with some from dryland rivers in Africa. Our aim is to identify the important sources of organic carbon that ultimately support aquatic food webs in dryland rivers and to highlight the anthropogenic factors that may disrupt important processes and lead to a decline in ecosystem health.

IN-STREAM PRIMARY PRODUCTION

In small forest stream ecosystems, in-stream primary production is often limited by shading from the dense riparian canopy (Feminella *et al.*, 1989; Boston & Hill, 1991) and contributes little to the stream food web. In sparsely vegetated biomes, direct riparian regulation of in-stream primary production is often markedly reduced and algae can provide an important source of organic carbon for consumers (Minshall, 1978; Finlay, 2001). Shading from the steep walls of narrow canyons or gorges may, however, have a similar effect in regulating in-stream production in some arid rivers (see, for example, Fig. 4.1a).

Arid-zone streams and rivers are much more metabolically active than their temperate counterparts, with gross primary production often one to two orders of magnitude greater (Fisher, 1995; Lamberti & Steinman, 1997; see Table 4.1). High rates of benthic respiration are also a feature (Table 4.1) and tend to be associated with autotrophic

Figure 4.1. Desert streams and rivers in Australia. (a) Shading by canyon walls: Standley Chasm, McDonnell Ranges, Northern Territory (Photo S. Bunn). (b) Sparse riparian vegetation and clear water in the Prince Regent River, north-western Australia (Photo R. Stone). (c) Turbid waterhole, Kyabra Creek, Queensland (Photo S. Bunn).

processes (i.e. autorespiration) rather than with the decomposition of terrestrial organic matter typical of many forest streams (Lamberti & Steinman, 1997). High rates of aquatic primary production in desert streams have been attributed to high light intensity, low current velocity, high temperatures and intensive internal recycling of nutrients (see, for example, Busch & Fisher, 1981; Velasco *et al.*, 2003). In these shallow, clear-water streams, aquatic photosynthesis can quickly become light-saturated (Busch & Fisher, 1981). In the absence of light limitation, nitrogen is the most commonly limiting element of streams in the arid and semi-arid south-west of the USA (Grimm *et al.*, 1981). Little additional information is available on nutrient limitation in other river systems, although the relatively high stable nitrogen isotope values of benthic algae recorded in Cooper Creek waterholes suggest

Table 4.1. *Rates of gross primary production (GPP) and respiration (R_{24}) in 14 desert streams and rivers with data for two temperate forest streams included for comparison*

Errors where given are ± 1 SE

River	Comment	Rate (gC m^2 d^{-1})		Source
		GPP	R_{24}	
Deep Creek, Idaho, USA	Cool-desert stream, production dominated by periphyton and macrophytes	3.2	2.67	Minshall (1978)[a]
Rattlesnake Springs, Washington, USA	Cool-desert spring stream	7.4	6.2	Cushing & Wolf (1984)[a]
Mojave Desert, California, USA	Thermal spring stream	3.25	2.56	Naiman (1976)[a]
Pinto Creek, Arizona USA	Desert-pristine	1.86	1.50	Lewis & Gerking (1979)
Sycamore Creek, Arizona USA	Warm desert	2.98	1.78	Busch & Fisher (1981)
Salmon River, Idaho, USA	Fourth-order, semi-arid. Seasonal mean	0.19–0.77	0.18–0.42	Bott et al. (1985)
Vaal River, South Africa	Phytoplankton production (^{14}C, light/dark bottle method). Turbid river, highly perturbed	0.15–2.05	(Nov–Aug)	Pieterse & Roos (1987)
		2.10	P : R = 1.18	Roos & Pieterse (1989)

Site	Description			Reference
White Nile River, Khartoum, Egypt		2.4		Payne (1986)
Chicamo stream, Spain.	Semi-arid stream: *Chara* (5%), epipelic algae (88%), epilithon (7%). (Average over six occasions.)	13.70	6.88	Velasco et al. (2003)
Cooper Creek waterholes (12), Queensland, Australia	Benthic littoral metabolism only	2.02 ± 0.25	1.36 ± 0.18	Bunn et al. (2003)
Warrego River waterholes (15), Queensland, Australia	Benthic littoral metabolism, Oct 2001	0.16 ± 0.02	0.25 ± 0.02	Fellows, C. S. and Bunn, S. E., unpublished data
	Apr 2002	0.14 ± 0.04	0.37 ± 0.09	
Ord River, Western Australia	Regulated sites, seasonal means	0.34 ± 0.04	0.30 ± 0.04	Davies, P. M., unpublished data
Ord River tributaries (3), Australia	Unregulated sites, seasonal means	0.28 ± 0.03	0.35 ± 0.04	Davies, P. M., unpublished data
Robe River, Western Australia	Eight permanent pools, late dry season	0.67 ± 0.11	0.74 ± 0.09	Davies, P. M., unpublished data
Augusta Creek, Michigan, USA	Deciduous forest stream	0.09	0.23	Bott et al. (1985)[b]
Mack Creek, Oregon, USA	Montane coniferous forest	0.10	0.14	Bott et al. (1985)[b]

[a] From Fisher (1995).
[b] From Webster & Meyer (1997).

that there is little N-fixation (Bunn et al., 2003). This is also the case in arid, clear-water systems in north-western Australia in the Pilbara and Kimberley regions (Fig. 4.1b) (P. M. Davies, unpublished data).

In some desert rivers, high turbidity due to fine clays in suspension markedly influences gross primary productivity. For example, in the rivers of western Queensland, Australia (Fig. 4.1c), turbidity remains high in waterholes even during the long periods between flows (up to 24 months) (Bailey, 2001; Bunn et al., 2003). Mean photic zone depth (i.e. 1% ambient light) in 30 waterholes in Cooper Creek and the Warrego River in western Queensland was less than 23 cm (Table 4.2). Few aquatic macrophytes of any kind have been recorded in these waterholes. However, despite this high natural turbidity, permanent river waterholes in Cooper Creek often feature a highly productive 'bath-tub ring' of algae, restricted to the shallow littoral margins (Bunn & Davies, 1999; Bunn et al., 2003; (Fig. 4.2a). Similar littoral bands of benthic algae occur in waterholes in other desert rivers in Australia (Fig. 4.2b). Rates of primary production in this zone are among the highest recorded for streams and rivers in Australia and remain high even during winter (Table 4.1). As would be expected, rates of benthic primary production and respiration below the photic zone are extremely low, although these rivers are typically net producers of organic carbon at the waterhole scale (Bunn et al., 2003). Much of the spatial variation in benthic

Table 4.2. Mean (± 1 SE) light extinction coefficients and mean (± 1 SE), maximum and minimum photic zone depths measured with a Li-Cor quantum sensor in 30 turbid river waterholes in western Queensland, Australia, in 2000–02

	Mean extinction coecient (cm^{-1})	Photic zone depth (cm)		
		mean	maximum	minimum
Cooper Creek waterholes (15)				
April 2001	0.20 (0.02)	26.9 (2.5)	48	10
September 2001	0.23 (0.02)	24.8 (4.2)	75	12
pooled	0.22 (0.02)	25.9 (2.4)		
Warrego River waterholes (15)				
October 2001	0.38 (0.08)	16.2 (2.0)	30	3
April 2002	0.26 (0.03)	22.7 (3.4)	54	8
pooled	0.32 (0.04)	19.4 (2.0)		

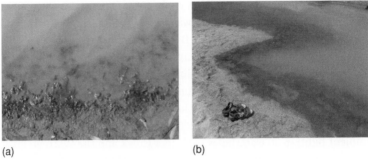

(a)   (b)

Figure 4.2. Bathtub ring' of benthic algae in desert rivers. (a) Yappi
waterhole, Cooper Creek, Queensland; (b) Simpsons Gap, McDonnell
Ranges, Northern Territory, Australia (Photos S. Bunn).

primary production in river waterholes can be explained by variations
in turbidity (C. S. Fellows and S. E., Bunn unpublished data) and this
in turn may be influenced by waterhole morphology, including fetch
length (Davis *et al.*, 2002). In contrast to Cooper Creek, rates of benthic
metabolism in the Warrego River catchment, in the upper Darling Basin,
are relatively low (Table 4.1) despite similarities in climate, turbidity and
nutrient status. Differences in waterhole morphology (steeper slopes
and narrower littoral zone in the Warrego), bio-perturbation by introduced carp *Cyprinus carpio* (absent in the Cooper) or more frequent flow
pulses in the Warrego may contribute to these differences.

Phytoplankton production is also occasionally high in the surface
waters of these turbid systems in Australia during periods of no flow, as
indicated by significant diel variations in dissolved oxygen (Bunn *et al.*,
2003). Rates of water column production, measured by using light and
dark bottle chambers during extended periods of no flow in the same
waterholes, range from 1.5 mgC $l^{-1}d^{-1}$ to 500 mgC $l^{-1}d^{-1}$ (P. M. Davies,
unpublished data). Similarly, high phytoplankton production observed
in the Vaal River in South Africa (Table 4.1) was generally restricted to
the top one metre; the river behaved more like a lentic waterbody in
this regard (Pieterse & Roos, 1987).

In-stream processes in desert streams can show considerable temporal variability in response to flow events, although they typically
recover rapidly after flood or drought (Fisher *et al.*, 1982). For example,
flood disturbances in a Sonoran desert stream decreased algal biomass
and gross primary production, but algal standing stocks returned to

50% of maximum levels within ten days (Grimm, 1987) and gross primary production (GPP) increased to approximately 4.6 gC m^{-2} d^{-1} within 28 days (Jones et al., 1997). Similarly, the flood regime in a semi-arid Spanish stream had little long-term effect on epipelic algae, as the availability of algal propagules and rapid growth rates allowed biomass and production values to return to pre-disturbance levels in less than a month (Velasco et al., 2003).

Flow pulses (i.e. flows confined to the channel) in turbid river systems are likely to have a significant influence on aquatic primary production. Although these events may top up previously isolated water-holes, bring in new nutrients and enhance the connectivity of popula-tions of aquatic biota, increases in depth of only 20 cm can submerge once productive littoral bands of benthic algae below the photic zone (Table 4.2). Flow pulse events lasting days to weeks will affect consumers dependent on algal food resources, especially if benthic algae are unable to track relatively rapid fluctuations in water depth.

FLOODPLAIN PRODUCTIVITY

The high productivity of floodplains favoured the development of an-cient cultures in arid and semi-arid regions, such as those along the Nile and Euphrates (Tockner & Stanford, 2002; Tockner et al., 2006). As in other floodplain systems, the duration of inundation of dryland river floodplains undoubtedly affects decomposition, nutrient cycling and the biomass and productivity of plants and animals (Davies et al., 1994; Chapters 5-7, this volume). Floods in Namibian rivers carry vast quantities of organic matter, which are deposited in the lower reaches and greatly contribute to the productivity of floodplain soils (Jacobson et al., 1995, 2000a,b). However, there is little published information available on aquatic production on inundated floodplains of desert rivers. Vast areas of shallow, warm, nutrient-rich water on floodplains will stimulate high productivity. For example, the lakes area north of the semi-arid Central Delta of the Niger River is known for its abundant phytoplankton blooms (principally the diatom *Melosira*), which can be traced by using satellite imagery (Welcomme, 1986a).

We measured rates of benthic and pelagic metabolism on the inundated floodplain of Cooper Creek in Australia from late February to mid-April 2000. At the height of this flood (return frequency of about 1 : 14 years), nearly 14 000 km^2 of floodplain was inundated (Fig. 4.3). We monitored dissolved oxygen within *in situ* Perspex chambers over 24 h

Figure 4.3. Cooper Creek floodplain, Australia, in March 2000 (Photo R. Ashdown). Approximately 14 000 km^2 was inundated at this time.

(see Bunn *et al.*, 2003). Open-bottomed chambers (diameter 29.5 cm, height 35 cm) were sealed by pushing at least 10 cm into the substrate. Open-water measurements were made with floodplain water enclosed in the same chambers with a plastic base, anchored to a fixed station near the water surface. All chambers had a central port for the polarographic oxygen sensor (YSI 5739, USA) and side ports for a 12 V recirculating pump. Dissolved oxygen and temperature within each chamber were measured electronically over at least 24 h at 10 min intervals and recorded by using a portable data logger (TPS Model 601). These data were converted into units of carbon, assuming a photosynthetic quotient of unity (Lambert, 1984; Bender *et al.*, 1987). After the measurement period, the volume of water enclosed by each chamber was measured *in situ* to determine absolute rates of metabolism.

Rates of benthic and pelagic gross primary production were low in the early phase of floodplain inundation (< 4 days), although there was an initial high rate of benthic respiration (Table 4.3). After 30 days, high rates of benthic metabolism were recorded as the flood waters began to recede. Although rates were not as high as those observed in waterholes during prolonged dry periods (Table 4.1), we estimated that the amount of algal carbon produced on the floodplain during a single

Table 4.3. *Mean (± 1 SE) rates of gross primary production (GPP) and respiration (R_{24}) from the Cooper Creek floodplain, Australia, during a major flood, February–April 2000*

Time since inundation	Benthic metabolism (gC m² d⁻¹)			Pelagic metabolism (gC m² d⁻¹)		
	N	GPP	R_{24}	N	GPP	R_{24}
< 4 days	27	0.015 (0.005)	0.284 (0.038)	7	0.003 (0.009)	0.005 (0.013)
16 days	23	0.036 (0.004)	0.131 (0.014)	8	0.008 (0.024)	0.021 (0.023)
30 days	21	1.366 (0.293)	0.696 (0.141)	8	0.093 (0.056)	0.036 (0.018)

(b)

(a)

Figure 4.4. Algae on floodplains (a) from Cooper Creek, March 2000 flood (Photo R. Ashdown) and (b) grown from dry sediment samples collected from the floodplain and experimentally inundated (Photo S. Hamilton).

day of inundation was equivalent to over 80 years of aquatic production in the permanent waterholes during the dry. Floating algae (mainly *Anabaena*) were observed associated with emergent floodplain plants during the early phases of this flood (Fig. 4.4a). Algal scums also quickly developed when samples of floodplain soils were experimentally inundated in the laboratory (Fig. 4.4b). The presence of algae in floodplain soils and their rapid response to inundation appears to be characteristic of these dryland river systems. The resulting 'boom' in primary production on the floodplain undoubtedly contributes to the proliferation of aquatic invertebrates, especially small crustaceans (Chapter 6, this volume). As flood waters recede, plant growth is stimulated and leads to a substantial increase in above-ground plant biomass (Capon, 2004). Longer flood peaks with slow-moving water on the floodplain result in more water being absorbed by the soil. This leads to a deeper soil moisture profile, a larger area flooded and a longer period in which plants maintain growth (Edmonston, 2001).

TERRESTRIAL SOURCES OF ORGANIC CARBON

The riparian vegetation of desert rivers is often markedly distinct from that of the surrounding catchment (see, for example, Fig. 4.5). Distinctive riparian forests, such as those of the western catchments of Namibia, are often referred to as linear oases (Jacobson *et al.*, 1995). Stream and river channels provide water to support trees and shrubs; in many desert systems, channels are at least partly shaded by overhanging

Figure 4.5. 'Ecological arteries of the landscape': Robe River, north-western Australia (Photo P. Davies).

Table 4.4. *Rates of terrestrial leaf-litter inputs in four desert streams and rivers, compared with two deciduous and one tropical forest stream*

River	Comment	Rate (g m^2 yr^{-1})	Source
Deep Creek, Idaho, USA	Great Basin Desert; sagebrush	2.4[a]	Minshall (1978)
Rattlesnake Springs, Washington, USA	Cold desert, shrub steppe	242	Cushing (1997)
Sycamore Creek, Arizona, USA	Sonoran Desert scrub	16.5	Jones et al. (1997)
Oued Zegzel, Morocco	Semi-arid, temporary stream	59–218	Chergui et al. (1999)
Augusta Creek, Michigan, USA	Deciduous forest	448[a]	Triska et al. (1984)
Mack Creek, Oregon, USA	Montane coniferous forest	730[a]	Cummins et al. (1983)
Rio Icacos, Puerto Rico	Tropical forest	400[a]	McDowell & Ashbury (1994)

[a] From Webster & Meyer (1997).

vegetation. However, others (e.g. the Karoo, a semi-desert vegetation biome in southern Africa) have little canopy cover (Davies et al., 1994).

Substantial variation (44%) in litter fall in stream ecosystems among different biomes is explained by precipitation, with arid lands, tundra and boreal forests having the lowest values (Benfield, 1997). Riparian inputs (leaves and invertebrates) represent a potentially important source of organic carbon, although annual rates are considerably less than those in more temperate or tropical systems (Table 4.4).

In intermittently flowing streams, and on floodplains, terrestrial breakdown of leaf litter may influence the dynamics of organic matter in streams. Microbial enrichment of leaf material may occur during the dry period but does not necessarily enhance decomposition (Herbst & Reice, 1982). Biotic fragmentation by invertebrate shredders is important in temperate streams (see, for example, Irons et al., 1994) but shredder numbers are low or zero in arid-zone streams (see, for example, Davis et al., 1993; Schade & Fisher, 1997; Pomeroy et al., 2000), suggesting little influence on leaf breakdown.

Streams in arid and semi-arid regions also typically have low levels of organic matter storage (fine and coarse benthic organic matter and wood) compared with temperate systems (Jones, 1997). The lack of wood and debris dams is a feature of many desert rivers in southern Africa (Davies *et al.*, 1995). Wood loads in Cooper Creek in western Queensland are also low relative to others in Australia, reflecting sparse riparian tree cover (Marsh *et al.*, 2001). Riparian vegetation along dryland river systems is often structured and maintained by flooding (Stromberg *et al.*, 1991; Jacobson *et al.*, 1995; Pettit *et al.*, 2001; Stromberg, 2001; Capon, 2004; Chapter 5, this volume). Massive episodic floods in some dryland rivers have a long-lasting impact on riparian zones and can demolish whole reaches of riparian forest (Jacobson *et al.*, 1995). In turn, this can influence terrestrial inputs (leaves and invertebrates) as well as the supply of wood to the channel.

FOOD WEBS IN DESERT STREAMS AND RIVERS

Riparian vegetation inputs are important in mesic systems, contributing up to 99% of the organic carbon available in the food web (Pomeroy *et al.*, 2000). In contrast, algal biomass and primary production contributed 99% of the total organic input to Sycamore Creek in the Sonoran Desert (Jones *et al.*, 1997). Even in a cold desert stream, most organic matter in transport is autochthonous in origin (Minshall, 1978). Perhaps not surprisingly, allochthonous inputs may not be such an important source of carbon for consumers in arid stream ecosystems (see, for example, Grimm, 1987; Jones *et al.*, 1997; Vidal-Abarca *et al.*, 2001; Bunn *et al.*, 2003).

There are several reasons why terrestrial inputs may not be important in arid stream ecosystems. Most of the sites studied have had little or no riparian vegetation (see, for example, Schade & Fisher, 1997; Velasco *et al.*, 2003). Extreme flooding can significantly reduce storage of leaf litter and its availability to consumers (Schade & Fisher, 1997; Vidal-Abarca *et al*, 2001). Furthermore, riparian species in arid zones tend to produce litter with relatively low nutritional quality (see, for example, Francis & Sheldon, 2002) and may make the leaves unpalatable to invertebrates. Shredder densities in arid and semi-arid stream systems are typically low (Ward *et al.*, 1986; Davies *et al.*, 1994; Martinez *et al.*, 1998) and leaching, microbial respiration and physical breakdown are likely to be the most important processing agents of coarse organic matter. Perhaps not surprisingly, macroinvertebrate abundance and biomass can be significantly correlated (positively) with chlorophyll *a* (algae) rather than with leaf litter (Schade & Fisher, 1997).

Few studies have been undertaken on the diets of fish in arid or semi-arid river systems. Dryland river fish communities appear to be less diverse than their temperate counterparts and show few examples of specialised feeding niches (Skelton, 1986; Welcomme, 1986b; Chapter 7, this volume; Balcombe *et al.*, 2006). Fish of the Orange River have a broad spectrum of feeding habits and most would be considered to be omnivorous (Skelton, 1986). The cyprinid species *Oreoleuciscus humilis* inhabits small desert rivers in closed desert watersheds of Mongolia and feeds mainly on insect larvae and on plants (Dgebuadze, 1995). Most species of fish in the Niger River show marked feeding patterns associated with flooding; feeding is either reduced or suspended during the dry season (Welcomme, 1986b). Exceptions are zooplanktivorous fish, which feed during slack water when their food is concentrated. There is a stepped growth of some fish in the Central Delta of the Niger associated with annual flooding, and interannual variation in growth is associated with flood intensity and duration (Welcomme, 1986b). As in many other floodplain river systems, fish catches in the Niger at the reach scale are a function of floodplain area (Welcomme, 1986b).

The diets of ten species of fish from isolated river waterholes in the Cooper Creek system in arid Australia were also found to be simple (Balcombe *et al.*, 2006). Zooplankton (mostly calanoid copepods) was a major component (> 50%) of the diet of all but one species during this no-flow period. Rainbow fish *Melanotaenia splendida* was the notable exception, with a relatively high terrestrial contribution to the diet (average of 80%). In contrast, at the beginning of a large flood in March 2000, seven fish species had broad diets, feeding on a variety of aquatic and terrestrial sources. However, late in this flood, most species fed only on aquatic resources (< 3% terrestrial). Again, rainbow fish was the notable exception and fed mostly on terrestrial insects, in both the early flood (58%) and the late flood (31%) (Balcombe *et al.*, 2006).

Stable isotope analysis has confirmed that benthic algal sources of carbon are the major source of energy supporting large populations of snails, crustaceans and fish in Cooper Creek (Bunn & Davies, 1999; Bunn *et al.*, 2003;) (Fig. 4.6). Spatial and temporal variation in the stable carbon and nitrogen isotope signatures of consumers suggested that phytoplankton and/or zooplankton was the other likely major source. However, with the exception of juvenile bony bream *Nematalosa erebi*, no

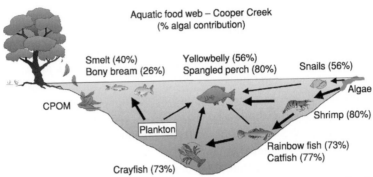

Figure 4.6. Food web structure in turbid river waterholes, Cooper Creek, Australia, based on stable isotope data (modified from Bunn & Davies, 1999). Percentage of biomass of consumers derived from benthic algae is given in parentheses (from Bunn *et al.*, 2003).

species of fish had a stable isotope signature indicative of a substantive contribution from a phytoplankton or zooplankton source. Similarly, in the Ord River in north-western Australia, stable isotope analyses showed that algal material made up the majority (> 50%) of the biomass carbon of native fish (P. M. Davies, unpublished data). The incorporation of algae into consumers increased during the wet season, corresponding with a reach-scale elevation in aquatic primary production.

Although ecosystem models of large rivers emphasise the importance of longitudinal or lateral inputs of terrestrial organic matter as a source of organic carbon for aquatic consumers (see, for example, Vannote *et al.*, 1980; Junk *et al.*, 1989), stable isotope data suggest that this is unlikely in desert river food webs. This is despite extensive floodplains fed by a vast network of anastomosing channels and distributaries that provide a far greater terrestrial–water interface than would occur with a single river channel (Walker *et al.*, 1995). Only chironomid larvae collected from benthic leaf packs in Cooper Creek showed evidence of a terrestrial carbon diet, although the extremely low carbon isotope value ($\delta^{13}C = -54.7‰$) suggests that this is derived via methanotrophic bacteria (Bunn *et al.*, 2003). This cannot be a major microbial pathway, however, because no high-order consumers showed evidence of ^{13}C-depletion. Similar stable isotope studies of large floodplain systems in tropical and temperate environments suggest that the dependence of aquatic food webs on algal carbon may be a feature of many large rivers (Lewis *et al.*, 2001; Thorp & Delong, 2002; Winemiller, 2006).

AQUATIC SUBSIDIES OF RIPARIAN FOOD WEBS

Desert streams show some of the highest rates of secondary production recorded for lotic systems (Jackson & Fisher, 1986; Gaines, 1987), attributed to the ample supply of food and high turnover of small, multivoltine fauna (Fisher, 1995). High secondary production of insects in desert streams may contribute substantially to the food supply of insectivores, including birds, spiders and reptiles (see, for example, Jackson & Fisher, 1986; Lynch *et al.*, 2002; Sabo & Power, 2002). For example, riparian spiders along a Sonoran Desert stream obtained most of their biomass carbon and a significant proportion of their nitrogen from in-stream sources (Sanzone *et al.*, 2003). The high abundance and diversity of spiders in this riparian zone was also attributed to aquatic subsidies of emergent insects. In such productive desert streams, the net flux of energy and nutrients is likely to be from the stream to the riparian zone, rather than in the reverse direction (Martí *et al.*, 2000).

Aquatic subsidies may extend beyond the biota of riparian zones in desert river systems. For example, bald eagles *Haliaeetus leucocephalus* in Arizona foraged primarily near shore in shallow river waters and most of their prey items (76%) were fish (Grubb, 1995). Australian waterbirds use floodplain wetlands flexibly in semi-arid and arid areas of Australia, shifting their distribution and abundance to productive habitat and breeding when flooding triggers sufficient food production (Kingsford *et al.*, 1999; Dorfman & Kingsford, 2001; Roshier *et al.*, 2002). The response of floodplain pastures to flooding is also a significant aquatic subsidy that underpins the viability of the pastoral industry in many dryland river catchments (Brock, 1999; Kingsford, 1999).

THREATS TO ECOSYSTEM PROCESSES IN DESERT RIVERS

As with most floodplain river systems of the world, water resource development undoubtedly poses the most significant threat to ecosystem processes in dryland river systems (Kingsford, 2000; Tockner & Stanford, 2002; Chapters 8 and 9, this volume). For example, river regulation and deliberate draining of wetlands has led to complete collapse of the ecosystem complex of the Mesopotamian wetlands in the middle and lower basin of the Tigris and Euphrates rivers and the disappearance of the social, cultural and economic base of the Marsh Arabs (Tockner *et al.*, 2006).

Changing the frequency, duration and areal extent of inundation of flood waters through upstream regulation, water harvesting or levee construction can alter productivity at the landscape scale. Given the vast areas (tens of thousands of kilometres for some desert floodplain rivers) and the relatively high rates of aquatic production compared with terrestrial sources, such impacts will have significant cascading effects on the vast numbers of waterbirds that capitalise on this epi-sodic food resource (Kingsford, 2000; Roshier *et al.*, 2002). Terrestrial fauna also receive significant subsidies from this aquatic production (Chapter 7, this volume); the long-term persistence of populations may well be threatened by reductions caused by flow regulation and water abstraction.

Given the overriding importance of algae in desert river food webs, factors that influence the production and composition of aquatic plants will seriously affect populations of consumers. For example, clearing of streamside vegetation in chaparral habitats in Arizona to enhance streamflow (Ingebo, 1971), together with cattle grazing, flow regulation and water diversion for agriculture, has accounted for large areas of riparian loss in the western USA (Fisher, 1995). This has led to an in-creased tendency for flash flooding and enhanced sediment transport to occur. Both of these factors are likely to reduce algal productivity, through scouring of bed materials (see, for example, Grimm, 1987) or increased turbidity, respectively. The effects of agricultural herbicides on aquatic algae are poorly understood, even though several chemicals (e.g. atrazine) are routinely found in dryland rivers (Fairweather, 1999).

In turbid desert river systems (e.g. Cooper Creek, Australia), factors influencing the distribution and productivity of the 'bathtub ring' of algae (Fig. 4.6) will have a pronounced effect on ecosystem function. For example, rapid drawdown of water in river waterholes (e.g. by pumping for irrigation) will expose the shallow band of algae. Littoral algae may be tolerant to desiccation but repeated exposure will limit primary production and reduce the availability of this food re-source to aquatic grazers. Similarly, uncontrolled access of stock and feral animals to the margins of river waterholes can physically disturb the algal zone, affecting aquatic primary production and threatening the food base of snails, crustaceans and fish. Even a moderate level of disturbance significantly lowers algal production and recovery to pre-disturbance levels takes many days in Cooper Creek waterholes (P. M. Davies and S. E. Bunn unpublished data).

Salinisation, either associated with changes in catchment vegetation or from irrigation, also significantly threatens some desert river systems (Tockner et al., 2006; Chapter 10, this volume). Although salinity is often a natural feature in these systems, increased salinity can markedly affect turbidity (through flocculation of fine particles) and affect the composition and production of aquatic plants. High salinity can also prevent bacterial and fungal growth on leaf detritus and decrease decomposition rates (Reice & Herbst, 1982).

Invasive species, both plant and animal, can also affect aquatic ecosystem processes in desert rivers. For example, introduction of riparian tamarisk trees (Tamarix spp.) along streams of the American southwest has led to the narrowing of active channels and an increased incidence of overbank flooding (Graf, 1978). Introduced carp may also affect benthic algal production in Australian dryland rivers, either through bio-perturbation of the littoral zone or through increased turbidity associated with feeding activity (King et al., 1997).

Desert rivers truly represent the ecological arteries of dryland landscapes, a significant proportion of the earth's surface. They are characterised by high productivity, an episodic 'boom and bust' nature and their capacity to exert an enormous influence on the biota of associated riparian and floodplain ecosystems. Competition for water, especially for agriculture, and other anthropogenic disturbances, are likely to disrupt the key ecosystem processes that sustain aquatic and terrestrial biota. Water resource managers need to have an improved recognition and understanding of these processes to ensure that the health of dryland rivers and their associated floodplain ecosystems is protected and, if necessary, restored.

ACKNOWLEDGEMENTS

We acknowledge funding support from the Cooperative Research Centre for Freshwater Ecology, Land and Water Australia (National Riparian Lands Program), Environment Australia (Environmental Flows Initiative) and the Water and Rivers Commission of Western Australia for our research on Australian dryland rivers. We also thank the many supportive landholders in the Lake Eyre and Murray–Darling Basins for their hospitality and friendship. Stephen Hamilton and Richard Kingsford are thanked for their constructive comments on the manuscript.

REFERENCES

Bailey, V. (2001). *Western Streams Water Quality Monitoring Project.* (ISBN 0 7345 1798 X.) Brisbane, Australia: Queensland Department of Natural Resources and Mines.

Balcombe, S. R., Bunn, S. E., McKenzie-Smith, F. J. and Davies, P. M. (2006). Variability of fish diets between dry and flood periods in an arid zone floodplain river. *Journal of Fish Biology* (in press).

Bender, M., Grande, K., Johnson, K., *et al.* (1987). A comparison of four methods for determining planktonic community production. *Limnology and Oceanography,* **32**, 1085–98.

Benfield, E. F. (1997). Comparison of litter fall input to streams. *Journal of the North American Benthological Society,* **16**, 104–8.

Boston, H. L. and Hill, W. R. (1991). Photosynthesis–light relations of stream periphyton communities. *Limnology and Oceanography,* **36**, 644–6.

Bott, T. L., Brock, J. T., Dunn, C. S., *et al.* (1985). Benthic community metabolism in four temperate stream systems: an interbiome comparison and evaluation of the river continuum concept. *Hydrobiologia,* **123**, 3–45.

Brock, M. A. (1999). Are aquatic plant seed banks resilient to water regime alteration? Implications for the Paroo river system. In *Free-flowing River: the Ecology of the Paroo River,* ed. R. T. Kingsford, pp. 129–37. Sydney: New South Wales National Parks and Wildlife Service.

Bunn, S. E. and Davies, P. M. (1999). Aquatic food webs in turbid, arid zone rivers: Preliminary data from Cooper Creek, Queensland. In *Free-flowing River: the Ecology of the Paroo River,* ed. R. T. Kingsford, pp. 67–76. Sydney: New South Wales National Parks and Wildlife Service.

Bunn, S. E., Davies, P. M. and Winning, M. (2003). Sources of organic carbon supporting the food web of an arid zone floodplain river. *Freshwater Biology,* **48**, 619–35.

Busch, D. E. and Fisher, S. G. (1981). Metabolism of a desert stream. *Freshwater Biology,* **11**, 301–7.

Capon, S. (2004). *Flow variability and vegetation dynamics in a large arid floodplain.* Ph.D. Thesis. Faculty of Environmental Sciences, Griffith University, Brisbane, Australia.

Chergui, H., Pattee, E., Essafi, K. and Mhamdi, M. A. (1999). Moroccan limnology. In *Limnology in Developing Countries, Vol. 2,* ed. R. G. Wetzel and B. Gopal, pp. 235–330. New Delhi, India: International Association for Theoretical and Applied Limnology, International Scientific Publications.

Cummins, K. W. (1974). Structure and function of stream ecosystems. *BioScience,* **24**, 631–641.

Cummins, K. W., Sedell, J. R., Swanson, F. J., Minshall, G. W., Fisher, S. G., Cushing, C. E., Petersen, R. C. and Vannote, R. L. (1983). Organic matter budgets for stream ecosystems: problems and their evaluation. In *Stream Ecology,* eds. J. R. Barnes and G. W. Minshall, pp. 299–353. New York: Plenum Press.

Cushing, C. E. (1997). Organic matter dynamics in Rattlesnake Springs, Washington, USA. *Journal of the North American Benthological Society,* **16**, 39–43.

Cushing, C. E. and Wolf, E. G. (1984). Primary production in Rattlesnake Springs, a cold desert spring-stream. *Hydrobiologia*, **114**, 229–236.

Davies, B. R., O'Keefe, J. H. and Snaddon, C. D. (1995). River and stream ecosystems in southern Africa: predictably unpredictable. In *River and Stream Ecosystems*, eds. C. E. Cushing and K. W. Cummins, pp. 537–600. The Netherlands: Elsevier Science BV.

Davies, B. R., Thoms, M. C., Walker, K. F., O'Keefe, J. H. and Gore, J. A. (1994). Dryland rivers: their ecology, conservation and management. In *The Rivers Handbook, Vol. 2*, eds. P. Calow and G. E. Petts, pp. 484–511. Oxford: Blackwell Scientific.

Davis, J. A., Harrington, S. A. and Friend, J. A. (1993). Invertebrate communities of relict streams in the arid zone: the George Gill Range, Central Australia. *Australian Journal of Marine and Freshwater Research*, **44**, 483–505.

Davis, L., Thoms, M., Fellows, C. and Bunn, S. (2002). Physical and ecological associations in dryland refugia: waterholes of the Cooper Creek, Australia. In *The Structure, Function and Management Implications of Fluvial Sedimentary Systems*. Proceedings of an international symposium, Alice Springs, Australia, September 2002. IAHS Publication No. **276**, 77–84.

Dgebuadze, Y. Y. (1995). The land inland-water ecotones and fish population of Lake Valley (West Mongolia). *Hydrobiologia*, **303**, 235–245.

Dorfman, E. J. and Kingsford, R. T. (2001). Scale-dependent patterns of abundance and habitat use by cormorants in arid Australia and the importance of nomadism. *Journal of Arid Environments*, **49**, 677–694.

Edmonston, V. (2001). *Managing the channel country sustainably*. Brisbane, Australia: Queensland Department of Primary Industries.

Fairweather, P. G. (1999). Pesticide contamination and irrigation schemes: What have we learnt so far? In *Free-flowing river: the ecology of the Paroo River*, ed. R. T. Kingsford, pp. 223–232. Sydney: New South Wales National Parks and Wildlife Service.

Feminella, J. W., Power, M. E. and Resh, V. H. (1989). Periphyton responses to invertebrate grazing and riparian canopy in three northern California coastal streams. *Freshwater Biology*, **22**, 445–457.

Finlay, J. C. (2001). Stable-carbon-isotope ratios of river biota: implications for energy flow in lotic food webs. *Ecology*, **82**, 1052–1064.

Fisher, S. G. (1995). Stream ecosystems of the western United States. In *River and Stream Ecosystems*, eds. C. E. Cushing and K. W. Cummins, pp. 61–88. The Netherlands: Elsevier Science BV.

Fisher, S. G., Gray, L. J., Grimm, N. B. and Busch, D. E. (1982). Temporal succession in a desert stream ecosystem following flash flooding. *Ecological Monographs*, **52**, 93–110.

Francis, C. and Sheldon, F. (2002). River Red Gum (*Eucalyptus camaldulensis* Dehnh.) organic matter as a carbon source in the lower Darling River, Australia. *Hydrobiologia*, **481**, 113–124.

Gaines, W. L. (1987). *Secondary production of benthic insects in three cold-desert streams*. Pacific Northwest Laboratory Publication, PNB 6286, Richland, Washington, USA. 39pp. (cited in Fisher, 1995).

Graf, W. F. (1978). Fluvial adjustments to the spread of tamarisk in the Colorado Plateau region. *Bulletin of the Geological Society of America*, **89**, 1491–1501.

Gregory, S. V., Swanson, F. J., McKee, W. A. and Cummins, K. W. (1991). An ecosystem perspective of riparian zones. *Bioscience*, **41**, 540–551.

Grimm, N. B. (1987). Nitrogen dynamics during succession in a desert stream. *Ecology*, **68**, 1157–1170.

Grimm, N. B., Fisher, S. G. and Minckley, W. L. (1981). Nitrogen and phosphorus dynamics in hot desert streams of Southwestern USA. *Hydrobiologia*, **83**, 303–312.

Grubb, T. G. (1995). Food-habits of bald eagles breeding in the Arizona desert. *Wilson Bulletin*, **107**, 258–274.

Herbst, G. and Reice, S. R. (1982). Comparative leaf litter decomposition in temporary and permanent streams in semi-arid regions of Israel. *Journal of Arid Environments*, **5**, 305–318.

Ingebo, P. A. (1971). Suppression of channelside chaparral cover increases streamflow. *Journal of Soil and Water Conservation*, **26**, 79–81.

Irons, J. G. I., Oswood, M. W., Stout, R. J. and Pringle, C. M. (1994). Latitudinal patterns in leaf litter breakdown: is temperature really important? *Freshwater Biology*, **32**, 401–411.

Jackson, J. K. and Fisher, S. G. (1986). Secondary production, emergence, and export of aquatic insects of a Sonoran Desert stream. *Ecology*, **67**, 629–638.

Jacobson, P. J., Jacobson, K. M. and Seely, M. K. (1995). *Ephemeral rivers and their catchments: sustaining people and development in Western Namibia.* Windhoek: Desert Research Foundation of Namibia.

Jacobson, P. J., Jacobson, K. M., Angermeier, P. L. and Cherry, D. S. (2000a). Variation in material transport and water chemistry along a large ephemeral river in the Namib Desert. *Freshwater Biology*, **44**, 481–491.

Jacobson, P. J., Jacobson, K. M., Angermeier, P. L. and Cherry, D. S. (2000b). Hydrologic influences on soil properties along ephemeral rivers in the Namib Desert. *Journal of Arid Environments*, **45**, 21–34.

Jones, J. B. (1997). Benthic organic matter storage in streams: influence of detrital import and export, retention mechanisms and climate. *Journal of the North American Benthological Society*, **16**, 109–119.

Jones, J. B., Schade, J. D., Fisher, S. G. and Grimm, N. B. (1997). Organic matter dynamics in Sycamore Creek, a desert stream in Arizona, USA. *Journal of the North American Benthological Society*, **16**, 78–82.

Junk, W. J., Bayley, P. B. and Sparks, R. E. (1989). The flood pulse concept in river-floodplain systems. In *Proceedings of the international large river symposium.* ed. D. P. Dodge, *Canadian Special Publications Fisheries and Aquatic Sciences*, **106**, 110–127.

King, A. J., Robertson, A. I., and Healey, M. R. (1997). Experimental manipulations of the biomass of introduced carp (*Cyprinus carpio*) in billabongs.1. Impacts on water-column properties. *Marine and Freshwater Research*, **48**, 435–443.

Kingsford, R. T. (1999). The potential impact of water extraction on the Paroo and Warrego rivers. In *Free-flowing river: the ecology of the Paroo River*, ed. R. T. Kingsford, pp. 257–277. Sydney: New South Wales National Parks and Wildlife Service.

Kingsford, R. T. (2000). Ecological impacts of dams, water diversions and river management on floodplain wetlands in Australia. *Austral Ecology*, **25**, 109–127.

Kingsford, R. T., Curtin, A. L. and Porter, J. L. (1999). Water flows on Cooper Creek determine 'boom' and 'bust' periods for waterbirds. *Biological Conservation*, **88**, 231–248.

Lambert, W. (1984). The measurement of respiration. In *A Manual on Methods for the Assessment of Secondary Productivity in Fresh Waters. Second Edition.* eds. J. A. Downing and F. H. Rigler, pp. 413–468. IBP Handbook 17. Oxford: Blackwell Scientific Publications.

Lamberti, G. A. and Steinman, A. D. (1997). A comparison of primary production in stream ecosystems. *Journal of the North American Benthological Society*, **16**, 95–104.

Lewis, M. A. and Gerking, S. D. (1979). Primary production in a polluted intermittent desert stream. *American Midland Naturalist*, **102**, 172–174.

Lewis, W. M., Hamilton, S. K., Rodríguez, M. A., Saunders, III J. F. and Lasi, M. A. (2001). Foodweb analysis of the Orinoco floodplain based on production estimates and stable isotope data. *Journal of the North American Benthological Society*, **20**, 241–254.

Lynch, R. J., Bunn S. E. and Catterall, C. P. (2002). Adult aquatic insects: potential contributors to riparian food webs in Australia's wet-dry tropics. *Austral Ecology*, **27**, 515–526.

Marsh, N., Rutherfurd, I. and Jerie, K. (2001). Predicting pre-disturbance loading and distribution of large woody debris. In *Proceedings of the Third Australian Stream Management Conference*, eds. I. Rutherfurd, F. Sheldon, G. Brierley and C. Kenyon, pp. 391–397. Brisbane, Australia: Cooperative Research Centre for Catchment Hydrology.

Martí, E., Fisher, S. G., Schade, J. D. and Grimm, N. B. (2000). Flood frequency and stream-riparian linkages in arid lands. In *Stream and groundwaters* eds. J. B. Jones and P. J. Mulholland, pp. 111–136. New York: Academic Press.

Martinez, B., Velasco, J., Suarez, M. L. and Vidal-Abarca, M. R. (1998). Benthic organic matter dynamics in an intermittent stream in South-East Spain. *Archiv für Hydrobiologie*, **141**, 303–320.

McDowell, W. H. and Ashbury, C. E. (1994). Export of carbon, nitrogen, and major ions from three tropical montane watersheds. *Limnology and Oceanography*, **39**, 111–125.

Middleton, N. J. and Thomas, D. S. G. (1997). *World Atlas of Desertification (2nd Edn)*. London: United Nations Environment Programme, Edward Arnold.

Minshall, G. W. (1978). Autotrophy in stream ecosystems. *BioScience*, **28**, 767–771.

Naiman, R. J. (1976). Primary production, standing stock, and export of organic matter in a Mohave Desert thermal stream. *Limnology and Oceanography*, **21**, 60–73.

Payne, A. I. (1986). *The Ecology of Tropical Lakes and Rivers*. New York: Wiley.

Pettit, N. E., Froend, R. H. and Davies, P. M. (2001). Identifying the natural flow regime and the relationship with riparian vegetation for two contrasting Western Australian rivers. *Regulated Rivers Research and Management*, **17**, 201–215.

Pieterse, A. J. H. and Roos, J. C. (1987). Preliminary observations on primary productivity and phytoplankton associations in the Vaal River at Balkfontein, South Africa. *Archiv für Hydrobiologie*, **110**, 499–518.

Pomeroy, K. E., Shannon, J. P. and Blinn, D. W. (2000). Leaf breakdown in a regulated desert river: Colorado River, Arizona, U. S. A. *Hydrobiologia*, **434**, 3–199.

Puckridge, J. T., Sheldon, F., Walker, K. F. and Boulton, A. J. (1998). Flow variability and the ecology of large rivers. *Marine and Freshwater Research*, **49**, 55–72.

Reice, S. R. and Herbst, G. (1982). The role of salinity in decomposition of leaves of *Phragmites australis* in desert streams. *Journal of Arid Environments*, **5**, 361–368.

Roos, J. C. and Pieterse, A. J. H. (1989). Short-term changes in water quality parameters in the Vaal River. II Primary productivity and community metabolism. In *Proceedings of the Fourth South African National Hydrological*

Symposium, eds. S. Kienzle and H. Maaren, pp. 226–234. Pretoria: University of Pretoria (cited in Davies *et al.*, 1995).

Roshier, D. A., Robertson, A. I. and Kingsford, R. T. (2002). Responses of waterbirds to flooding in an arid region of Australia and implications for conservation. *Biological Conservation*, **106**, 399–411.

Sabo, J. L. and Power, M. E. (2002). River-watershed exchange: effects of riverine subsidies on riparian lizards and their terrestrial prey. *Ecology*, **83**, 1860–1869.

Sanzone, D. M., Meyer, J. L., Martí, E., Gardiner, E. P., Tank, J. L. and Grimm, N. B. (2003). Carbon and nitrogen transfer from a desert stream to riparian predators. *Oecologia*, **134**, 238–250.

Schade, J. D. and Fisher, S. G. (1997). Leaf litter in a Sonoran Desert stream ecosystem. *Journal of the North American Benthological Society*, **16**, 612–626.

Skelton, P. H. (1986). Fish of the Orange-Vaal system. In *The ecology of river systems*. eds. B. R. Davies and K. F. Walker, pp. 143–161. Dordrecht, The Netherlands. Dr W. Junk Publishers.

Stromberg, J. C. (2001). Restoration of riparian vegetation in the south-western United States: importance of flow regimes and fluvial dynamism. *Journal of Arid Environments*, **49**, 17–34.

Stromberg, J. C., Patten, D. T. and Richter, B. D. (1991). Flood flows and dynamics of Sonoran riparian forests. *Rivers*, **2**, 221–235.

Thoms, M. C. and Sheldon, F. (2000). Lowland rivers: an Australian perspective. *Regulated Rivers Research and Management*, **16**, 375–383.

Thorp, J. H. and Delong, A. D. (2002). Dominance of autochthonous autotrophic carbon in food webs of heterotrophic rivers. *Oikos*, **96**, 543–550.

Tockner, K. and Stanford, J. A. (2002). Riverine floodplains: present state and future trends. *Environmental Conservation*, **29**, 308–330.

Tockner, K., Bunn, S. E., Gordon, C., Naiman, R. J., Quinn, G. P. and Stanford, J. A. (in press). Floodplains: Critically threatened ecosystems. In *Future of aquatic ecosystems*. ed. N. Polunin. Cambridge: Cambridge University Press.

Triska, F. J., Sedell, J. R., Cromack, K., Gregory, S. V. and McCorison, F. M. (1984). Nitrogen budget for a small coniferous forest stream. *Ecological Monographs*, **54**, 119–140.

Vannote, R. L., Minshall, G. W., Cummins, K. W., Sedell, J. R. and Cushing, C. E. (1980). The river continuum concept. *Canadian Journal of Fisheries and Aquatic Sciences*, **37**, 130–137.

Velasco, J., Millan, A., Vidal-Abarca, M. R., Suarez, M. L., Guerrero, C. and Ortega, M. (2003). Macrophytic, epipelic and epilithic primary production in a semiarid Mediterranean stream. *Freshwater Biology*, **48**, 1408–1420.

Vidal-Abarca, M. R., Suarez, M. L., Guerrero, C., Velasco, J., Moreno, J. L., Millan, A. and Peran, A. (2001). Dynamics of dissolved and particulate organic carbon in a saline and semiarid stream of southeast Spain (Chicamo stream). *Hydrobiologia*, **455**, 71–78.

Walker, K. F., Sheldon, F. and Puckridge, J. T. (1995). An ecological perspective on large dryland rivers. *Regulated Rivers: Research and Management*, **11**, 85–104.

Ward, J. V., Zimmermann, H. J. and Cline, L. D. (1986). Lotic zoobenthos of the Colorado system. In *The ecology of river systems*. eds. B. R. Davies and K. F. Walker, pp. 403–423. Dordrecht, The Netherlands. Dr W. Junk Publishers.

Webster, J. R. and Meyer, J. L. (1997). Stream organic matter budgets – introduction. *Journal of the North American Benthological Society*, **16**, 5–13.

Welcomme, R. L. (1986a). The Niger River system. In *The ecology of river systems*. eds. B. R. Davies and K. F. Walker, pp. 9–24. Dordrecht, The Netherlands. Dr W. Junk Publishers.

Welcomme, R. L. (1986b). Fish of the Niger system. In *The ecology of river systems*. eds. B. R. Davies and K. F. Walker, pp. 25–48. Dordrecht, The Netherlands. Dr W. Junk Publishers.

Winemiller, K. O. (2006). Floodplain river food webs: generalizations and implications for fisheries management. *River Research and Applications* (in press).

5

Disturbance of plant communities dependent on desert rivers

MARGARET A. BROCK, SAMANTHA J. CAPON AND JOHN L. PORTER

INTRODUCTION

Plant habitats in desert river ecosystems throughout the world are characterised by extreme variability in natural hydrological regimes. Floods and droughts occur unpredictably (Chapter 2, this volume), expanding and contracting across the landscape and altering the size, shape and connectivity of aquatic habitats (Naiman *et al.*,1993; Stanley *et al.*, 1997; Puckridge *et al.*,1998; Ward, 1998) and influencing resource availability and other factors limiting to plant life. Plant species persisting in desert river habitats exhibit a range of responses to the disturbances of flooding and drying. Temporal and spatial patterns in plant community composition and structure reflect these hydrological disturbances (Auble *et al.*, 1994; Stanley *et al.*, 1997; Bornette *et al.*, 2001).

In many desert landscapes, anthropogenic activities have altered natural hydrological regimes, changing the diversity and dynamics of desert river plant communities. Although plant communities are resilient to natural disturbance regimes they are sensitive to the altered flooding and reduced connectivity caused by river regulation (Auble *et al.*, 1994; Nilsson *et al.*, 1997; Andersson *et al.*, 2000; Nilsson & Berggren, 2000). Understanding relationships between natural and anthropogenic disturbances and plant communities is therefore critical to efforts of conservation and restoration in desert rivers.

Ecology of Desert Rivers, ed. R. T. Kingsford. Published by Cambridge University Press.
© Cambridge University Press 2006.

In this chapter we consider plant diversity and community dynamics in desert river ecosystems, focussing on the role of natural hydrological disturbances. The effects of flooding and drying on plant habitat characteristics are discussed and we review the diversity of plants in desert river habitats, their responses to these disturbances and their survival mechanisms. These are discussed within the temporal and spatial patterns of plant community dynamics and how they respond to natural and anthropogenic disturbances of desert rivers. We exemplify this with case studies from two Australian desert rivers.

PLANT HABITATS IN DESERT RIVER ECOSYSTEMS

Defining characteristics

Plant habitats in desert river ecosystems are either permanently or periodically connected by surface water to the rivers that flow through them. By definition, all deserts receive low average annual precipitation (Chapter 1, this volume). However, with a distribution throughout the world's inhabited continents, much diversity exists among the world's deserts. Hyper-arid, arid and semi-arid deserts respectively receive < 25 mm, 25–200 mm and 200–500 mm mean annual precipitation (Davies et al., 1994). Also, climatically, deserts differ in their temperature regimes from hot (e.g. Simpson Desert, Australia), to cold (e.g. China's central plains). Biotic and anthropogenic characteristics, including herbivory and land use, also vary widely among deserts. In spite of these differences, desert rivers of the world have extremely variable hydrological regimes (Davies et al., 1994; Puckridge et al., 1998; Chapter 2, this volume). Surface water flows in desert river ecosystems typically fluctuate unpredictably between periods of drought and flood events of varying magnitude, duration and timing (Chapter 2, this volume). Both floods and droughts constitute major natural disturbances that considerably influence the physical, chemical and biotic functioning of river ecosystems (Junk et al., 1989; Walker et al., 1995; Lake, 2000; Kingsford et al., 2004). Consequently, the salient feature of desert river ecosystems, and the plant habitats within them, is the spatial and temporal variability of hydrological patterns.

Plant habitats are generally described in terms of resources available to plants (e.g. water, light and nutrients) and factors limiting to plants (e.g. oxygen deficiency, light limitation and salinity). These limiting factors, which may impede plant growth or other life-history stages such as reproduction, are the 'stresses'. In desert river ecosystems, plant habitat characteristics are related to temporal and spatial patterns in the natural hydrological disturbance regime. Flooding and drying result in large temporal fluctuations in the resources in and stresses on plant habitats (e.g. soil moisture). Spatial patterns in plant habitat types also result from interactions between topography and variable frequencies and durations of flooding and drying (Chapter 3, this volume). This hydrological variability produces highly spatially and temporally variable plant habitats in deserts (Pollock *et al.*, 1998; Puckridge *et al.*, 2000; Stromberg, 2001).

Temporal patterns

Floods are 'pulse' disturbances and comprise discrete, short-term events that alter abiotic characteristics of riverine habitats (Fisher *et al.*, 1982; Lake, 2000). In many large rivers, including those in deserts, floods essentially 'reset' the ecosystem (Townsend, 1989; Junk *et al.*, 1989; Walker *et al.*, 1995). During flooding, resource availability and physical stresses change for plants. Flooding increases water and nutrient availability and alters the light and temperature conditions for plants in submerged habitats (Blom & Voesenek, 1996). Anoxia and increased toxicity of soils also result from inundation (Blom & Voesenek, 1996). Fast-flowing flood waters can damage plant stems (Menges, 1986; Young *et al.*, 2001) and displace plants adapted to slow flow conditions, but can also transport propagules (Nilsson *et al.*, 1991). Flood waters may also flush salts and litter material from plant habitats and create new sites of freshly deposited sediments for plant colonisation (Stromberg, 2001).

Droughts are 'ramp' disturbances, gradually increasing in intensity over time (Stanley *et al.*, 1997; Lake, 2000). In addition to reducing the area of aquatic habitat, droughts (or less severe drying periods) can increase salinity by evapoconcentration of salts (Boulton & Brock, 1999). Droughts also reduce connectivity (Stanley *et al.*, 1997; Ward *et al.*, 1999), alter the turbidity of subsequent floods (van der Wielen, 2002), facilitate the release of sediment-bound nutrients (Baldwin & Mitchell, 2000) and stimulate changes in seed dormancy in plant propagules (Bonis *et al.*, 1995; Brock & Casanova, 1997; Baskin & Baskin, 1998).

The scale of plant habitat changes induced by either flood or drought disturbance depends on the disturbance's magnitude, frequency, duration and timing. Flood magnitude affects the extent of plant habitat as well as the severity of the impact. For example, water depth will determine the amount of light reaching submerged plants.

Similarly, drought duration influences the accumulation of litter material and its flammability, increasing the probability of fire (Stromberg, 2001). Frequency influences the potential for habitats to change. The effects of a flood following a long drought, for instance, may differ greatly from those of a similar flood soon after the previous flood. Differences in flood (and drought) frequency at a landscape scale produce spatial patterns in plants.

Spatial patterns

The distribution of riverine plant communities generally reflects habitat types related to specific landforms (e.g. active channels, terraces) (Hupp & Osterkamp, 1996). In desert river ecosystems, the boundary between aquatic and terrestrial habitats is indistinct, with some poorly developed landforms found in less variable river ecosystems (Hupp & Osterkamp, 1996). In addition, plant habitats in desert riverine ecosystems exhibit considerable spatial heterogeneity as a result of microtopographic variation in elevation affecting flooding (Pollock et al., 1998; Stromberg, 2001). Nevertheless, four major plant habitat types, channel, channel bank, floodplain and wetland, can be identified (Fig. 5.1). The distinction among these habitats varies within different desert rivers and related regional characteristics such as rainfall, geomorphology and topography.

Habitat characteristics

Channel habitats (in-stream) are the most frequently flooded plant habitats of desert rivers (Fig. 5.1). In desert rivers, flows may be perennial, intermittent or ephemeral and channel habitats can be devoid of surface water for significant periods of time (Chapter 2, this volume). The geomorphologic characteristics of channel habitats vary widely among desert rivers (Chapter 3, this volume) and may be as different as the disconnected strings of highly turbid waterholes found in Australia's Cooper Creek (see, for example, Bunn & Davies, 1999; Chapter 4, this volume) and the ephemeral arroyos of North America's Chihuahuan Desert (Killingbeck & Whitford, 2001). Some desert river channel habitats may also have an estuarine influence, such as the channels of the Colorado River Delta (Glenn et al., 2001). Abiotic factors that structure plant communities in channel habitats include

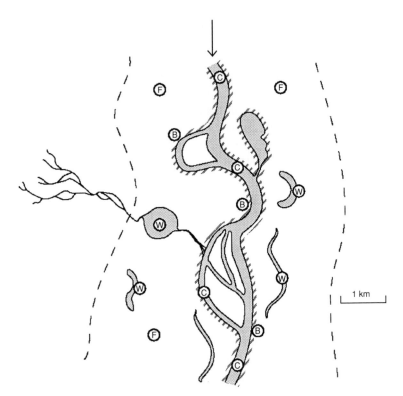

Figure 5.1. Plant habitats in desert rivers. Aerial view of river channel (C), channel bank (diagonal lines; B), floodplain (F) and wetland (W) habitats. Dashed lines indicate edges of the floodplain; the arrow shows direction of flow.

hydrological (frequency, duration, depth and timing of surface water flows), hydraulic (flow velocity) and water quality factors (salinity, nutrient and sediment loads).

Channel banks, including naturally raised levees, are referred to in geomorphologic terms as riparian habitats (Fig. 5.1) (Hupp & Osterkamp, 1996). Plant communities in channel bank habitats can access surface water present in channels or periodically from overbank floods. Many large desert rivers also have extensive floodplains beyond their banks (Fig. 5.1), as 'areas of low lying land that are subject to inundation by lateral overflow water from rivers with which they are associated' (Junk & Welcomme, 1990). Flood disturbance occurs less often in floodplain habitats than in channel banks (Hupp & Osterkamp,

1996) and its frequency decreases with increasing distance or elevation from channels. Floodplain habitats in some desert river ecosystems may be tens of kilometres away from channels and inundated less than once a decade (Capon, 2002). Flood frequency, in addition to rainfall, depth to groundwater (Scott et al., 2000; Snyder & Williams, 2000) and soil properties (e.g. salinity) (Akeroyd et al., 1998; Stromberg 2001), can determine plant community distributions in riparian and floodplain habitats. In many desert river ecosystems, grazing and fire regimes may also be influential (Carr, 1998; Ali et al., 2000; Robertson & Rowling, 2000; Stromberg, 2001).

Various wetland habitats also occur within many desert rivers, including shallow swamps, deflation basins, claypans and lakes (e.g. oxbows (billabongs), waterholes and salt lakes) (Fig. 5.1). These may occur within floodplain and channel bank habitats or within channels themselves when rivers cease to flow. Wetland habitats are inundated sporadically by overbank floods or by channel flows in terminal systems such as Lake Eyre or the Aral Sea. They may also receive water from rainfall or upwelling of groundwater (see, for example, Alvarez-Cobelas et al., 2001). Water quality attributes (salinity and turbidity) are important determinants of plant community composition in wetland habitats (Porter, 2002).

Plant diversity and distribution within channel, channel bank, floodplain and wetland habitats

Plant communities in desert river habitats include the visible flora of flowering plants, ferns, fern allies, bryophytes and algae that either depend on water (i.e. submerged, floating-leaved and emergent aquatic plants) or tolerate water (i.e. semi-terrestrial plants) for all or part of their life cycles. Large algae are represented predominantly by submerged charophytes from the green algae; records of bryophyte distributions are lacking. The microscopic photosynthetic components of these communities are not considered further in this chapter, but their functional role is important for desert rivers (Chapter 4, this volume). Most visible plants in these systems are flowering plants with limited representation of ferns and fern allies such as *Marsilea* spp. and *Equisetum* spp. (Zimmerman et al., 1999; Capon, 2003). The plants of these communities have a variety of life forms (i.e. trees, shrubs, graminoids (including all grass-like plants such as grasses, sedges and rushes) and forbs) and life histories (i.e. perennial, biennial, annual or ephemeral). Plant types are generally distributed according to main habitats, primarily reflecting hydrology.

CHANNELS

Large plants such as trees are usually sparse in channel habitats of dryland rivers (Fig. 5.2) (Timms, 1998, 1999). The paucity of records for plant communities in desert river channels may reflect the difficulties of collecting fragile aquatic plants that only persist in the seed bank, following the recession of floodwaters. Aquatic plant communities are often presumed to be absent in arid rivers (Beadle, 1981; Cowlishaw & Davies, 1997; Casanova, 1999). In semi-arid rivers, submerged, free-floating and emergent aquatic species occur (Roberts & Sainty, 1996); some of these may be recent arrivals that have colonised in response to the more stable conditions created by river regulation (Walker *et al.*, 1994).

Free-floating plants (e.g. *Azolla* spp.) occur only in areas of slow flow, whereas submerged plants are generally more abundant where turbidity is low (Beadle, 1981). The margins of channel habitats in semi-arid rivers can also support emergent graminoid species such as the cosmopolitan *Typha* spp., *Cyperus* spp., *Eragrostis* spp. and *Phragmites australis* (see, for example, Carr, 1998; Glenn *et al.*, 2001).

CHANNEL BANKS

Plant communities in channel bank habitats (Fig. 5.2) are well described for many arid and semi-arid rivers, possibly owing to research effort on commercially valuable riverbank trees. Riparian plant communities usually represent the most diverse and structurally complex plant communities in desert river ecosystems and provide critical habitat for fauna (Poff *et al.*, 1997; Powell & Steidl, 2000; Woinarski *et al.*, 2000; Chapter 7, this volume). Trees and shrubs account for the majority of plant cover in these habitats (Fig. 5.3); structural associations range from closed forest to open shrubland (see, for example, Carr, 1998; Zimmerman *et al.*, 1999; Ali *et al.*, 2000).

Several genera of desert river tree and shrub are remarkably cosmopolitan. Cottonwoods *Populus* spp. are dominant native tree species in channel bank habitats of the Colorado and San Pedro Rivers in the south-west USA (Snyder & Williams, 2000; Zamora-Arroyo *et al.*, 2001) and also occur frequently along the Ulungur and Ertis Rivers in the cold desert region of central China (An *et al.*, 2002). Willows

Figure 5.2. Highly variable flows, including powerful floods and periods of
no flow, and dynamic sediments in the main channel of the Shingwedzi
River, South Africa, make it difficult for vegetation to establish, but
riparian areas on the channel banks support dense vegetation, including
large trees (Photo K. Rogers).

Salix spp. are widely distributed shrubs found along channel banks of
China's cold desert river systems, semi-arid Mediterranean systems of
Spain (Salinas *et al.*, 2000) and throughout the south-west USA, including the banks of ephemeral canyon streams in Arizona (Zimmerman *et
al.*, 1999). In the semi-arid Murray–Darling Basin of Australia, exotic
willows have also become widespread. Salt-tolerant tamarisks *Tamarix*
spp. are recorded from channel bank habitats in several African desert
river systems (Cowlishaw & Davies, 1997; Ali *et al.*, 2000), semi-arid river
systems of Spain (Salinas *et al.*, 2000) and in central China (An *et al.*,
2002). Exotic species belonging to this genus also occupy large areas
of channel bank habitat in desert river systems of the south-west USA
(e.g. the Colorado River) (Stromberg, 2001; Zamora-Arroyo *et al.*, 2001).
Mesquite *Prosopis* spp. occurs in the Colorado River (Snyder & Williams,
2000) and as an exotic in some Australian and African desert rivers
e.g. the Turkwel River, Kenya (Stave *et al.*, 2003); the Swakop River,
Namibia (Cowlishaw & Davies, 1997). Other common, but less widespread, trees and shrubs include *Acacia* spp. in African and Australian
desert river systems and *Eucalyptus* spp. in Australian systems (Fig. 5.3).

Figure 5.3. River red gum *Eucalyptus camaldulensis*, black box *E. largiflorens*, and coolibah *E. coolabah* trees on channel banks of the Paroo River, Australia, withstand dry periods by accessing groundwater, whereas the shrubby lignum *Muehlenbeckia florulenta* has dormant stems during dry periods (Photo J. L. Porter). Recruitment of these species occurs during unpredictable flood periods.

Channel bank woodlands in desert river systems have grassy understoreys, which fluctuate in composition with rainfall and over-bank flooding. Graminoid species also dominate channel bank habitats where trees and shrubs are sparse or absent (see, for example, Higgins *et al.*, 1997; Carr, 1998). Graminoid and forb species in channel bank habitats of desert rivers are less well known, although these may account for a greater proportion of species diversity than trees and shrubs (see, for example, Zimmerman *et al.*, 1999). Common grasses on channel banks of desert rivers throughout the world include *Phragmites australis* and *Echinochloa* spp. Graminoid sedges and rushes such as *Cyperus* spp., *Eleocharis* spp., *Carex* spp., *Scirpus* spp. and *Juncus* spp. are also widely distributed (see, for example, Carr, 1998; Capon, 2002; An *et al.*, 2002; Fig. 5.4). Members of the family Fabaceae (*Indigofera* spp., *Cullen* spp., *Tephrosia* spp.) are often conspicuous in the understorey of channel bank woodlands in African and Australian desert river systems (Carr, 1998; Capon, 2002; Ali *et al.*, 2000). In saline channel bank habitats, salt-tolerant species may occur (Stromberg, 2001), such as the salt grass

Figure 5.4. During flood periods, many plants germinate from the seed bank, such as floating-leaved or emergent *Eleocharis* sp., *Nymphoides crenata*, *Ludwidgia peploides*, *Goodenia* sp., *Ranunculus* sp., *Damasonium minus* and *Marsilea angustifolium* and submergent *Nitella partita*, *Chara australis*, *Najas* sp. and *Myriophyllum verrucosum* in the open water of Pied Stilt Swamp, a fresh temporary wetland near the Paroo River, Australia. Poplar box *Eucalyptus populnea* and black box trees *E. largiflorens* are able to withstand dry times with deep root systems and wax-coated sclerophyllous leaves (Photo J. L. Porter).

Distichlis palmeri that dominates the estuarine reaches of the Colorado River delta (Glenn *et al.*, 2001).

FLOODPLAINS

Plant communities in desert river floodplains are usually transitional between those of channel bank and upland terrestrial habitats (Carr, 1998). In some desert rivers, woodland and forest communities extend from the channel bank onto the floodplain and are gradually replaced by terrestrial species with increasing distance or elevation from the channel (see, for example, An *et al.*, 2002; Stave *et al.*, 2003) (Fig. 5.5). In other rivers, such as Australia's Lake Eyre Basin and the San Pedro River Basin in Arizona, trees are mostly restricted to narrow woodland corridors on raised levee banks with only scattered distributions on floodplains (Scott *et al.*, 2000; Capon, 2003). Many large desert river floodplains in Africa, Australia and North America are dominated by treeless graminoid and

Figure 5.5. Floods and droughts are critically important disturbance factors in the structure of vegetation communities on desert rivers. The vegetation of the floodplain of the Nyl River, Africa, moves from species more tolerant of flooding near river channels to species least tolerant of flooding on the edge of the floodplain. This includes graminoid species that germinate during floods; perennial riparian vegetation grows on the channel bank of the river (Photo K. Rogers).

forb communities (Hughes, 1988; Higgins *et al.*, 1997; Carr, 1998; Scott *et al.*, 2000; Capon, 2003) (Figs. 5.4 and 5.5). Commonly occurring grasses include *Sporobolus* spp., which dominate floodplain grasslands in North America (Scott *et al.*, 2000; Stromberg, 2001), Australia (Capon, 2003) and Africa (Carr, 1998), *Eragrostis* spp. and *Aristida* spp. Patchy or clumped distributions of shrubs may also be present. *Acacia* shrubs occur frequently in African systems (Carr, 1998; Stave *et al.*, 2003) and in the south-west USA (Scott *et al.*, 2000); members of the family Chenopodiaceae (e.g. *Atriplex* spp., *Chenopodium* spp.) occur in desert river floodplains of central Australia (Capon, 2002) and the south-west USA (Stromberg, 2001). In saline areas, salt-tolerant shrubs such as *Tamarix* spp. and *Atriplex* spp. often dominate (Stromberg, 2001; Zamora-Arroyo *et al.*, 2001). Desert river floodplains may also support a high diversity of ephemeral plants, including obligate and facultative annual and ephemeral graminoids and forbs that only appear following rainfall or flooding (Cowlishaw & Davies, 1997; Ali *et al.*, 2000; Capon, 2003).

WETLANDS

A wide variety of wetlands may be present within the channels, channel banks and floodplains of desert rivers (Fig. 5.1) and each of these may support distinctive plant communities. Floodplain wetlands, including swamps and marshes in North America, Asia, Africa and Australia, are frequently dominated by emergent graminoid plants, particularly *Typha domingensis* and *Phragmites australis* (Alvarez-Cobelas *et al.*, 2001; Glenn *et al.*, 2001; Stromberg, 2001; An *et al.*, 2002). *Carex* spp., *Scirpus* spp. and *Juncus* spp. are also widely distributed emergent plants in desert river swamps (Stromberg, 2001; An *et al.*, 2002). Shrub thickets are prevalent in wetlands of some desert river floodplains (Kingsford & Porter, 1999). Lignum *Muehlenbeckia florulenta*, for example, occurs commonly in such habitats throughout arid and semi-arid Australia (Craig *et al.*, 1991). Oxbow lakes and relict channels in floodplains may also support shrub, grassland and aquatic communities (see, for example, Carr, 1998).

Freshwater and saline lakes occur naturally on desert river floodplains; plant communities differ between these. Plants of freshwater permanent, semi-permanent or temporary wetlands often include members of cosmopolitan genera such as *Myriophyllum*, *Potamogeton*, *Callitriche*, *Marsilea*, *Scirpus*, *Cyperus* and *Ludwigia*, although many species are locally endemic (Fig. 5.4). Charophytes from the cosmopolitan genera *Chara* and *Nitella* may also be present in freshwater lakes (Fig. 5.4) (Alvarez-Cobelas *et al.*, 2001; Porter, 2002). In saline lakes, plant communities may be dominated by submerged plants from cosmopolitan salt-tolerant genera of charophytes (e.g. *Lamprothamnium*) and angiosperms such as *Lepilaena* spp. and *Ruppia* spp. (Brock, 1981; Alvarez-Cobelas *et al.*, 2001; Porter, 2002). Where these wetlands are temporary, other graminoid and forb species will germinate and colonise wet sediments when floodwaters recede. In addition, trees and shrubs present in channel bank and floodplain habitats often fringe desert river wetlands and lakes.

SURVIVAL MECHANISMS AND RESPONSES TO DISTURBANCE

Plants inhabiting desert river habitats persist through natural disturbances of floods and droughts and their temporal and spatial variability. A wide variety of survival mechanisms and responses to flooding and drying occur among plants in river ecosystems (Brock & Casanova, 1997; Casanova, 1999). These can be divided into physiological and morphological traits that enable tolerance and life history traits which allow avoidance. Plants tolerate drought with succulence and the production of underground water storage organs, such as tubers, rhizomes and bulbs (Larcher, 2003). Physiological and morphological traits for flood tolerance include the rapid elongation of stems, development of adventitious roots and aerenchyma (intercellular spaces allowing improved gas exchange) and the use of alternative metabolic pathways (Blom *et al.*, 1990; Blom & Voesenek, 1996). Some wetland plants also exhibit variable morphologies in response to fluctuating water levels, such as leaf heterophylly in *Callitriche* spp., *Ludwigia* spp. and *Myriophyllum* spp. (Brock, 1991; Cronk & Fennessy, 2000) and different growth forms in charophytes and angiosperms (Brock, 1991; Brock & Casanova, 1991). Alternatively, life-history patterns allow some plants to escape in time, by coinciding seed dispersal with the recession of seasonal floods (Blom *et al.*, 1990) or producing dormant seeds that germinate with triggers such as light, temperature or flooding and drying (Baskin & Baskin, 1998; Leck & Brock, 2000; Brock *et al.*, 2003). Plants in desert river habitats may possess a combination of different survival mechanisms for tolerance of flood or drought. Tolerance of disturbances may also vary with life-history stage: for example, mature trees or seedlings will differ in their response to flood depth.

Many trees and shrubs in channel bank and floodplain habitats of desert rivers can tolerate floods and droughts as mature adults (Fig. 5.3). For example, channel bank trees in arid zones can flexibly use water sources, switching between groundwater, rainfall and surface water, depending on the prevailing conditions (Snyder & Williams, 2000). Succulent-stemmed shrubs may shed their leaves during drought to reduce water loss (Larcher, 2003). Lignum, a common shrub in desert river floodplain and swamp habitats in Australia, persists through drought as dormant stems and produces leaves and flowers in response to flooding (Roberts & Marston, 2000). The survival of seedlings of trees and shrubs is probably more affected by disturbance than survival of adults, often depending on groundwater accessibility and appropriate flooding, following germination (Dexter, 1967; Stromberg, 2001). Usually, the seeds of tree species are short-lived and germination relies on propagule dispersal coinciding with suitable conditions, such as the recession of floodwaters (Blom *et al.*, 1990). Consequently, recruitment of trees in desert river ecosystems is usually closely related to flooding, and population structures often reflect the history of flood events (Roberts, 1993; Pettit, 2000; Pettit, *et al.*, 2001; Stromberg, 2001; Capon, 2002; Chapter 3, this volume).

In contrast to desert river trees and shrubs, graminoid and forb species often persist through disturbances as dormant propagules, rather than as mature plants. Large persistent soil seed banks are common in desert river floodplain and wetland habitats (Brock & Rogers, 1998; Capon, 2002; Porter, 2002) and many seeds are viable for a decade or more (Leck & Brock, 2000). Germination is usually triggered by cues related to rainfall or flooding and associated changes in temperature, light or water conditions (Baskin & Baskin, 1998; Leck, 1989; Leck & Brock 2000). Some species, such as *Echinochloa turneriana*, can germinate in anoxic conditions during flooding, whereas others germinate in moist sediments, following floodwater recession (Conover & Geiger, 1984; Baskin & Baskin, 1998). Seeds of submerged and free-floating aquatic plants usually germinate exclusively in submerged conditions (van der Valk & Davis, 1978; Gerritsen & Greening, 1989; Boedeltje *et al.*, 2002). Aquatic graminoids and forbs in desert rivers often have extremely rapid life cycles, replenishing soil seed banks before the onset of drought. For submerged aquatic species the entire reproductive cycle occurs under water (e.g. charophytes) or with only pollination at the water surface (e.g. angiosperms such as *Ruppia* spp. and *Lepilaena* spp.) (Fig. 5.6). Other plants (e.g. *Glossostigma* spp.) can grow rapidly under water but only reproduce once flood waters recede in a narrow window of opportunity to flower and set seed before the sediment dries out. In many desert floodplain and wetland species, only a fraction of the soil seed bank germinates in a single germination event, ensuring the survival of future generations should recruitment fail (Brock & Rogers, 1998; Leck & Brock, 2000).

Given the diversity of survival mechanisms, plant species in river habitats can be classified into groups on the basis of their responses to environmental conditions (e.g. hydrology). Naiman & DeCamps (1997), identify four groups: the invaders that produce large numbers of wind- and water-dispersed propagules that colonise river habitats; the endurers that resprout after breakage or burial of either the stem or the roots; the resisters that withstand flooding for extended periods; and the avoiders that lack adaptations to specific disturbances. Alternatively, Brock & Casanova (1997) categorise plants by their responses to hydrological change. So submerged plants germinate, grow and reproduce in flooded conditions; amphibious plants may alter morphology in response to water presence or absence; and semi-terrestrial plants do

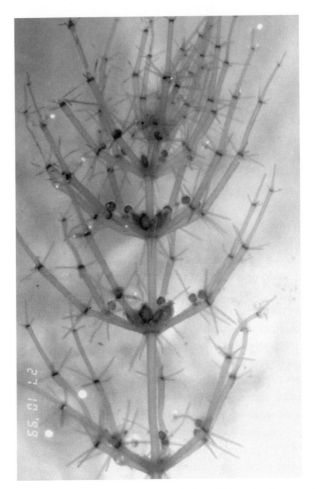

Figure 5.6. Floods provide a narrow window of time for the reproductive cycles of some submerged aquatic species, such as *Lamprothamnium macropogon* (pictured, height *c.* 4 cm), *Lepilaena* spp. and *Ruppia tuberosa* from saline wetlands of the Paroo and Bulloo River basins, Australia. After flooding, a pulse of germination occurs and plants grow rapidly before setting seed that can persist in sediment seed banks during long dry periods.

not tolerate flooding for prolonged periods. In desert rivers where little is known of plant attributes, plants have been grouped by life form (e.g. tree, graminoid, forb) and life history (e.g. annual, perennial) (see, for example, Higgins *et al.*, 1997; Capon, 2003).

COMMUNITY DYNAMICS

As a result of variable hydrological disturbance regimes, plant communities in desert rivers are highly dynamic, both temporally and spatially. Shifts in community composition and structure over time reflect variation in how different plant groups tolerate flooding and drying. Floods kill flood-intolerant species and stop the germination and growth of hydrophytic species; extended drying leads to the replacement of these species by more mesic or xeric species (Capon, 2003). The duration and timing of disturbance often determine plant community responses (Townsend, 1989). For example, mortality of shrubs and the magnitude of germination events on arid floodplains may depend on flood duration (Capon, 2003). In the Lake Eyre Basin of Australia, the seasonal timing of flooding also determines the germination response of arid floodplain plant communities: summer flooding promotes the growth of graminoids and winter floods induce germination of forbs (Edmonston, 2001).

Differences in the frequency and duration of hydrological disturbances can also change spatial patterns of plant community composition and structure, among and within habitats. Plant communities of channel bank, floodplain and wetland habitats in desert rivers often show zonation along gradients of flood frequency, duration or depth (Bren, 1992; Hupp & Osterkamp, 1996; Higgins et al., 1997, Bendix & Hupp, 2000; Capon, 2002, 2003; Chapter 3, this volume). In frequently flooded areas, plant community composition is determined primarily by abiotic factors, with domination by flood-tolerant perennials or annuals that rapidly complete their life cycles between flood events (Blom et al., 1990). Biotic factors such as competition may be more important in rarely flooded areas (Lenssen et al., 1999) where drought-tolerant shrubs and perennial graminoids may dominate (Capon, 2003). Soil seed banks in desert river floodplain and wetland habitats can also exhibit spatial patterns in response to flooding, with higher diversity and density of seeds occurring in more frequently flooded areas (Capon, 2004).

PLANT COMMUNITIES IN TWO AUSTRALIAN DESERT
RIVER HABITATS

Australia is the driest inhabited continent, with almost two thirds of its land mass either arid or semi-arid (Williams, 1995). About 60% of the

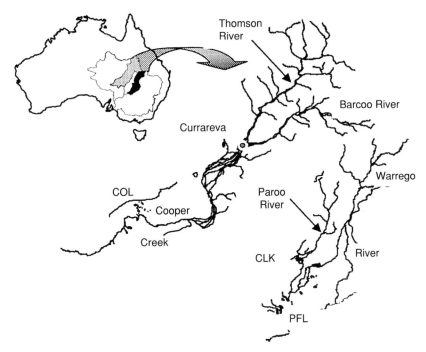

Figure 5.7. Cooper Creek (Lake Eyre Basin) and Paroo–Warrego (Murray–Darling Basin) river catchments in arid Australia.

continent is drained by two major inland river basins, the Lake Eyre Basin and the Murray–Darling Basin, where there is an enormous variety of desert river ecosystems. We demonstrate characteristics of plant communities in desert river habitats with examples of floodplain and wetland plant communities from two Australian desert river systems, the Paroo–Warrego River and Cooper Creek (Fig. 5.7). Our two case studies from these basins illustrate the plant habitats and communities and their response to highly variable hydrological regimes.

Wetland plant communities of the Paroo–Warrego River

The Paroo and Warrego Rivers are the last largely free-flowing rivers remaining in the Murray–Darling Basin; their catchments contain some of the most extensive and important wetlands in the Australian arid zone (Kingsford, 1995; Kingsford *et al.*, 2001, 2004). Located on the north-western fringe of the Murray–Darling Basin (Fig. 5.7), their 137 670 km^2 catchment area has a mean annual rainfall of approximately 300 mm (Gehrke, 2001). Erratic rainfall produces highly variable flows (Puckridge et al., 1998; Young, 1999; Gehrke, 2001; Chapter 2, this volume), which expand and contract water through complex anastomosing channels to extensive overflows and lakes (Gehrke & Harris, 2001). The Warrego has a particularly variable flow (Chapter 2, this volume).

Wetlands of the Paroo–Warrego are unusually diverse because of enormous variation in water regime, water quality, size, depth and geomorphology (Timms, 1993; Kingsford & Porter, 1994, 1999; Timms & Boulton, 2001). Permanent, semi-permanent and temporary freshwater lakes are present; temporary saline lakes also occur. Plant communities in these wetlands undergo considerable changes with flooding and drying, as aquatic plants and annual grasses, herbs and forbs become abundant during and after inundation before declining again. During prolonged dry periods, ground cover may be virtually absent, with woody perennial shrubs and trees as the main above-ground components. A large proportion of the overall species diversity remains hidden below the surface, persisting only as propagules in seed banks or underground storage organs (e.g. rhizomes or tubers).

Freshwater wetlands in channel and channel bank habitats include active river channels, remnant or blocked channel systems, overflow lakes, waterholes and oxbow lakes ('billabongs') that are permanently or semi-permanently flooded. Salinity is low in these habitats (total dissolved solutes (TDS) < 1 g l^{-1}) and light penetration is limited by high turbidity (Secchi transparency < 6 cm). Channel habitats have little vegetative cover in open-water zones, but species such as water primrose Ludwigia peploides ssp. montevidensis, knotweed Persicaria attenuata and duckweed Lemna spp. occur occasionally in sheltered waters. Channel bank habitats typically have a narrow longitudinal band of river red gum Eucalyptus camaldulensis with yapunyah E. ochrophloia and river cooba Acacia stenophylla more common with increasing distance from river channels. An understorey of woody shrubs (e.g. Eremophila spp. and lignum) is often present and, following flooding, a diverse array of grasses and annual forbs also appear. Sedges and rushes (e.g. Eleocharis spp., Juncus spp.) are common.

Overflow lakes (Fig. 5.7), formed from remnant or blocked channel systems, differ from river channel wetlands in that they can have extensive catchments that receive significant local runoff as well as river flows (Kingsford *et al.*, 2001). Plant communities in the channel bank habitats of these are similar to those of active channels but often have a dense fringing band of spiny sedge *Cyperus gymnocaulos* and scattered grasses. Within the shelter of littoral sedges, herbs (e.g. *Myriophyllum* spp. and *Marsilea* spp.) and submerged plants (e.g. *Nitella* spp.) may occur, but plants are usually absent from large areas of deep open water. Despite the high turbidity and paucity of above-ground vegetation, persistent seed banks, including those of submerged charophytes, come from open-water habitats in these wetlands (J. Porter, unpublished data).

Floodplain wetlands occur mostly on the fringes of floodplains and fill predominantly from local runoff, only connecting with river channels during large floods. They are generally small (< 100 ha) and shallow (< 2 m) and hold water temporarily (< 25% of the time). Flooding duration varies from 1 to 6 months, salinity from 0.06 to 0.44 g l^{-1} mean TDS, and turbidity from 5 to 45 cm mean Secchi transparency. Water quality within a particular wetland varies widely, depending on antecedent rainfall, ground cover, soil moisture and the relative contributions of local runoff and river flows to flooding (J. Porter, unpublished. data). These wetlands are fringed by a narrow band of trees, mostly black box *Eucalyptus largiflorens* and bimble box *E. populnea* ssp. *bimbil*. Trees are sparse or absent within the wetlands, but dense shrub thickets of lignum and nitre goosefoot *Chenopodium nitrariaceum* may be present. At the margins of standing water, there are often sedges and rushes (e.g. *Eleocharis* spp., *Juncus* spp. and *Cyperus* spp.). Floating-leaved species such as nardoo *Marsilea angustifolia* (Fig. 5.8), water lily *Ottelia ovalifolia*, wavy marshwort *Nymphoides crenata* and milfoil *Myriophyllum verrucosum* are frequent in open water. If the water is clear, submerged plants, including water nymph *Najas* sp., *Chara* spp. and *Nitella* spp., may also occur. As water levels recede, amphibious herbs (e.g. waterwort *Elatine gratioloides*, mudwort *Glossostigma diandrum*) appear and on higher ground or, after further drying, semi-terrestrial herbs and forbs may establish. Seed densities of up to 62 650 m^{-2} with high species diversity have been recorded in these floodplain wetlands (J. Porter, unpublished data).

Saline wetlands also occur within the Paroo–Warrego and are mostly shallow, wind-formed deflation basins that fill predominantly from local runoff but sometimes receive water from river flows (Timms, 1999). Salinity varies greatly (5–350 g l^{-1} TDS), depending on geomorphology, time since flooding and source of flood water. Turbidity is usually

Figure 5.8. During flood periods, floating-leaved species such as nardoo *Marsilea angustifolia* on a temporary wetland (near the Paroo River, Australia) spread across the wetland and floodplain and reproduce as floodwaters recede. Perennial emergent species like this sedge *Eleocharis acuta* produce new growth and seeds (Photo J. L. Porter).

low because dissolved salts cause flocculation of clay particles, allowing light to penetrate (Porter, 2002). Species diversity is generally low and inversely related to salinity. When flooded, saline wetlands typically have a dense cover of submerged macrophytes including *Ruppia* spp., *Lepilaena* spp. and charophytes (e.g. *Nitella* spp., *Lamprothamnium* spp. and *Chara* spp.), forming productive habitat for aquatic fauna (e.g. waterbirds) (Kingsford & Porter, 1994). Fringing plant communities include salt-tolerant shrubs (e.g. samphire *Halosarcia* spp. and saltbush *Atriplex* spp.) and, after flooding, sparse cover of sedges, grasses and forbs. Wetlands with extremes of salinity and drying disturbance have significantly lower seed bank density and diversity (J. Porter, unpublished data).

Floodplain plant communities of Cooper Creek

Cooper Creek, in the Lake Eyre Basin of central Australia, is one of the world's largest remaining wild rivers (Fig. 5.7). Over 70% of its 296 000 km^2 catchment area receives less than 400 mm average annual rainfall and the river usually exists as a chain of disconnected waterholes. Like many other desert rivers, the Cooper Creek is characterised by extreme variability in rainfall and surface-water discharge (Walker *et al.*, 1995) and is considered to be one of the most hydrologically variable rivers in the world (Puckridge *et al.*, 1998). Large floods supply highly productive flooded habitat for native fish species (Puckridge *et al.*, 2000; Chapter 4, this volume) and waterbirds (Kingsford *et al.*, 1999, 2004). When large flood events occur, a network of braided and anastomosing channels, combined with a low topographic gradient, produce widespread flooding. Extensive floodplains, stretching up to 60 km in width, form almost one third of the catchment (Graetz, 1980; Roshier *et al.*, 2001).

The flooding regime is the principal factor structuring plant community composition on the floodplain (Capon, 2004). Woody vegetation, dominated by river red gum and coolibah, *Eucalyptus coolabah* is mostly restricted to the channel bank habitats of the permanent waterholes of the channel. Open shrubland of Queensland bluebush *Chenopodium auricomum* and lignum occurs in swampy wetland areas in the more frequently flooded areas of the floodplain. Most of the floodplain has short grass and forb communities that change in composition with hydrology (Boyland, 1984).

Only a few species on the floodplain exhibit true flood tolerance. For example, the perennial fern nardoo *Marsilea drummondii* grows in shallow aquatic habitats during flooding and spreads across the floodplain as water recedes, where it survives if there is enough moisture (Cunningham *et al.*, 1992) (Fig. 5.8). Reproduction of nardoo occurs with floodwater recession when sporocarps develop and detach after they have dried out, remaining in the soil until the next flood releases their spores for germination. The dominant shrub species, Queensland bluebush and lignum, are also flood-tolerant (Cunningham *et al.*, 1992). They survive vegetatively through short periods of inundation but are killed by extended flood events (Capon, 2003). Many ephemeral graminoids (e.g. *Cyperus difformis, C. bifax, Eleocharis* spp., *Echinochloa* spp.) also exhibit some flood tolerance, but do not survive long dry periods as adults.

Given the high natural hydrological variability of the floodplain, most plant species escape, rather than tolerate, the stresses associated with flooding. Many annual species maintain large persistent soil seed banks that germinate following flooding (Capon, 2002). Channel millet *Echinochloa turneriana* germinates during inundation whereas the annual legume Cooper clover *Trigonella suavissima* requires sustained flooding for mass germination (Boyland, 1984). Such species have rapid growth rates and reproduction, so seeds enter the soil seed bank before drought or further inundation. Consequently, plant communities with many ephemeral and annual grass and forb species develop with flooding (Capon, 2003). Even these large germination events have little effect on the overall abundance of seeds stored in the soil seed bank (Capon, 2004).

The spatial variation in flooding across the Cooper Creek floodplain creates a complex gradient along which plant habitat resources and stresses vary (Huston, 1994). The composition of plant communities shifts along this gradient. In frequently flooded areas, flood-tolerant species (e.g. nardoo and *Cyperus* spp.) dominate, whereas perennial grasses and sub-shrubs (e.g. *Sclerolaena* spp. and *Atriplex* spp.) dominate the rarely inundated edges of the floodplain (Capon, 2002). The distribution of shrub and sub-shrub species is related to flood history. Lignum occupies patches inundated more frequently and for a longer duration than patches of Queensland bluebush, producing a sharp zonation, whereas drought-tolerant perennial sub-shrubs are prevalent in infrequently flooded areas (Capon, 2002). Large-scale flooding homogenises vegetation across the floodplain landscape; plant communities at the edges of the floodplain exhibit responses to inundation similiar to those of areas close to channels (Capon, 2003). As drying progresses, local factors such as soil type and rainfall influence plant community composition, causing changes throughout the landscape. High spatial and temporal variability in flooding and drying patterns across the Cooper Creek floodplain consequently maintains a heterogeneous mosaic of plant communities differing in composition and structure over large and small temporal and spatial scales.

Anthropogenic disturbances in the Paroo–Warrego and Cooper Creek desert river systems

The pastoral industry, predominantly for sheep and cattle, is the main land use in the catchments of the Paroo–Warrego and Cooper Creek in Australia (Fig. 5.7). In these catchments, grazing and invasive species are the main anthropogenic disturbances, as rivers have little water resource development.

Livestock grazing in the Cooper Creek and Paroo catchments has some impact on plant community composition and structure, particularly in channel bank woodlands (Edmonston, 2001; Pettit, 2002; Robertson & Rowling, 2000), but stock levels at present are generally low, limiting severity of impacts. Increased numbers of artificial watering points increases grazing pressure across isolated wetland habitats, also providing reliable water for pest species such as feral goats *Capris hirca* and pigs *Sus scrofa* (Landsberg *et al.*, 1997). Floodplain grass and forb communities of the Cooper Creek floodplain shift in composition with selective grazing, during drying, but this has little impact on the massive germination responses to inundation (Edmonston, 2001).

Honey from the European honeybee *Apis mellifera* from Yapunyah is an important industry on the Paroo, Warrego and Bulloo floodplains. The introduced honeybees, which may outcompete native pollinators, may reduce populations of native pollinator species, reducing seed set and recruitment of some plant species (Anderson, 1989).

Invasion by plant species is also a result of agricultural activity in both catchments. Species introduced for stock fodder or by grazing are now widespread on the floodplains (e.g. *Medicago* spp. and grasses such as barley grass *Hordeum* spp., barnyard grass *Echinochloa* spp. and buffel grass *Cenchrus ciliaris*). Weed species such as noogoora burr *Xanthium occidentale* and water lettuce *Pistia stratioides* are spread by water, and could reach high densities and potentially threaten floodplain communities. Animal pest species such as feral goats and pigs may affect channel bank vegetation with palatable species, such as the nationally rare aquatic plant cooboree *Aponogeton queenslandicus* with its fleshy tuber, probably most at risk.

The potential for alterations to natural flooding patterns poses the most serious threat to plant communities of Cooper Creek and the Paroo–Warrego systems (Kingsford *et al.*, 1998). Extraction or diversion of river flows is one of the major threats facing these unregulated river systems (Kingsford, 2000; Chapter 8, this volume). Currently, little water resource development has occurred within the catchments, and there are restrictions to limit future levels of water extraction in the short term (Queensland Department of Natural Resources, 1998; Queensland State Government, 2000). Floodwaters may also be influenced by developments that alter topography (e.g. roads or levee banks). Potential changes to flood extent, duration and frequency from increased water extraction for irrigation or other developments would predictably affect plant community composition and structure (for example, aquatic species may be replaced by mesic and xeric species shifting the position of vegetation zones, giving invasive species a competitive advantage).

ANTHROPOGENIC DISTURBANCES TO PLANT COMMUNITIES
IN DESERT RIVERS

The range of anthropogenic disturbances in desert rivers throughout the world results from agricultural, urban and industrial activities. Threats to desert river plant communities include alteration of flows, eutrophication, salinisation, land clearing, grazing of domestic animals and the introduction of exotic species. The alteration of surface water flow regimes is recognised as the major threat to plant communities in river habitats (Nilsson & Svedmark, 2002; Tockner & Stanford, 2002; Chapter 8, this volume). Riparian vegetation is especially sensitive to modified flow patterns that produce substantial changes without changing mean annual flows (Auble *et al.*, 1994). Extraction and diversion of surface water, in addition to river regulation through the construction of dams, weirs and levee banks, have dramatic impacts on natural hydrological disturbance regimes in desert rivers (Davies *et al.*, 1994; Walker *et al.*, 1995; Thoms & Sheldon, 2000; Chapters 8 and 9, this volume) and their plant communities. In addition, the effects of other anthropogenic disturbances, such as eutrophication and salinisation, are often exacerbated by the changes to river flows (see, for example, Alvarez-Cobelas *et al.*, 2001; Stromberg, 2001; Chapter 10, this volume).

Changes to surface water flow regimes in desert river systems usually involve a reduction of mean annual flows and an increase in median flows (Walker *et al.*, 1995; Puckridge *et al.*, 1998). The frequency and duration of flood events are consequently diminished, as are periods of low flow or drought (Walker *et al.*, 1995). Channel bank vegetation in desert rivers is especially sensitive to changes in minimum and maximum flows (Auble *et al.*, 1994). In the rivers of the Murray–Darling Basin, Australia, the seasonal timing of floods may also be altered (Maheshwari *et al.*, 1995; Thoms & Sheldon, 2000). Overall, anthropogenic alterations to surface water flows decrease the natural variability of hydrological disturbances (Walker *et al.*, 1995; Thoms & Sheldon, 2000). In channel habitats, this often increases duration of surface water flows, and sometimes creates new habitats (e.g. weir pools) (Chapter 9, this volume). In contrast, channel bank, floodplain and wetland habitats usually experience less frequent flooding and more prolonged drying. This significantly alters characteristics of plant habitats, temporally and spatially, affecting plant life history.

The alteration of flows in desert rivers affects dispersal, germination and growth (see, for example, Blanch *et al.*, 2000). River regulation and the suppression of overbank floods reduced recruitment of native trees in the Colorado River of North America (Zamora-Arroyo *et al.*, 2001) and the Turkwel River of Kenya (Stave *et al.*, 2003). Plant community composition and structure shifts as different groups of plants either become established or are extirpated in response to the new regime. For example, some aquatic macrophytes have increased in weir pools since river regulation has stabilised flows of the Murray–Darling Basin, Australia (Walker *et al.*, 1994; Blanch *et al.*, 2000) and the Yobe River, Nigeria (Goes, 2002). In the semi-arid Las Tablas de Daimiel wetland of Spain, increased drying and eutrophication have occurred as a result of river regulation (Alvarez-Cobelas *et al.*, 2001). Free-floating aquatic plants have subsequently disappeared from the community and previously large areas of cut sedge *Cladium mariscus* have been reduced and replaced by species more tolerant of long dry periods and high nutrient levels, such as the reed *Phragmites australis*.

Changes to spatial patterns in plant community composition and structure can also result from alterations of flood frequency gradients. In channel bank and floodplain habitats, vegetation zones may migrate towards the channel (Hughes & Cass, 1997). For example, river red gum has invaded hydrophytic grass communities in floodplains of the River Murray since the reduction of flood frequency by flow regulation (Bren, 1992). Similarly, upland desert species invaded forests reducing native tree diversity in the cold desert of China's Altai Plain, following alterations to surface water hydrology (An *et al.*, 2002). Exotic species, favoured by new habitats, can also invade native plant communities with changes to flooding. In channel bank habitats of the Colorado River, native cottonwood and willow communities have been replaced by thickets of the invasive saltcedar *Tamarix* spp. This shrub is more tolerant of drought, salinity and burning, all of which have increased with reduced surface water (Glenn *et al.*, 2001; Stromberg, 2001). In the Murray–Darling Basin, cumbungi *Typha* spp. has invaded many channels because of regular low flows.

Overall, changes to natural hydrological regimes in desert rivers reduce temporal and spatial heterogeneity of plant habitats, resulting in the loss of biodiversity and homogenisation of plant community composition and structure. Given the ecological importance of plant communities in desert rivers (e.g. channel bank stabilisation, wildlife habitat), there may be significant secondary impacts. Fortunately,

there is some evidence to suggest that restoration of natural hydrological regimes in desert rivers may be sufficient to partly reverse such
deleterious changes in plant communities (see, for example, Glenn *et
al.*, 2001; Stromberg, 2001). However, increasing human population
pressure drives unprecedented demand for water resource developments in many deserts around the world (Davies *et al.*, 1994; Chapters
8 and 11, this volume). Plant communities in river, floodplain and
wetland habitats are some of the most degraded and threatened on
the planet (see, for example, Tockner & Stanford, 2002) and those
in desert river habitats will continue to be affected by recent
anthropogenic disturbances.

CONCLUSIONS

The variability of the hydrological regime is the key determinant of
both plant community structure in time and space and the types of
plant present in desert rivers. This variability in flow has selected for
many flexible and opportunistic species that respond to favourable
conditions through rapid germination, growth, reproduction and seed
addition to large persistent soil seed banks. Each plant species is different; the combination of spatially and temporally variable species is a
function of hydrology in each of the channel, channel bank, floodplain
and wetland habitats in a river. This dynamism in the face of hydrological variability indicates natural resilience. However, anthropogenic
changes to flow reduce the variability and extent of flooding, changing
composition of vegetation communities by favouring species adapted to
stability. The synergistic effects of flow alteration with other threats
(invasives, grazing, salinity, pollution) on plant communities may
exacerbate the changes to plant communities.

ACKNOWLEDGEMENTS

We thank two anonymous referees for their comments, which greatly
improved the structure and content of this manuscript. Thanks are also
due to Dr Richard Kingsford for editorial comments, suggestions and
encouragement, which also enhanced the manuscript. Kevin Rogers
generously provided photographs of African rivers, and Stuart McVicar
assisted with Fig. 5.1.

REFERENCES

Akeroyd, M. D., Tyerman, S. D., Walker, G. R. and Jolly, I. D. (1998). Impact of flooding on the water use of semi-arid riparian eucalypts. *Journal of Hydrology*, 206, 104–7.

Ali, M. M., Dickinson, G. and Murphys, K. J. (2000). Predictors of plant diversity in a hyperarid desert wadi ecosystem. *Journal of Arid Environments*, 45, 215–30.

Alvarez-Cobelas, M., Cirujano, S. and Sanchez-Carillo, S. (2001). Hydrological and botanical man-made changes in the Spanish wetland of Las Tablas de Daimiel. *Biological Conservation*, 97, 90–8.

An, S., Cheng, X., Sun, S., Wang, Y. and Li, J. (2002). Composition change and vegetation degradation of riparian forests in the Altai plain, NW China. *Plant Ecology*, 164, 75–84.

Anderson, J. M. E. (1989). Honeybees in natural ecosystems. In *Mediterranean Landscapes in Australia: Mallee Ecosystems and their Management*, ed. J. C Noble and R. Bradstock pp. 300–4. Melbourne: CSIRO Publishing.

Andersson, E., Nilsson, C. and Johansson, M. E. (2000). Effects of river fragmentation on plant dispersal and riparian flora. *Regulated Rivers: Research and Management*, 16, 83–9.

Auble, G. T., Friedman, J. M. and Scott, M. L. (1994). Relating riparian vegetation to present and future streamflows. *Ecological Applications*, 4, 544–54.

Baldwin, D. S. and Mitchell, A. M. (2000). The effects of drying and re-flooding on the sediment and soil nutrient dynamics of lowland river-floodplain systems: a synthesis. *Regulated Rivers: Research and Management*, 16, 457–67.

Baskin, C. C. and Baskin, J. M. (1998). *Seeds: Ecology, Biogeography and Evolution of Dormancy and Germination*. San Diego: Academic Press.

Beadle, N. C. W. (1981). *The Vegetation of Australia*. Cambridge: Cambridge University Press.

Bendix, J. and Hupp, C. R. (2000). Hydrological and geomorphological impacts on riparian plant communities. *Hydrological Processes*, 14, 2977–90.

Blanch, S., Walker, K. F. and Ganf, G. G. (2000). Water regimes and littoral plants in four weir pools of the River Murray, Australia. *Regulated Rivers: Research and Management*, 16, 445–56.

Blom, C. W. P. M., Bogemann, G. M., Laan, P. *et al.* (1990). Adaptation to flooding in plants from river areas. *Aquatic Botany*, 38, 29–47.

Blom, C. W. P. M. and Voesenek, L. A. C. J. (1996). Flooding: the survival strategies of plants. *Trends in Ecology and Evolution*, 11, 290–5.

Boedeltje, G., ter Heerdt, G. N. J. and Bakker, J. P. (2002). Applying the seedling-emergence method under waterlogged conditions to detect the seed bank of aquatic plants in submerged sediments. *Aquatic Botany*, 72, 121–8.

Bonis, A., Lepart, J. and Grillas, P. (1995). Seed bank dynamics and coexistence of annual macrophytes in a temporary and variable habitat. *Oikos*, 74, 81–92.

Bornette, G., Piegay, H., Citterio, A., Amoros, C. and Godreau, V. (2001). Aquatic plant diversity in four river floodplains: a comparison at two hierarchical levels. *Biodiversity and Conservation* 10, 1683–701.

Boulton, A. J. and Brock, M. A. (1999). *Australian Freshwater Ecology: Processes and Management*. South Australia: Gleneagles Publishing.

Boyland, D. E. (1984). *Vegetation Survey of Queensland: South Western Queensland*. Brisbane: Queensland Department of Primary Industries.

Bren, L. J. (1992). Tree invasion of an intermittent wetland in relation to changes in the flooding frequency of the River Murray, Australia. *Australian Journal of Ecology*, 17, 395–408.

Brock, M. A. (1981). The ecology of halophytes in salt lakes in the south-east of South Australia. *Hydrobiologia*, **81**, 23–32.

Brock, M. A. (1991). Mechanisms for maintaining persistent populations of *Myriophyllum variifolium* in a fluctuating Australian shallow lake. *Aquatic Botany*, **39**, 211–9.

Brock, M. A. and Casanova, M. T. (1991). Plant survival in temporary waters: a comparison of charophytes and angiosperms. *Verhandlungen der Internationale Vereinigung für Theoretische und Angewandte Limnologie*, **24**, 2668–72.

Brock, M. A. and Casanova, M. T. (1997). Plant life at edge of wetlands: ecological responses to wetting and drying patterns. In *Frontiers in Ecology: Building the Links*, ed. N. Klomp and I. Lunt, pp. 181–92. Oxford: Elsevier Science.

Brock, M. A. and Rogers, K. H. (1998). The regeneration potential of the seed bank of an ephemeral floodplain in South Africa. *Aquatic Botany*, **61**, 123–35.

Brock, M. A., Nielsen, D. L, Shiel, R. J., Green, J. D. and Langley, J. D. (2003). Drought and aquatic community resilience: the role of eggs and seeds in sediments of temporary wetlands. *Freshwater Biology*, **48**, 1027–218.

Bunn, S. E. and Davies, P. M. (1999). Aquatic food webs in turbid, arid zone rivers: preliminary data from Cooper Creek, western Queensland. In *A Free-Flowing River: the Ecology of the Paroo River*, ed. R. T. Kingsford, pp. 67–76. Hurstville: NSW National Parks and Wildlife Service.

Capon, S. J. (2002). The effects of flood frequency on floodplain vegetation in a variable, arid catchment: Cooper Creek, south-west Queensland. In *Landscape Health of Queensland*, ed. A. J. Franks, J. Playford and A. Shapcott, pp. 223–6. St. Lucia, Queensland: The Royal Society of Queensland.

Capon, S. J. (2003). Plant community responses to wetting and drying in a large arid floodplain. *River Research and Applications*, **19**, 509–20.

Capon, S. J. (2004). Flow variability and vegetation dynamics in a large arid floodplain: Cooper Creek, Australia. Ph. D. thesis,: Griffith University, Brisbane.

Carr, C. J. (1998). Patterns of vegetation along the Omo river in southwest Ethiopia. *Plant Ecology*, **135**, 135–63.

Casanova, M. T. (1999). Plant establishment in Paroo wetlands: the importance of water regime. In *A Free-Flowing River: the Ecology of the Paroo River*, ed. R. T. Kingsford, pp. 138–48. Hurstville: NSW National Parks and Wildlife Service.

Conover, D. G. and Geiger, D. R. (1984). Germination of Australian channel millet [*Echinochloa turnerana* (Domin) J. M. Black] seeds. II Effects of anaerobic conditions, continuous flooding and low water potential. *Australian Journal of Plant Physiology*, **11**, 409–17.

Cowlishaw, G. and Davies, J. G. (1997). Flora of the Pro-Namib Desert Swakop River catchment, Namibia: community classification and implications for desert vegetation sampling. *Journal of Arid Environments*, **36**, 271–90.

Craig, A. E., Walker, K. F. and Boulton, A. J. (1991). Effects of edaphic factors and flood frequency on the abundance of lignum (*Muehlenbeckia florulenta* Meissner, Polygonaceae) on the River Murray Floodplain, South Australia. *Australian Journal of Botany*, **39**, 431–43.

Cronk, J. K. and Fennessy, M. S. (2000). *Wetland Plants: Biology and Ecology*. Florida: Lewis Publishers.

Cunningham, G. M., Mulham, W. E., Milthorpe, P. L. and Leigh, J. H. (1992). *Plants of Western New South Wales*. Melbourne: Inkata Press.

Davies, B. R., Thoms, M. C., Walker, K. F., O'Keefe, J. H. and Gore, J. A. (1994). Dryland rivers: their ecology, conservation and management. In *The Rivers Handbook*, vol. 2 ed. P. Calow and G. E. Petts, pp. 484–511. Oxford: Blackwell Scientific.

Dexter, B. D. (1967). Flooding and regeneration of river red gum, *Eucalyptus camaldulensis*, Dehn. *Forestry Commission of Victoria Research Bulletin*, **20**, 1–35.

Edmonston, V. (2001). *Managing the Channel Country Sustainably: Producers' Experiences*. Brisbane: Queensland Department of Primary Industries.

Fisher, S. G., Gray, L. J., Grimm, N. B. and Busch, D. E. (1982). Temporal succession in a desert stream ecosystem following flash flooding. *Ecological Monographs*, **52**, 93–110.

Gehrke, P. C. (2001). The Paroo River. In *Rivers as ecological systems: the Murray-Darling Basin*, ed. W. J. Young, pp. 119–31. Canberra: Murray Darling Basin Commission.

Gehrke, P. C. and Harris, J. H. (2001). Regional-scale effects of flow regulation on lowland riverine fish communities in New South Wales, Australia. *Regulated Rivers Research and Management*, **17**, 369–91.

Gerritsen, J. and Greening, H. S. (1989). Marsh seed banks of the Okefenokee swamp: effects of hydrologic regime and nutrients. *Ecology*, **70**, 750–763.

Glenn, E. P., Zamora-Arroyo, F., Nagler, P. L. *et al.* (2001). Ecology and conservation biology of the Colorado River Delta, Mexico. *Journal of Arid Environments*, **49**, 5–15.

Goes, B. J. M. (2002). Effects of river regulation on aquatic macrophyte growth and floods in the Hadejia-Nguru wetlands and flow in the Yobe River, northern Nigeria; implications for future water management. *River Research and Applications*, **18**, 81–95.

Graetz, R. D. (1980). *The Potential Application of LANDSAT Imagery to Land Resource Management in the Channel Country*. Canberra: CSIRO Division of Land Resource Management.

Higgins, S. I., Rogers, K. H. and Kemper, J. (1997). A description of the functional vegetation pattern of a semi-arid floodplain, South Africa. *Plant Ecology*, **129**, 95–101.

Hughes, F. M. R. (1988). The ecology of African floodplain forests in semi-arid and arid zones: a review. *Journal of Biogeography*, **15**, 127–40.

Hughes, J. W. and Cass, W. B. (1997). Pattern and process of a floodplain forest, Vermont, USA: predicted responses of vegetation to perturbation. *Journal of Applied Ecology*, **34**, 594–612.

Hupp, C. R. and Osterkamp, W. R. (1996) Riparian vegetation and fluvial geomorphic processes. *Geomorphology*, **14**, 277–95.

Huston, M. A. (1994). *Biological Diversity: the Coexistence of Species in Changing Landscapes*. Cambridge: Cambridge University Press.

Junk, W. J. and Welcomme, R. L. (1990). Floodplains. In *Wetlands and Shallow Continental Water Bodies*, ed. B. C. Patten, S. E. Jorgensen and H. Dumont, pp. 491–524. The Hague: SPB Academic Publishers.

Junk, W. J., Bayley, P. B. and Sparks, R. E. (1989). The flood pulse concept in river-floodplain systems. *Canadian Special Publication of Fisheries and Aquatic Sciences*, **106**, 110–27.

Killingbeck, K. T. and Whitford, W. G. (2001). Nutrient resorption in shrubs growing by design, and by default in Chichuahuan Desert arroyos. *Oecologia*, **128**, 351–9.

Kingsford, R. T. (1995). Occurrence of high concentrations of waterbirds in arid Australia. *Journal of Arid Environments*, **29**, 421–5.

Kingsford, R. T. (2000). Protecting rivers in arid regions or pumping them dry? *Hydrobiologia*, **427**, 1–11.

Kingsford, R. T. and Porter, J. (1994). Waterbirds on an adjacent freshwater and salt lake in arid Australia. *Biological Conservation*, **69**, 219–28.

Kingsford, R. T. & Porter, J. (1999). Wetlands and waterbirds of the Paroo and Warrego Rivers. In *A Free-Flowing River: the Ecology of the Paroo River*, ed. R. T. Kingsford, pp. 23–50. Sydney: NSW National Parks and Wildlife Service.

Kingsford, R. T., Boulton, A. J. and Puckridge, J. M. (1998). Challenges in managing dryland rivers crossing political boundaries: lessons from Cooper Creek and the Paroo River, central Australia. *Aquatic Conservation: Marine and Freshwater Ecosystems*, **8**, 361–78.

Kingsford, R. T., Brandis, K., Thomas, R. *et al.* (2004). Classifying landform at broad spatial scales: the distribution and conservation of wetlands in New South Wales, Australia. *Marine and Freshwater Research*, **55**, 17–31.

Kingsford, R. T., Curtin, A. L. and Porter, J. (1999). Water flows on Cooper Creek in arid Australia determine 'boom' and 'bust' periods for waterbirds. *Biological Conservation*, **88**, 231–48.

Kingsford, R. T., Jenkins, K. M. and Porter, J. L. (2004). Imposed hydrological stability on lakes in arid Australia and effects on waterbirds. *Ecology*, **85**, 2478–92.

Kingsford, R. T., Thomas, R. F. & Curtin, A. L. (2001). Conservation of wetlands in the Paroo and Warrego catchments in arid Australia. *Pacific Conservation Biology*, **7**, 21–33.

Lake, P. S. (2000). Disturbance, patchiness, and diversity in streams. *Journal of the North American Benthological Society*, **19**, 573–92.

Landsberg, J., James, C. D. and Morton, S. (1997). Assessing the effects of grazing on biodiversity in Australia's rangelands. *Australian Biologist*, **10**, 153–62.

Larcher, W. (2003). *Physiological Plant Ecology: Ecophysiology and Stress Physiology of Functional Groups*. Berlin: Springer-Verlag.

Leck, M. A. (1989). Wetland seed banks. In *Ecology of Soil Seed Banks*, ed. M. A. Leck, V. T. Parker and R. L. Simpson, pp. 283–305. San Diego: Academic Press.

Leck, M. A. and Brock, M. A. (2000). Ecological and evolutionary trends in wetlands: evidence from seeds and seed banks in New South Wales, Australia and New Jersey, USA. *Plant Species Biology*, **15**, 97–112.

Lenssen, J., Menting, F., van der Putten, W. and Blom, K. (1999). Control of plant species richness and zonation of functional groups along a freshwater flooding gradient. *Oikos*, **86**, 523–34.

Maheshwari, B. L., Walker, K. F., and McMahon, T. A. (1995). Effects of regulation on the flow regime of the River Murray, Australia. *Regulated Rivers* **10**, 15–38.

Menges, E. S. (1986). Environmental correlates of herb species composition in five southern Wisconsin floodplain forests. *The American Midland Naturalist*, **115**, 106–17.

Naiman, R. J. and DeCamps, H. (1997). The ecology of interfaces: Riparian zones. *Annual Review of Ecology and Systematics*, **28**, 621–658.

Naiman, R. J., Decamps, H. and Pollock, M. (1993). The role of riparian corridors in maintaining regional biodiversity. *Ecological Applications*, **3**, 209–12.

Nilsson, C. and Berggren, K. (2000). Alterations of riparian ecosystems caused by river regulation. *BioScience*, **50**, 783–92.

Nilsson, C. and Svedmark, M. (2002). Basic principles and ecological consequences of changing water regimes: riparian plant communities. *Environmental Management*, **30**, 468–80.

Nilsson, C., Gardfjell, M. and Grelsson, G. (1991). Importance of hydrochory in structuring plant communities along rivers. *Canadian Journal of Botany*, **69**, 2631–3.

Nilsson, C., Jansson, R. and Zinko, U. (1997). Long-term responses of river-margin vegetation to water-level regulation. *Science*, **276**, 798–800.

Pettit, N. E. (2000). Factors affecting the recruitment of riparian vegetation on the Ord and Blackwood Rivers in Western Australia. Ph. D. Thesis, Edith Cowan University, Perth.

Pettit, N. E. (2002). Riparian vegetation of a permanent waterhole on Cooper Creek, southwest Queensland. *Proceedings of the Royal Society of Queensland*, **110**, 15–25.

Pettit, N. E., Froend, R. H. and Davies, P. M. (2001). Identifying the natural flow regime and the relationship with riparian vegetation for two contrasting Western Australian Rivers. *Regulated Rivers: Research and Management*, **17**, 201–15.

Poff, N. L., Allan, J. D., Bain, M. B. *et al.* (1997). The natural flow regime. *Bioscience*, **47**, 769–84.

Pollock, M. M., Naiman, R. J. and Hanley, T. A. (1998). Plant species richness in riparian wetlands – a test of biodiversity theory. *Ecology*, **79**, 94–105.

Porter, J. (2002). Effects of salinity, turbidity and water regime on arid zone wetland seed banks. *Internationale Vereinigung für Theoretische und Angewandte Limnologie*, **28**, 1468–71.

Powell, B. F. and Steidl, R. J. (2000). Nesting habitat and reproductive success of southwestern riparian birds. *Condor*, **102**, 823–31.

Puckridge, J. T., Sheldon, F., Walker, K. F. and Boulton, A. J. (1998). Flow variability and the ecology of large rivers. *Marine and Freshwater Research*, **49**, 55–72.

Puckridge, J. T., Walker, K. F. and Costelloe, J. F. (2000). Hydrological persistence and the ecology of dryland rivers. *Regulated Rivers: Research and Management*, **16**, 385–402.

Queensland Department of Natural Resources (1998). *Draft Water Management Plan for Cooper Creek: Information Paper*. Brisbane: Queensland Department of Natural Resources.

Queensland State Government (2000). *Water Resource (Cooper Creek) Plan 2000*. Brisbane: Queensland State Government.

Roberts, J. (1993). Regeneration and growth of coolabah, *Eucalyptus coolabah* ssp. *arida*, a riparian tree in the Cooper Creek region of South Australia. *Australian Journal of Ecology*, **18**, 345–50.

Roberts, J. and Marston, F. (2000). *Water regime of wetland and floodplain plants in the Murray-Darling Basin: a source book of ecological knowledge*. Report No 30/00. Canberra: CSIRO Land and Water.

Roberts, J. and Sainty, G. (1996). *Listening to the Lachlan*. Sydney: Sainty and Associates.

Robertson, A. I. and Rowling, R. W. (2000). Effects of livestock on riparian zone vegetation in an Australian dryland river. *Regulated Rivers: Research and Management*, **16**, 527–41.

Roshier, D. A., Whetton, P. H., Allan, R. J. and Robertson, A. I. (2001). Distribution and persistence of temporary wetland habitats in arid Australia in relation to climate. *Austral Ecology*, **26**, 371–84.

Salinas, M. J., Blanca, G. & Romero, A. T. (2000). Riparian vegetation and water chemistry in a basin under semiarid mediterranean climate, Andarax River, Spain. *Environmental Management*, **26**, 539–52.

Scott, R. L., Shuttleworth, W. J., Goodrich, D. C. and Maddock, T. (2000). The water use of two dominant vegetation communities in a semiarid riparian ecosystem. *Agricultural and Forest Meteorology*, **105**, 241–56.

Snyder, K. A. and Williams, D. G. (2000). Water sources used by riparian trees varies among stream types on the San Pedro River, Arizona. *Agricultural and Forest Meteorology*, **105**, 227–40.

Stanley, E. H., Fisher, S. G. and Grimm, N. B. (1997). Ecosystem expansion and contraction in streams. *BioScience*, **47**, 427–35.

Stave, J., Oba, G., Bjora, C. S. *et al.* (2003). Spatial and temporal woodland patterns along the lower Turkwel River, Kenya. *African Journal of Ecology*, **41**, 224–36.

Stromberg, J. C. (2001). Restoration of riparian vegetation in the south-western USA: importance of flow regimes and fluvial dynamism. *Journal of Arid Environments*, **49**, 17–34.

Thoms, M. C. and Sheldon, F. (2000). Water resource development and hydrological change in a large dryland river: the Barwon-Darling River, Australia. *Journal of Hydrology*, **228**, 10–21.

Timms, B. V. (1993). Saline lakes of the Paroo, inland New South Wales, Australia. *Hydrobiologia*, **267**, 1–21.

Timms, B. V. (1998). Spatial and temporal variation in wetlands of the Paroo catchment of the Murray Darling Basin and implications for management. In *Wetlands in a Dry Land: Understanding for Management*, ed. W. D. Williams, pp. 123–9. Canberra: Environment Australia.

Timms, B. V. (1999). Local runoff, Paroo floods and water extraction impacts on wetlands of Currawinya National Park. In *A Free-Flowing River: the Ecology of the Paroo River*, ed. R. T. Kingsford, pp. 51–66. Hurstville: New South Wales National Parks and Wildlife Service.

Timms, B. V. and Boulton, A. J. (2001). Typology of arid-zone floodplain wetlands of the Paroo River (inland Australia) and the influence of water regime, turbidity, and salinity on their aquatic invertebrate assemblages. *Archiv für Hydrobiologie*, **153**, 1–27.

Tockner, K. and Stanford, J. A. (2002). Riverine flood plains: present state and future trends. *Environmental Conservation*, **29**, 308–30.

Townsend, C. R. (1989). The patch dynamics concept of stream community ecology. *Journal of the North American Benthological Society*, **8**, 36–50.

van der Valk, A. G. and Davis, C. B. (1978). The role of seed banks in the vegetation dynamics of prairie glacial marshes. *Ecology*, **59**, 322–35.

van der Wielen, M. (2002). Drying cycles as a switch between alternative stable states in wetlands. *Internationale Vereinigung für Theoretische und Angewandte Limnologie*, **28**, 1448–51.

Walker, K. F., Boulton, A. J., Thoms, M. C. and Sheldon, F. (1994). Effects of water-level changes induced by weirs on the distribution of littoral plants along the River Murray, South Australia. *Australian Journal of Marine and Freshwater Research*, **45**, 1421–38.

Walker, K. F., Sheldon, F. and Puckridge, J. T. (1995). A perspective on dryland river ecosystems. *Regulated Rivers: Research and Management*, **11**, 85–104.

Ward, J. V. (1998). Riverine landscapes: biodiversity patterns, disturbance regimes, and aquatic conservation. *Biological Conservation*, **83**, 269–78.

Ward, J. V., Tockner, K. and Schiemer, F. (1999). Biodiversity of floodplain river ecosystems: ecotones and connectivity. *Regulated Rivers: Research and Management*, **15**, 125–39.

Williams, W. D. (1995). Dryland rivers: a global review. *Australian Biologist*, **8**, 175–80.

Woinarski, J. C. Z., Brock, C., Armstrong, M. *et al.* (2000). Bird distribution in riparian vegetation in the extensive natural landscape of Australia's tropical savanna: a broad-scale survey and analysis of a distributional data base. *Journal of Biogeography*, **27**, 843–68.

Young, W. J. (1999). Hydrologic descriptions of semi-arid rivers: an ecological perspective. In *A Free-Flowing River: the Ecology of the Paroo River*, ed. R. T. Kingsford, pp. 77–96. Hurstville: New South Wales National Parks and Wildlife Service.

Young, W. J., Schiller, C. B., Harris, J. H., Roberts, J. and Hillman, T. J. (2001). River flow, processes, habitats and river life. In *Rivers as Ecological Systems: the Murray–Darling Basin*, ed. W. J. Young, pp. 45–99. Canberra: Murray Darling Basin Commission.

Zamora-Arroyo, F., Nagler, P. L., Briggs, N. *et al.* (2001). Regeneration of native trees in response to flood releases from the United States into the delta of the Colorado River, Mexico. *Journal of Arid Environments*, **49**, 49–64.

Zimmerman, J. C., DeWald, L. E. and Rowlands, P. G. (1999). Vegetation diversity in an interconnected ephemeral riparian system of north-central Arizona, USA. *Biological Conservation*, **90**, 217–28.

6

Natural disturbance and aquatic invertebrates in desert rivers

A. J. BOULTON, F. SHELDON AND K. M. JENKINS

INTRODUCTION

Flowing waters in deserts vary from ephemeral rills that carry water only after irregular and episodic downpours to the lowland stretches of perennial rivers whose headwaters are fed by groundwater interflow, snowmelt, or monsoonal rains. Many deserts have uncoordinated (arheic) drainage patterns. Here, flow may depend as much on where in the desert rain fell as on the weak gradients of the poorly defined channels. In other desert areas, meandering endorheic channels end in internal basins that can contain water for long periods of time. For example, the Lake Eyre Basin is a large endorheic drainage system in Australia that fills irregularly in response to erratic incursions of moist tropical air from the north. Major floods can occur, associated with La Niña phases of the El Niño Southern Oscillation (Puckridge *et al.*, 2000), triggering 'booms' in productivity of waterbirds, fish, and invertebrates (Kingsford *et al.*, 1999; Timms, 1999; Chapters 2, 4, and 7, this volume).

In a few desert areas, springs emerge but usually only flow a short distance before percolating into the alluvium (Jacobson *et al.*, 1995; Boulton & Williams, 1996). These desert springs have relatively consistent physicochemical characteristics but support fewer invertebrate

Ecology of Desert Rivers, ed. R. T. Kingsford. Published by Cambridge University Press.
© Cambridge University Press 2006.

species compared with nearby intermittent desert streams (Meffe & Marsh, 1983). None the less, many of these desert springs have high conservation value, supporting relictual invertebrate communities (see, for example, Davis *et al.*, 1993). One unique spring habitat is the mound spring, a natural outlet for subsurface waters rich in carbonate that precipitates out on evaporation to form 'mounds'. Although not strictly rivers, mound springs provide a significant permanent and mostly flowing water resource in an otherwise arid environment. Many mound springs in arid central Australia harbour endemic invertebrates, dominated by hydrobiid molluscs and crustaceans with limited mobility and dispersal potential (Harris, 1992). Springs also occur in 'cold' desert areas (e.g. much of Oregon and Nevada in the western USA) and, like their hot-desert counterparts, support relatively few aquatic invertebrate taxa (Cushing, 1996).

In this chapter, we briefly review the roles played by aquatic invertebrates in desert rivers before exploring how their community composition and abundance in desert streams are governed by the twin natural disturbances of flooding and drying. Recovery by aquatic macroinvertebrates after scouring flash floods in an Arizonan desert stream is compared with invertebrate responses to the more gradual inundation across immense Australian desert river floodplains. Impacts of drying on aquatic invertebrates are assessed in the context of deteriorating water quality in the drying pools of desert streams and the broader disruptions of hydrological and ecological connectivity in floodplain desert river systems. Fundamental to invertebrate responses to both these physical disturbances are their recolonisation mechanisms, ranging from aerial adults to resting eggs. We conclude that, in nearly every desert stream and river, invertebrates play key roles in many ecosystem processes whose rates, significance and timing are governed by the alternating natural hydrological disturbances of flooding and drying.

DIVERSITY, DENSITY AND COMMUNITY COMPOSITION OF INVERTEBRATES IN DESERT RIVERS

Aquatic invertebrate species richness in desert rivers is typically lower than in equivalent-sized permanent rivers and tends to be correlated with the duration of flow and water permanence, the availability of suitable substrate and microhabitat complexity, and proximity to permanent refuges (reviews in Comín & Williams, 1994; Boulton *et al.*,

2000; Sheldon et al., 2002). Species richness and composition of some taxonomic groups (e.g. anostracan crustaceans) are usually higher and more variable in desert river pools and connected wetlands than in nearby discrete claypans (Timms, 2002). Regional biodiversity of these groups may be especially high, owing to the presence of many distinct types of wetland, each of which harbours characteristic species (Madden et al., 2002; Timms, 2002).

Recent repeated sampling in desert rivers in the Lake Eyre Basin, central Australia, has revealed surprisingly high diversities of rotifers over time: although at any one particular time the community may be dominated by only a few taxa, these change according to the hydrological connections among the rivers and their floodplain wetlands (ARIDFLO unpublished data, Dr R. Shiel, personal communication, 2003). A similar process probably accounts for the reasonably high diversity of macroinvertebrate taxa collected across the Lake Eyre Basin (Madden et al., 2002); this spatial spectrum of hydrological isolation interspersed with periods of floodplain inundation and connection is considered necessary for the maintenance of the ecological integrity of arid-zone rivers (Choy et al., 2002).

Although invertebrate diversity might be low in many desert streams and rivers, densities can be extremely high. For example, in Sycamore Creek, a desert stream in Arizona, densities of macroinvertebrates reached 146 000 individuals per square metre (Grimm & Fisher, 1989). In another Arizonan desert stream, Aravaipa Creek, maximum macroinvertebrate densities averaged 167 000 m^{-2} during winter 1976 (calculated from Table 6 in Meffe & Minckley, 1987). Similar maximum densities (130 000–155 000 m^{-2}) were collected in unshaded riffles and runs of several desert streams draining the Flinders Ranges, South Australia (A. J. Boulton, unpublished data), (Fig. 6.1).

Desert river macroinvertebrate communities are dominated by insects, especially mobile bugs (Hemiptera), beetles (Coleoptera) and true flies (Diptera). This last group, represented by chironomid midges, simuliid blackflies and biting ceratopogonids, typically dominates secondary production; most of these species have rapid life cycles (days to months) (Gray, 1981). Although diversity is low, rates of secondary production are extremely high (see later) when the systems contain water. The microinvertebrate community composition of lowland desert rivers and their inundated floodplains is typically dominated by rotifers and microcrustaceans (Jenkins & Boulton, 2003). As with macroinvertebrates, secondary production rates are high whereas

Figure 6.1. Arkaba Creek, Flinders Ranges, South Australia. This dryland creek flows for only a few months but densities of macroinvertebrates can reach 150 000 individuals m^{-2} (Photo A. Boulton).

community composition tends to be dominated by relatively few taxa at any one time (Jenkins, 2002).

Patterns in invertebrate diversity in desert rivers appear to be controlled by fine-scale variables such as microhabitat complexity and by broad-scale factors including river size, drainage pattern, and flooding and drying disturbances. The interactions between microhabitat availability and physical disturbance have hardly been explored in desert rivers but this is a promising line of research. For example, in the Colorado River, driftwood supports high numbers of taxa by providing a richer source of biofilm than cobbles, a refuge during floods, and habitat for secondary consumers such as dragonflies and caddis flies (Haden *et al.*, 1999). Many desert rivers are turbid and have limited amounts of stable substratum so that hard surfaces such as driftwood become disproportionately important microhabitats for aquatic invertebrates, especially during floods. Another environmental factor likely to be exacerbated in desert rivers draining sparsely vegetated catchments is the abrasive effect on benthic invertebrates exerted by the high suspended sediment loads; here again, driftwood appears to play a key role as a refuge (Haden *et al.*, 1999).

THE FUNCTIONAL ROLES OF INVERTEBRATES IN DESERT RIVERS

In desert streams, invertebrates play much the same roles as they do in temperate systems, mediating functional processes such as organic matter breakdown, feeding on algae and microbial biofilms, transforming nutrients, and providing an important food source for fish, waterbirds, amphibians and other vertebrates (Boulton, 1999). However, the relative importance of some of these invertebrate roles deviates significantly from the predictions by models of river function based on temperate systems. For example, terrestrially derived carbon in some desert streams appears to be less important to aquatic food webs than in comparable temperate streams. In one turbid desert river, Cooper Creek in central Australia, filamentous algae in the shallow littoral zone is the major source of energy for aquatic consumers, ultimately supporting high densities of fish and crustaceans such as shrimps (Bunn et al., 2003). In another desert river (Colorado River, Arizona), dual stable isotope analysis indicated that leaf material was not the primary food for invertebrates associated with leaf packs, and cottonwood *Populus fremontii* leaf breakdown rates were low, probably owing to low densities of shredding invertebrates (Pomeroy et al., 2000). In contrast, *P. fremontii* leaf litter broke down rapidly in Sycamore Creek, an Arizonan desert stream where shredders were also rare but water temperatures were much higher (Schade & Fisher, 1997). More data from other rivers are needed to determine the range of variability in organic matter decomposition processes in these ecosystems but it seems that, unlike most temperate streams, desert stream invertebrates are less involved in leaf decomposition and do not rely heavily on terrestrial sources of organic matter (Davis et al., 1993; Sheldon et al., 2002; Bunn et al., 2003; Chapter 4, this volume).

From the few detailed studies on the invertebrate ecology of desert streams, it appears that where surface water persists for longer than a few weeks, aquatic invertebrates rapidly become abundant enough to contribute substantially to secondary production. This can have significant repercussions for associated riparian environments. For example, secondary production of aquatic macroinvertebrates in Sycamore Creek exceeded 120 g m^{-2} y^{-1}; although standing biomass was similar to that of similar temperate streams, turnover was much higher (Jackson & Fisher, 1986). Much of this invertebrate production transfers to adjacent terrestrial ecosystems. A tracer study indicated that orb-weaving spiders (Araneidae and Tetragnathidae) in the riparian vegetation of Sycamore Creek obtained nearly 100% of their carbon from eating emerging aquatic insects (Sanzone *et al.*, 2003). In a desert river in south-east Spain, the mayfly *Caenis luctuosa* reproduced continuously throughout the year despite the considerable spatial intermittency of flow, with an annual estimated dry mass production of 6.35 g m^{-2} y^{-1}, far higher than previous measurements for this genus in temperate streams (Perán *et al.*, 1999).

In some of the larger desert rivers, secondary production of zooplankton can also be substantial, driven by high water temperatures and the ready availability of pulsed nutrients and carbon associated with floodplain inundation (Boulton & Jenkins, 1998). These zooplankton, many derived from resting stages in the sediments, underpin a complex food web culminating in fish and waterbirds that feed on the high productivity derived from newly inundated floodplains in these dryland rivers (Kingsford *et al.* 1999; Timms, 1999; Jenkins & Boulton, 2003). Again, the rapid rates of secondary production by aquatic invertebrates play a key role in the 'boom and bust' cycles that typify most desert ecosystems, driven by the irregular alternating patterns of water availability and drawdown.

THE CONTRASTING DISTURBANCES OF FLOODING AND DRYING

Desert ecosystems, by definition, are water-limited. As terrestrial vegetation is usually sparse, catchments have little storage capacity so runoff in desert streams is swift, flashy, and often occurs as sheetflow across the largely impermeable desert landscape (Jacobson *et al.*, 1995). In hilly country, this causes sudden scouring flash floods capable of removing up to 98% of the standing crop of invertebrate fauna (Fisher *et al.*, 1982) and resetting successional sequences of invertebrate community composition (Meffe & Minckley, 1987; Grimm & Fisher, 1989; Boulton *et al.*, 1992a). At the other end of the hydrological spectrum lies the prolonged disturbance of drying. In most desert streams, the bed is dry for far longer than it holds water. Few aquatic macroinvertebrates can physiologically tolerate drying or water loss for longer than ten days, and their persistence in desert streams largely relies on an ability to recolonise rapidly from nearby refuges (Carl, 1989; Stanley *et al.*, 1994; Pires *et al.* 2000). Once established, many of them have extremely short life cycles (e.g. less than 3 weeks from egg to adult) (Gray, 1981) and populations build up swiftly.

The twin disturbances of flooding and drying change aquatic invertebrate community composition over time. Depending on the severity and type of disturbance, the composition of the assemblage that either resists the disturbance or recolonises soon after influences many ecosystem processes and governs subsequent succession trajectories. One of the best examples of this is the variable community response to differently sized flash floods in Sycamore Creek (Grimm & Fisher, 1989). Although recovery after each flash flood was swift, the community composition of the newly arrived colonists differed seasonally. For eight post-flood sequences in Sycamore Creek, chironomid larvae, mayfly nymphs (*Tricorythodes* and *Fallceon*) and oligochaetes dominated spring post-flood colonists whereas after summer floods early recolonisation was primarily by the caddis fly larvae *Cheumatopsyche* and *Helicopsyche* and the snail *Physella* (Boulton *et al.*, 1992a). The spring recolonists are primarily collector–gatherer invertebrates feeding on benthic fine particulate organic matter, whereas the summer colonists are filter-feeders or graze benthic algae. The implications of this difference in post-flood dominance by different functional feeding groups has not been explored but grazing would be expected to influence summer primary productivity more than it does in spring. In turn, this could have repercussions on nutrient uptake in this nitrogen-limited stream, potentially altering algal succession and subsequent resistance to later floods (Grimm & Fisher, 1989).

Relative densities of aquatic invertebrates in Sycamore Creek also varied within seasons among years. For example, the three mayfly taxa that were common in spring 1985 and 1986 virtually disappeared in spring 1987, whereas densities of snails and chironomids increased over the three years (Boulton *et al.*, 1992a). Such annual differences in response trajectories have not been studied in other desert streams so we must be cautious about speculating on the effects of disturbance on functional roles of invertebrates on the basis of such limited datasets. Longer-term cycles of drought are superimposed on the short, irregular flow regimes of desert streams; responses to this ramp disturbance (Lake, 2000, 2003) are likely to be lagged and complex.

INVERTEBRATE RESPONSES TO FLASH FLOODS IN DESERT
STREAMS

In upland desert streams, heavy and intense rainfall on relatively impervious arid catchments results in flash floods capable of virtually
eliminating all benthic algae and removing up to 98% of the invertebrate standing crop (Fisher *et al.*, 1982). However, the effects of these
flash floods are not simple functions of the magnitude of the flood or
the time since last flood. Grimm & Fisher (1989) assessed aquatic invertebrate responses to 13 flash floods varying in maximum discharge from
0.3 to 58.4 m^3 s^{-1} in Sycamore Creek, an Arizonan desert stream.
Although the decline in macroinvertebrate densities was related
broadly to the magnitude of maximum discharge, the trend was not
always consistent; later reanalysis of these data (Boulton *et al.*, 1992a)
showed that some common taxa displayed variable degrees of resistance to similarly sized floods. For example, a flash flood of 2.4 m^3 s^{-1}
reduced densities of water mites, chironomid midges and stratiomyid
dipteran larvae, but two other events of similar size (3.3 and 2.4 m^3 s^{-1})
did not have any significant effect. This differential resistance to
flooding appears to be related to temporal variability in the resistance
of benthic algal mats in Sycamore Creek as well as a strong seasonal
component to assemblage composition in this desert stream (Boulton
et al., 1992a).

Resilience, assessed as rate of recovery, was consistently high
in Sycamore Creek. Densities of aquatic macroinvertebrates reached
20 000 individuals m^{-2} in 8–36 days and 40 000 m^{-2} in 13–52 d, with
recovery being significantly faster in warm, low-flow seasons (Grimm &
Fisher, 1989). Such high resilience by aquatic macroinvertebrates to
flash floods has been reported in other desert streams (Iraq: Carl,
1989; Australia: Boulton & Williams, 1996; Oman: Victor & Al-Mahrouqi,
1996; Arizona: Lytle, 2000; Portugal: Pires *et al.*, 2000) and largely results
from aerial recolonisation. Recolonisation patterns in Sycamore Creek
are seasonal, with adult beetles (Coleoptera) and bugs (Hemiptera)
being the main colonists in winter whereas mayflies (Ephemeroptera)
and true flies (Diptera) dominate colonisation after summer floods
(Gray & Fisher, 1981). Although flash floods can cause high mortality
in the juvenile aquatic stages of many desert stream insects, these
ecological effects may be mitigated by the evolution of life-history
strategies that allow the terrestrial adult to avoid floods. For example,

86% of emergence by a stream caddis fly in the Chihuahuan Desert (Arizona) occurred before the long-term mean arrival date of the first seasonal flood, suggesting compensation for disturbance by flash floods by having emergence strategies synchronised with long-term flood dynamics (Lytle, 2002).

Flash floods also influence aquatic invertebrates living within the saturated sediments of desert streams in the hyporheic zone where surface water exchanges with subsurface groundwater. In Sycamore Creek, densities of most taxa of hyporheic invertebrates down to 30 cm declined after flash flooding even though the bed sediments were not scoured away (Boulton & Stanley, 1995). However, return to pre-flood densities by most taxa occurred within 5 d, indicating a similar high resilience to that observed for surface invertebrates. In this case, recolonisation appeared to come from deeper in the sediments (> 50 cm) where there was little change in community composition in response to flooding (Boulton & Stanley, 1995). The functional significance of hyporheic invertebrates to desert stream ecosystems is unknown but they probably play a substantial role in organic matter decomposition because many of the taxa are detritivorous (Boulton et al., 1992b). Perhaps their feeding activities mobilise and break down particulate organic matter in the stream sediments during flash floods, rendering nutrients and carbon available for surface processes located above upwelling zones where hyporheic water enters the stream.

INVERTEBRATE RESPONSES TO FLOODPLAIN INUNDATION IN DESERT RIVERS

In contrast to the flash floods described above, floodplain inundation in some lowland desert rivers is typically much slower, with floodwaters taking months to pass from upper to lower reaches. For example, in the Darling River, Australia, summer floods creep over hundreds of kilometres of dry sediments with gradients as little as 4 cm km^{-1}, and eventually inundate the downstream reaches (Jenkins & Boulton, 2003) (Fig. 6.2). When the vast expanses of floodplain sediments that have lain dry for up to several decades are inundated, the floodwaters can disappear beneath the surface for days, re-wetting the dark grey clay soils and filling the gilgais (holes) and cracks that formed during drying. Microcrustaceans and rotifers colonise many desert rivers from diapausing resting stages in cracks and dry floodplain sediments (Boulton

Figure 6.2. Floodwaters draining the Darling River along Malta Creek, New South Wales. These floodwaters typically spread slowly across the floodplain, initiating a 'boom' period of productivity as resting stages of plants and invertebrates hatch from dry sediments, terrestrial nutrients are released and dissolved, and floodplain wetlands are flushed and rejuvenated (Photo K. Jenkins).

& Lloyd, 1992; Jenkins & Boulton, 1998; Skinner *et al.*, 2001). Sediments collected from lakes dry for 6–20 years along the Darling River flood-plain were inundated in laboratory microcosms. Within 1–3 days, a community dominated by protozoans, nematodes and bdelloid rotifers developed (Jenkins, 2002). After a week, the appearance of crustacean nauplii and rotifers caused a marked shift in community composition, particularly in the more recently flooded sediments. If the loss in productivity and diversity observed in microcosms can be scaled up to microinvertebrate responses to floods in whole desert river ecosystems (Jenkins & Boulton, 2003), human activities that extend durations of drying beyond *c*. 20 years will affect ecosystem productivity and diversity in these desert rivers.

During the course of a gradual flood, invertebrate assemblage composition in the advancing floodwaters 'ages' owing to differences in colonisation strategies, availability of bacteria, algae and nutrients, the presence of predators, and seasonal differences in temperature and dissolved oxygen. For example, in the semi-arid Lachlan River in Australia, a productive band of *Chironomus tepperi*, the dominant chironomid midge present after flooding, moved with the advancing floodwater margin through a continual cycle of emergence and egg laying (Maher & Carpenter, 1984). Larvae of *C. tepperi* appeared in the water column and mud within 5–7 days of flooding; maximum densities of 10 922 m^{-2} occurred 7 weeks later (Maher & Carpenter, 1984). The leading edge of these floodwaters is a 'moving littoral zone', carrying masses of organic matter, microcrustaceans, rotifers and even chironomid larvae (Maher & Carpenter, 1984; Jenkins & Boulton, 2003). As the slowly creeping waters drown or dislodge terrestrial invertebrates in the gilgais and cracks, opportunistic predators such as ibis follow the flood front, picking off spiders, crickets and other prey (R. Kingsford, personal communication, 2003), representing a potentially important yet transient pathway for carbon in these ecosystems.

Many large desert rivers inundate immense areas of floodplain over prolonged periods of time. Over one third of the catchment of the Cooper Creek and Diamantina River in the arid Lake Eyre Basin comprises floodplain soils, suggesting that a large proportion of the total catchment can be inundated during large flood events (Graetz, 1980). The hydrology of large desert rivers is exceptionally variable (Chapter 2, this volume), with Cooper Creek and the Diamantina River being two of the most variable in the world (Puckridge *et al.*, 1998). These desert rivers have hydrographs characterised by exceptionally large floods of long duration separated by periods of smaller irregular flooding or, more commonly, no flow. So it is not only the large floods that influence the structure of biotic communities in these rivers but also the lengthy periods of zero flow where the watercourses are reduced to disconnected waterbodies (Fig. 6.3) or aquatic refugia in an otherwise terrestrial environment (Sheldon *et al.*, 2002).

In these desert rivers, the diversity of the invertebrate assemblage for any one sampling occasion appears to be low compared with other rivers worldwide. For example, 70 taxa were recorded from habitats in the Coongie Lakes wetland complex on the lower Cooper Creek, central Australia (Sheldon *et al.*, 2002), 54 taxa from the Goyder Lagoon complex on the nearby Diamantina River (Sheldon & Puckridge, 1998), 80 taxa from the Murray and Darling Rivers in the neighbouring Murray–Darling Basin (Sheldon & Walker, 1998), and 95 taxa from the Chowilla floodplain on the lower River Murray, South Australia (Boulton & Lloyd, 1991). Comparing this with temperate European rivers, 251 taxa were

Figure 6.3. Clifton Hills Waterhole, Diamantina River, South Australia. The lengthy dry periods reduce watercourses to disconnected refugia, like this waterhole, where aquatic plants and animals must persist until the next flood. These waterholes also act as a magnet for terrestrial fauna that must come to drink or seek shelter in the vegetated riparian zone (Photo. F. Sheldon).

recorded from the Rhône and Ain rivers in France (Castella *et al.*, 1991), 131 from the river Po in Italy (Battegazzore *et al.*, 1992) and 206 taxa from the Morovian floodplain in the Czech Republic (Adamek & Sukop, 1992). Such relatively low values of taxon richness observed in Australian desert rivers are probably underestimates. Given the high level of spatial and temporal variability in flooding, drying and waterhole connection, we predict that there would be increases in measured biodiversity with more intensive sampling effort at broader spatial and temporal scales. Recent basin-scale research programmes on desert rivers, spanning multiple rivers and years, reported high aquatic invertebrate biodiversity (ARIDFLO unpublished data, J. Reid, personal communication, 2003).

The extreme spatial and temporal variability of habitats in these large desert rivers creates a relatively 'harsh' environment that favours generalist taxa with high vagility and low habitat specificity. This is reflected in the typical composition of macroinvertebrate assemblages

Figure 6.4. Koonchera Dune Waterhole, Goyders Lagoon, Diamantina River, South Australia. The invertebrate food web of this water hole is dominated by 'collectors' that eat the fine particulate organic matter resting on and in the substrate (Photo F. Sheldon).

in desert rivers, where highly mobile taxa such as hemipteran bugs, coleopteran beetles and chironomid midge larvae dominate the species diversity in the vast majority of sites. In samples from Cooper Creek and the Diamantina River, insects were the dominant group, accounting for 70% and 76% of taxa, and 55% and 63% of individual abundance, respectively (Sheldon & Puckridge, 1998; Sheldon et al., 2002). More sedentary, specialist taxa such as gastropods and bivalves occur mostly in the more permanent channels and waterholes (Sheldon & Puckridge, 1998; Sheldon et al., 2002). In waterholes of the Goyder Lagoon on the Diamantina River (Fig. 6.4), invertebrate collectors were the dominant feeding group across all habitats (Sheldon & Puckridge, 1998). The absence of shredders suggests that the major allochthonous inputs may not be derived directly from litter fall in the riparian zone, and that collectors use fine particulate organic matter (of both allochthonous and autochthonous origin) present within the substrate.

Under low flow and drying conditions, waterholes and lakes along desert rivers become 'aquatic patches' (refugia) in a dry terrestrial environment (Davis et al., 1993). As individual aquatic species differ in their probability of colonising the aquatic patches, owing to their differential abilities to disperse and the effects of varying distances between the refugia (cf. Olden *et al.*, 2001), waterholes that appear environmentally similar may have dissimilar faunas because of differential chances of colonisation and extinction for each species. In the Lake Eyre Basin rivers, connection and disconnection of waterbodies along desert rivers influences the aquatic community in a sequential fashion associated with landscape-level fluctuations in hydrology (Sheldon *et al.*, 2002). As water levels fall and the river becomes increasingly fragmented into disconnected waterbodies, the community composition of aquatic invertebrates in each site diverges according to those species present at the time of disconnection and the prevailing biotic and abiotic selective forces at work since disconnection. In this way, each site behaves like a unique, natural mesocosm and differs from other sites in ways that are not explained readily by measured physical or water chemical variables.

INVERTEBRATE RESPONSES TO DRYING AND WATER LOSS

Most desert rivers dry for months to decades (Chapter 2, this volume). Water loss has led to a variety of physiological and behavioural adaptations by desert river invertebrates. For example, the Australian desert crab *Holthuisana transversa* can recover from the loss of up to 43% of its total body water, produces minimal amounts of urine, and during dry periods can reduce its metabolism by 70% of its resting rate (Greenaway, 1984). Many desert river microcrustaceans and rotifers produce desiccation-resistant resting stages that persist in the sediments for decades (Jenkins & Boulton, 1998); viable resting stages of some invertebrates have been aged at 332 years from museum collections (Hairston *et al.*, 1995). However, it is not known whether these resting stages can persist for centuries in the harsher environment of the beds and floodplains of desert rivers.

In general, few aquatic insects or molluscs can physiologically resist rapid drying. In Sycamore Creek, invertebrate mortality in drying pools was high, owing to a combination of increased predation and deteriorating water quality (Stanley *et al.*, 1994). Taxa such as mayflies with high oxygen requirements were notably absent during the final stages of drying, whereas invertebrates capable of using atmospheric oxygen (e.g. adult beetles, corixid bugs) or possessing haemoglobin (*Chironomus* midges) persisted as the water and sediments became hypoxic

(Stanley *et al.*, 1994). Drying is rapid in Sycamore Creek, surface sediments become hot ($>$ 60 °C), and deeper sediments lose moisture rapidly and are prone to scour during summer flash floods (Gray, 1981). As a result, virtually no invertebrates in this stream burrow into the sediments to avoid drying or have desiccation-resistant stages (Boulton *et al.*, 1992c). Similarly, in desert wadis in western Algeria that dry for 4–6 months, only a few dipteran fly larvae were found in the sediments during the dry season and this hyporheic biotope was not considered to be an important refuge (Gagneur & Chaoui-Boudghane, 1991).

In Sycamore Creek, some aquatic invertebrates chased receding water margins upstream but were often stranded because drying was too rapid (Stanley *et al.*, 1994). Other invertebrates such as the dytiscid beetles *Laccophilus maculosus* and *Stictotarsus roffii*, the hydrophilid beetle *Tropisternus ellipticus* and the corixid bug *Graptocorixa serrulata* dispersed by flying or crawling when periods of drought caused stream pools to contract (Velasco & Millan, 1998). The majority of invertebrates in this stream apparently persist during the dry period by flying away to seek nearby waterbodies or humid refuges (Boulton *et al.*, 1992c) because they have little resistance to desiccation (Velasco & Millan, 1998). They also have very short aquatic life cycles, enhancing the probability of completing their development before disturbance by flash floods or drying. For example, mayflies, small dipterans and the corixid bug *G. serrulata* develop within three weeks and reproduce continuously (Gray, 1981). Mortality among aquatic invertebrates that cannot evade drying is high (Stanley *et al.*, 1994); their dead and dying bodies represent a food bonanza for terrestrial predators in the riparian zone (Chapter 7, this volume).

The periodically dry beds of some desert streams can be inhabited by a diverse assemblage of terrestrial invertebrates. For example, some 19 different orders of terrestrial invertebrates were collected from three sites on the Kruis River in the Western Cape Province, South Africa, by using pitfall traps within the dry stream bed over summer (Wishart, 2000). The inability of a large portion of the biomass of these invertebrates to escape the onset of flow by flying away suggests that periodically inundated areas of the stream channel constitute an ecologically significant *input* of food into the stream ecosystem (Wishart, 2000), perhaps partly compensating for predation of aquatic invertebrates during drying.

During drying in many desert streams, a steady increase in predator densities has been reported in the shrinking pools (Stanley *et al.*,

1994; Boulton & Williams, 1996). As predation pressure increases, the extent of organic matter decomposition carried out by detritivores declines, and organic matter processing may virtually stall during drying. Working on desert streams in Israel, Herbst & Reice (1982) demonstrated that leaf litter decomposition was retarded by periodic drying. Total colonisation rates of invertebrates were more important to rates of leaf decomposition than were the kinds of colonisers or the diversity of species (Herbst & Reice, 1982) but the extent to which this was controlled by predation was not clear.

At a system level, drying leads to fragmentation of stream reaches and disruption of transport mechanisms of organic matter, nutrients and invertebrate drift. Although the implications of this for nutrient spiralling and organic matter have been reviewed in Sycamore Creek (Stanley *et al.*, 1997), there are few studies of the effects of such fragmentation on system-level ecosystem processes mediated by invertebrates. In confined reaches where water persists, grazing pressure and organic matter breakdown may be temporarily accelerated as faunal densities rise during drying, but these processes ultimately slow down when grazers and detritivores fall prey for the escalating predator population. Although drying and loss of flow will interrupt transport of organic matter into these areas, it is possible that this may not be significant to aquatic invertebrates where secondary productivity is derived from algae in the shallow littoral zone (see, for example, Bunn *et al.*, 2003; Chapter 4, this volume). However, rapid drying and drawdown of water in these pools may expose the shallow marginal band of algae, reducing its availability to grazers. Furthermore, access by livestock to the edges of these pools (cf. Choy *et al.*, 2002) with concomitant trampling, siltation, and physical disruption of the shallow littoral zone is also likely to impact on this productive region.

CONCLUSION: WHY NATURAL DISTURBANCES ARE ESSENTIAL TO INVERTEBRATES IN DESERT RIVERS

Although invertebrate diversity in desert rivers is lower than in their temperate counterparts, aquatic invertebrates still play central roles in many riverine ecosystem foodwebs by grazing algal biofilms and fungi while providing prey for other invertebrates, fish, waterbirds, and terrestrial predators. High water temperatures promote rapid turnover, recovery after flooding and drying is swift, and secondary production is substantial when water is in the system. These 'boom' periods are interspersed with long, unproductive 'bust' times. Many invertebrates have physiological and behavioural adaptations for dealing with the twin disturbances of drying and flooding, and resilience in most desert rivers is high.

The term 'disturbance' is a misnomer because the 'system reset-ting' that occurs with the fluctuating presence of water is fundamental to the boom and bust ecology of all desert rivers from small ephemeral streams to huge floodplain rivers. The 'real' disturbance occurs when the natural cycle of hydrological variation is destroyed by human activities such as the effects of river regulation on invertebrates of the Colorado River (see, for example, Stevens *et al.*, 1997) or excessive water abstrac-tion on mound springs (Harris, 1992) and desert floodplain wetlands (review in Boulton *et al.*, 2000; Chapter 8, this volume). The effects of other impacts on aquatic invertebrates, such as eutrophication (Bromley & Por, 1975), salinisation (Genxu & Guodong, 1999; Chapter 10, this volume) or riparian and bank damage by stock access (Choy *et al.*, 2002) tend to be exacerbated in desert rivers during periods of low or zero flow. Superimposed on these catchment-scale disruptions are the broader effects of climate change on desert aquatic ecosystems (Molles & Dahm, 1990; Grimm *et al.*, 1997), widespread trends of overpopulation in arid areas across the globe (Victor & Al-Mahrouqi, 1996; Davies & Day, 1998), and the uncertain outcomes of political tension and intense competi-tion for scarce water resources in socially volatile parts of the world (Wishart *et al.*, 2000; Chapter 11, this volume). Although we are unaware of the true value of 'ecosystem services' provided by aquatic inverte-brates in desert rivers, preservation of the naturally variable flow regime is essential so that the disturbances of flooding and drying can continue to drive the vital 'boom and bust' cycles of these threatened ecosystems.

ACKNOWLEDGEMENTS

We are grateful to Dr Richard Kingsford for the invitation to contribute to this book, and to a number of our colleagues and friends for useful discussions about aspects of invertebrate ecology covered in this paper, provision of references, or comments on early drafts of the manuscript: Dr Jim Puckridge, Professor Stuart Bunn, Mr John Burt, Dr Emily Stanley, Mr Derek 'Des' Buschman, Mr Julian Reid, and Dr

Satish Choy. We also thank two anonymous referees and Richard Kingsford for useful comments and suggestions that improved our paper.

REFERENCES

Adamek, Z. and Sukop, I. (1992). Invertebrate communities of former southern Moravian floodplains (Czechoslovakia) and impacts of regulation. *Regulated Rivers: Research and Management*, **7**, 181–92.

Battegazzore, M., Petersen, R. C., Moretti, G. and Rossaro, B. (1992). An evaluation of the environmental quality of the River Po using benthic macroinvertebrates. *Archiv für Hydrobiologie*, **125**, 175–206.

Boulton, A. J. (1999). Why variable flows are needed for invertebrates of semi-arid rivers. In *A Free-Flowing River: The Ecology of the Paroo River*, ed. R. T. Kingsford, pp. 113–28. Hurstville, New South Wales: NSW National Parks and Wildlife Service.

Boulton, A. J. and Jenkins, K. M. (1998). Flood regimes and invertebrate communities in floodplain wetlands. In *Wetlands in a Dry Land: Understanding for Management*, ed. W. D. Williams, pp. 137–48. Canberra, Australian Capital Territory: Environment Australia.

Boulton, A. J. and Lloyd, L. N. (1991). Macroinvertebrate assemblages in floodplain habitats of the lower River Murray, South Australia. *Regulated Rivers: Research and Management*, **6**, 183–201.

    (1992). Flooding frequency and invertebrate emergence from dry floodplain sediments of the River Murray, Australia. *Regulated Rivers: Research and Management*, **7**, 137–51.

Boulton, A. J. and Stanley, E. H. (1995). Hyporheic processes during flooding and drying in a Sonoran Desert stream. II. Faunal dynamics. *Archiv für Hydrobiologie*, **134**, 27–52.

Boulton, A. J. and Williams, W. D. (1996). Aquatic biota. In *The Natural History of the Flinders Ranges*, ed. C. R. Twidale, M. J. Tyler and M. Davies, pp. 102–12. Adelaide, South Australia: Royal Society of South Australia.

Boulton, A. J., Peterson, C. G., Grimm, N. B. and Fisher, S. G. (1992a). Stability of an aquatic macroinvertebrate community in a multi-year hydrologic disturbance regime. *Ecology*, **73**, 2192–207.

Boulton, A. J., Sheldon, F., Thoms, M. C. and Stanley, E. H. (2000). Problems and constraints in managing rivers with variable flow regimes. In *Global Perspectives on River Conservation: Science, Policy and Practice*, ed. P. J. Boon, B. R. Davies and G. E. Petts, pp. 411–26. London: John Wiley and Sons.

Boulton, A. J., Stanley, E. H., Fisher, S. G. and Lake, P. S. (1992c). Over-summering strategies of macroinvertebrates in intermittent streams in Australia and Arizona. In *Aquatic Ecosystems in Semi-arid Regions: Implications for Resource Management*, ed. R. D. Robarts and M. L. Bothwell, pp. 227–37. Saskatoon, Canada: Environment Canada.

Boulton, A. J., Valett, H. M. and Fisher, S. G. (1992b). Spatial distribution and taxonomic composition of the hyporheos of several Sonoran Desert streams. *Archiv für Hydrobiologie*, **125**, 37–61.

Bromley, H. J. and Por, F. R. (1975). The metazoan fauna of a sewage-carrying wadi Nahal Soreq (Judean Hills, Israel). *Freshwater Biology*, **5**, 121–33.

Bunn, S. E., Davies, P. M. and Winning, M. (2003). Sources of organic carbon supporting the food web of an arid zone floodplain river. *Freshwater Biology*, 48, 619–35.

Carl, M. (1989). The ecology of a wadi in Iraq with particular reference to colonization strategies of aquatic macroinvertebrates. *Archiv für Hydrobiologie*, 166, 499–515.

Castella, E., Richardot-Coulet, M., Roux, C. and Richoux, P. (1991). Aquatic macroinvertebrate assemblages of two contrasting floodplains: the Rhone and Ain Rivers, France. *Regulated Rivers: Research and Management*, 6, 289–300.

Choy, S. C., Thomson, C. B. and Marshall, J. C. (2002). Ecological condition of central Australian arid-zone rivers. *Water Science and Technology*, 45, 225–32.

Comín, F. A. & Williams, W. D. (1994). Parched continents: our common future? In *Limnology Now: a Paradigm of Planetary Problems*, ed. R. Margalef, pp. 473–527. Amsterdam: Elsevier Science.

Cushing, C. E. (1996). The ecology of cold desert spring-streams. *Archiv für Hydrobiologie*, 135, 499–522.

Davies, B. R. and Day, J. A. (1998). *Vanishing Waters*. Cape Town, South Africa: University of Cape Town Press and Juta Press.

Davis, J. A., Harrington, S. A. and Friend, J. A. (1993). Invertebrate communities of relict streams in the arid zone: the George Gill Range, central Australia. *Australian Journal of Marine and Freshwater Research*, 44, 483–505.

Fisher, S. G., Gray, L. J., Grimm, N. B. and Busch, D. E. (1982). Temporal succession in a desert stream following flash flooding. *Ecological Monographs*, 52, 93–110.

Gagneur, J. and Chaoui-Boudghane, C. (1991). Sur le rôle du milieu hyporhéique pendant l'assèchement des oueds de l'Oest Algérien. *Stygologia*, 6, 77–89.

Genxu, W. and Guodong, C. (1999). The ecological features and significance of hydrology within arid inland river basins of China. *Environmental Geology*, 37, 218–22.

Graetz, R. D. (1980). *The Potential Application of Landsat Imagery to Land Resource Management in the Channel Country*. Perth: CSIRO Division of Land Resources Management.

Gray, L. J. (1981). Species composition and life histories of aquatic invertebrates in a lowland Sonoran Desert stream. *American Midland Naturalist*, 106, 229–42.

Gray, L. J. and Fisher, S. G. (1981). Postflood recolonization pathways of macroinvertebrates in a lowland Sonoran Desert stream. *American Midland Naturalist*, 106, 249–57.

Greenaway, P. (1984). Survival strategies in desert crabs. In *Arid Australia*, ed. H. G. Cogger and E. E. Cameron, pp. 145–52. Chipping Norton, New South Wales: Surrey Beatty & Sons.

Grimm, N. B. and Fisher, S. G. (1989). Stability of periphyton and macroinvertebrates to disturbance by flash floods in a desert stream. *Journal of the North American Benthological Society*, 8, 293–307.

Grimm, N. B., Chacón A., Dahm C. N. *et al.* (1997). Sensitivity of aquatic ecosystems to climatic and anthropogenic changes: The Basin and Range, American Southwest and Mexico. *Hydrological Processes*, 11, 1023–41.

Haden, G. A., Blinn, D. W., Shannon, J. P. and Wilson, K. P. (1999). Driftwood: an alternative habitat for macroinvertebrates in a large desert river. *Hydrobiologia*, 397, 179–86.

Hairston, N. G., Van Brunt, R. A. and Kearns, C. M. (1995). Age and survivorship of diapausing eggs in a sediment egg bank. *Ecology*, 76, 1706–11.

Harris, C. R. (1992). Mound springs: South Australian conservation initiatives. *Rangelands Journal*, **14**, 157–73.

Herbst, G. and Reice, S. R. (1982). Comparative leaf litter decomposition in temporary and permanent streams in semi-arid regions of Israel. *Journal of Arid Environments*, **5**, 305–18.

Jackson, J. K. and Fisher, S. G. (1986). Secondary production, emergence, and export of aquatic insects of a Sonoran Desert stream. *Ecology*, **67**, 629–38.

Jacobson, P. J., Jacobson, K. M. and Seely, M. K. (1995). *Ephemeral Rivers and their Catchments: Sustaining People and Development in Western Namibia.* Windhoek, Namibia: Desert Research Foundation of Namibia.

Jenkins, K. M. (2002). Dynamics of aquatic microinvertebrate assemblages after inundation of an Australian dryland river floodplain. Ph.D. Thesis, University of New England, New South Wales.

Jenkins, K. M. and Boulton, A. J. (1998). Community dynamics of invertebrates emerging from reflooded lake sediments: flood pulse and aeolian influences. *International Journal of Ecological and Environmental Science*, **24**, 179–92.

(2003). Connectivity in a dryland river: short-term aquatic microinvertebrate recruitment following floodplain inundation. *Ecology*, **84**, 2708–23.

Kingsford, R. T., Curtin, A. L. and Porter, J. (1999). Water flows on Cooper Creek in arid Australia determine 'boom' and 'bust' periods for waterbirds. *Biological Conservation*, **88**, 231–48.

Lake, P. S. (2000). Disturbance, patchiness, and diversity in streams. *Journal of the North American Benthological Society*, **19**, 573–92.

(2003). Ecological effects of perturbation by drought in flowing waters. *Freshwater Biology*, **48**, 1161–72.

Lytle, D. A. (2000). Biotic and abiotic effects of flash flooding in a montane desert stream. *Archiv für Hydrobiologie*, **150**, 85–100.

(2002). Flash floods and aquatic insect life-history evolution: evaluation of multiple models. *Ecology*, **83**, 370–85.

Madden, C. P., McEvoy, P., Taylor, J. D. *et al.* (2002). Macroinvertebrates of watercourses in the Lake Eyre Basin, South Australia. *Verhandlungen der Internationalen Vereinigung für Theoretische und Angewandte Limnologie*, **28**, 591–600.

Maher, M. and Carpenter, S. M. (1984). Benthic studies of waterfowl breeding habitat in south-western New South Wales. II. Chironomid populations. *Australian Journal of Marine and Freshwater Research*, **35**, 97–110.

Meffe, G. K. and Marsh, P. C. (1983). Distribution of aquatic macroinvertebrates in three Sonoran Desert springbrooks. *Journal of Arid Environments*, **6**, 363–71.

Meffe, G. K. and Minckley, W. L. (1987). Persistence and stability of fish and invertebrate assemblages in a repeatedly disturbed Sonoran Desert stream. *American Midland Naturalist*, **117**, 177–91.

Molles, M. C. and Dahm, C. N. (1990). A perspective on El Niño and La Niña: global implications for stream ecology. *Journal of the North American Benthological Society*, **9**, 68–76.

Olden, J. D., Jackson, D. A. and Peres-Neto, P. R. (2001). Spatial isolation and fish communities in drainage lakes. *Oecologia*, **127**, 572–85.

Perán, A., Velasco, J. and Millan, A. (1999). Life cycle and secondary production of *Caenis luctuosa* (Ephemeroptera) in a semiarid stream (Southeast Spain). *Hydrobiologia*, **400**, 187–94.

Pires, A. M., Cowx, I. G. and Coelho, M. M. (2000). Benthic macroinvertebrate communities of intermittent streams in the middle reaches of the Guadiana Basin (Portugal). *Hydrobiologia*, **435**, 167–75.

Pomeroy, K. E., Shannon, J. P. and Blinn, D. W. (2000). Leaf breakdown in a regulated desert river: Colorado River, Arizona, USA. *Hydrobiologia*, **434**, 193–9.

Puckridge, J. T., Sheldon, F., Walker, K. F. and Boulton, A. J. (1998). Flow variability and the ecology of large rivers. *Marine and Freshwater Research*, **49**, 55–72.

Puckridge, J. T., Walker, K. F. and Costelloe, J. F. (2000). Hydrological persistence and the ecology of dryland rivers. *Regulated Rivers: Research and Management*, **16**, 385–402.

Sanzone, D. M., Meyer, J. L., Marti, E. *et al.* (2003). Carbon and nitrogen transfer from a desert stream to riparian predators. *Oecologia*, **134**, 238–50.

Schade, J. D. and Fisher, S. G. (1997). Leaf litter in a Sonoran Desert ecosystem. *Journal of the North American Benthological Society*, **16**, 612–26.

Sheldon, F. and Puckridge, J. T. (1998). Macroinvertebrate assemblages of Goyder Lagoon, Diamantina River, South Australia, *Transactions of the Royal Society of South Australia*, **122**, 17–31.

Sheldon, F. and Walker, K. F. (1998). Spatial distribution of littoral invertebrates in the lower Murray-Darling River system, Australia. *Marine and Freshwater Research*, **49**, 171–82.

Sheldon, F., Boulton, A. J. and Puckridge, J. T. (2002). Conservation value of variable connectedness: aquatic invertebrate assemblages of channel and floodplain habitats of a central Australian arid-zone river, Cooper Creek. *Biological Conservation*, **103**, 13–31.

Skinner, R., Sheldon, F. and Walker, K. F. (2001). Propagules in dry wetland sediments as indicators of ecological health: effects of salinity. *Regulated Rivers: Research and Management*, **17**, 191–7.

Stanley, E. H., Buschman, D. L., Boulton, A. J., Grimm, N. B. and Fisher, S. G. (1994). Invertebrate resistance and resilience to intermittency in a desert stream. *American Midland Naturalist*, **131**, 288–300.

Stanley, E. H., Fisher, S. G. and Grimm, N. B. (1997). Ecosystem expansion and contraction in streams. *BioScience*, **47**, 427–36.

Stevens, L. E., Shannon, J. P. and Blinn, D. W. (1997). Colorado River benthic ecology in Grand Canyon, Arizona, USA: Dam, tributary and geomorphological influences. *Regulated Rivers: Research and Management*, **13**, 129–49.

Timms, B. V. (1999). Local runoff, Paroo floods and water extraction impacts on the wetlands of Currawinya National Park. In *A Free-Flowing River: The Ecology of the Paroo River*, ed. R. T. Kingsford, pp. 51–66. Hurstville, New South Wales: NSW National Parks and Wildlife Service.

(2002). Biogeography and ecology of Anostraca (Crustacea) in middle Paroo catchment of the Australian arid-zone. *Hydrobiologia*, **486**, 225–38.

Velasco, J. and Millan, A. (1998). Insect dispersal in a drying desert stream: effects of temperature and water loss. *Southwestern Naturalist*, **43**, 80–7.

Victor, R. and Al-Mahrouqi, A. I. S. (1996). Physical, chemical and faunal characteristics of a perennial stream in arid northern Oman. *Journal of Arid Environments*, **34**, 465–76.

Wishart, M. J. (2000). The terrestrial invertebrate fauna of a temporary stream in southern Africa. *African Zoology*, **35**, 193–200.

Wishart, M. J., Gagneur, J. and El-Zanfaly, H. T. (2000). River conservation in North Africa and the Middle East. In *Global Perspectives on River Conservation: Science, Policy and Practice*, ed. P. J. Boon, B. R. Davies and G. E. Petts, pp. 127–54. London: John Wiley and Sons.

7

Vertebrates of desert rivers: meeting the challenges of temporal and spatial unpredictability

R. T. KINGSFORD, A. GEORGES AND P. J. UNMACK

INTRODUCTION

Vertebrates, the 'charismatic megafauna' of desert rivers, form a minute part of the biodiversity dependent on flows. Despite their numerical insignificance, they are particularly important to humans for food (e.g. fishes) or conservation (e.g. mammals and waterbirds); given our impact on the ecology of these unique systems, this makes them an essential component of the ecology of desert rivers. They also play an important functional role as top-level predators, affecting prey abundance, or sometimes shaping habitat availability and extent (e.g. by enlarging waterholes). Vertebrates are often the focus for conservation effort and policies, with declines in waterbird and fish populations forcing changes, albeit belated, to river management (Lemly et al., 2000; see Chapter 8, this volume).

There are thousands of species over all the major groups of vertebrates (fishes, amphibians, reptiles, birds, mammals) dependent on desert rivers of the world. Aquatic vertebrates are obviously directly dependent on desert rivers but other more 'terrestrial' vertebrates, including elephants, birds of prey (e.g. fish eagles) and small mammals

Ecology of Desert Rivers, ed. R. T. Kingsford. Published by Cambridge University Press.

(Briggs, 1992), have life histories significantly affected by flooding and drying patterns of rivers. This chapter focusses primarily on aquatic vertebrates but also includes more terrestrial species dependent on riparian areas and water from rivers. Treatment of these more terrestrial groups is not exhaustive. The presence and availability of water for these species is essential for survival, particularly as a source for drinking and refuge from, or area for, predation. Survival and reproduction of many vertebrate species depend on the spatial and temporal disturbance patterns of desert rivers. From its source, the desert river forms habitats that range from riparian corridors along the river to the terminal lakes, estuaries and floodplains, sometimes extending more than 1000 km. In the lower parts of the river, the aquatic habitats provide complex food webs for vertebrates in an otherwise poorly resourced desert landscape. It is no accident that biodiversity around wetlands in the desert is spectacularly high (e.g. Lake Manyara, Tanzania; Okavango Swamp, Botswana; Macquarie Marshes, Australia).

One vertebrate species now dominates the ecology of most desert rivers. Our role is pervasive, often irrevocably changing a river's ecology and disturbance regimes with river regulation (see Chapters 8 and 11, this volume). Besides water resource development, fishing is probably the most important ecosystem service for people living on desert rivers (Thomas, 1995). In addition, vertebrates now underpin the economies of wildlife tourism in many desert regions of the world (e.g. Okavango) (Ellery & McCarthy, 1994). There is also a tradition for hunting water-fowl for sport, emanating from the northern hemisphere (Kear, 1990), which contributes to local economies. Albeit not at the same scale, other exotic species of vertebrate, particularly fishes, have also significantly changed the ecology of desert rivers.

Despite the human focus on vertebrates, our understanding of their ecology and interaction with desert rivers and disturbance regimes remains relatively superficial (Nilsson & Dynesius, 1994), even for fish ecology. This is also attributable to the lack of research effort on vertebrates in desert regions (Kingsford, 1995) until relatively recently. The combination of spatial and temporal variability of desert rivers and extensive movement patterns of vertebrates, laterally and longitudin-ally, has also probably hampered our understanding. In this chapter we review current knowledge of vertebrate groups and their interactions with disturbance flow regimes in desert rivers. Rather than adopting an evolutionary classification order, we present the groups from most aquatic to least aquatic (fishes, amphibians, birds, reptiles and mammals).

Most large desert rivers begin in higher-altitude and more mesic regions of the world before they flow across desert regions (< 500 mm rainfall per year; e.g. Nile, Colorado, Murray–Darling, Tigris–Euphrates, Tarim) (see Chapter 1, this volume). These rivers are often sediment-laden and often flood large broad floodplains, wetlands or estuaries with relatively low gradient (see Fig. 7.1). Some desert rivers originate in areas of higher elevation, with high gradients and variable rainfall producing unpredictable brief flash floods. A lateral dimension matches the longitudinal one, producing different habitats: channels, waterholes, riparian corridors, floodplains, swamps, lakes, springs, depressions and estuaries (Fig. 7.1; Table 7.1). These categories provide a reference point for a discussion of different dependencies of vertebrates in three main regions: the upper, mid and lowland parts of a desert river. Although no river can be divorced from its catchment, we confine our discussion to where rivers flow or flood temporarily or permanently across deserts.

Large desert rivers can begin as small permanent streams in a mesic upper catchment, providing habitat for amphibians and a source of water for terrestrial vertebrates. Networks of tributary creeks and streams in the upper catchments of all desert rivers funnel flow into the main stem of a river system. Distinctive riparian vegetation usually grows along the tributary creeks and the main river in the upper catchment, providing an important ecotone between terrestrial and aquatic systems (Table 7.1) (Malanson, 1993). In the upper catchment, vegetation diversity, distribution and structure usually reflects high rainfall and the aquatic parts of the river are primarily confined to the channels or narrow floodplains. These provide habitat for fish species (tolerant of low temperatures), some turtles and a few specialist mammals. Most vertebrates using this part of the river are terrestrial species, drinking, bathing or preying on other animals (Table 7.1). Exotic fish species (e.g. trout species) may thrive in low temperatures exacerbated by hypolimnetic releases from storages (King *et al.*, 1998). Large storages in the upper catchment of a desert river are usually deep and provide poor habitat for many vertebrate species, such as waterbirds or riverine native fishes (Minckley, 1973).

For most aquatic vertebrates (fish may be an exception) the river in the upper part of the catchment plays more of a functional role, transporting nutrients and organic matter from nutrient-rich and

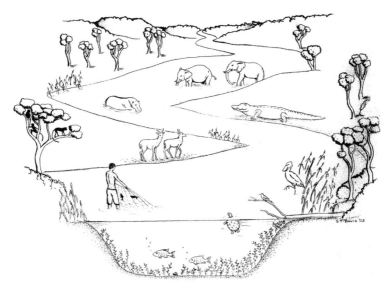

Figure 7.1. Schematic of vertebrates using a hypothetical African desert river system that starts in mountain ranges and ends in a large terminal wetland.

biodiverse vegetation communities to the mid and lower parts of the river. Most of the available aquatic habitats for vertebrates, in terms of diversity and extent, lie in the lower parts of desert rivers (Fig. 7.1). Waterholes and channels in the lower parts of a desert river (Table 7.1) often provide permanent habitat (Knighton & Nanson, 1994) for totally aquatic species (e.g. turtles, crocodiles (Fig. 7.2), fishes) when evaporation dries floodplain and wetland habitats. Floodplains are the most extensive areas on desert rivers (Kingsford *et al.*, 2001). Although floodplains are usually inundated for less than a year, complex vegetation diversity and abundance allow many terrestrial and aquatic vertebrates to survive and reproduce.

Large deltas often occur in the lowland regions of a desert river and these form relatively permanent wetland systems dependent on flows from desert rivers (Table 7.1). They are most extensive in the terminal parts of large desert rivers (Lemly *et al.*, 2000). Deltas are made up of complex permanent, semi-permanent and temporary aquatic habitats, with different inundation and drying frequencies. These habitats usually include extensive swamps, reed beds and lagoons as well as

Figure 7.2. Nile crocodiles *Crocodylus niloticus* in the Sabie River, a semi-arid river in South Africa, prey on fishes and herbivores when they come to drink. Birds such as these hammerkops *Scopus umbretta* feed on frogs, fishes and aquatic invertebrates that live in desert rivers (Photo K. Rogers).

distinct river channels and woodland floodplains. This results in a complex array of different habitats for terrestrial and aquatic verte-brates (Table 7.1; Fig. 7.1). Freshwater and saline lakes form another distinctive feature of desert rivers (Table 7.1). Generally, salt lakes are connected to more temporary and smaller desert rivers or local catch-ments (Kingsford *et al.*, 2001), although some terminal basins of large rivers also become saline as they dry (e.g. Lake Eyre, Australia). Salinity in saline lakes varies with patterns of freshwater inflows. Some large freshwater lakes in desert regions often retain water for long periods, sometimes almost permanently (Timms, 1998, 2001). The remain-ing aquatic habitats are classified as claypans or depressions (Timms & Boulton, 2001) (Table 7.1). These predominantly obtain water from small creeks or local runoff. All these diverse habitats provide essential and sometimes distinctive habitat for a broad suite of vertebrate species.

FISHES

There are more than 10 000 freshwater fish species (Matthews, 1998), many more than any other group of vertebrates. Fishes are abundant in

Table 7.1. *Different habitats and their use by vertebrates dependent on desert rivers (see Fig. 7.1) in the upper, middle and lower parts of a desert river catchment*

Location in the river: upper (U), middle (M) and lower (L).

Habitat	Location	Description and use by vertebrates	Vertebrates
Storage or dam	U	Relatively little shallow water but may allow growth of small amounts of aquatic vegetation, forming habitat for aquatic vertebrates	Fish species typical of lentic conditions. Terrestrial vertebrates drink and bathe. Some waterbirds (e.g. piscivores) but generally poor habitat
Spring	U, M, L	Usually relatively small areas with distinctive ecological communities, including fish and amphibian communities. Terrestrial vertebrates drink and bathe in springs	Unique habitat for fish species and some frogs. Used by few other aquatic vertebrates but some terrestrial vertebrates (e.g. passerine birds)
Channel	U, M, L	Usually relative narrow band of flowing water in upper catchment. Used for drinking and bathing by most vertebrates, apart from fish and turtle species. Increasing volume and width of water in mid and lower catchment, structurally more complex with different instream habitats (e.g. benches, wood)	Fish species, forest-dwelling mammals (e.g. monkeys, antelope), snakes (tree snakes) and birds (parrots) and some specialist mammal species (e.g. platypus, otters). In mid to lower parts, fish number and diversity high, increasing numbers of turtles and mammals (e.g. water rats, elephants, waterbuck). Focus for fish-eating birds (e.g. pelicans, cormorants)

Table 7.1. (cont.)

Habitat	Location	Description and use by vertebrates	Vertebrates
Riparian corridor	U, M, L	Habitat formed between the main river channel and terrestrial areas for aquatic and terrestrial vertebrates. Vegetation attracts herbivorous or fruit-eating vertebrates. In mid to lower sections, area increases	Birds and mammals feeding on fruits, seeds and vegetation (e.g. fruit pigeons, bats). Increasing diversity and abundance of terrestrial species using the structural and diverse vegetation (e.g. honeyeaters, bee-eaters, koalas, tree frogs, elephants, monkeys)
Waterholes, billabongs, wadis	M, L	Deep parts of the channel that produce refuge areas during dry periods for aquatic fauna, as these areas seldom dry out	Fish and turtle abundance high. Provide nesting habitats for waterbird species, particularly piscivores.
Floodplain	M, L	Usually increases with longitudinal distance down the catchment. Frequency of inundation determines vegetation composition. During floods extensive habitat created for aquatic and some terrestrial vertebrates, particularly for breeding	High diversity and abundance of fishes, frogs, waterbirds, aquatic reptiles and mammals in feeding and breeding areas. Terrestrial vertebrates also breed (e.g. budgerigars, raptors) during floods. Dry floodplains provide habitat for terrestrial vertebrates: birds, small and large mammals and reptiles
Swamps	M, L	Generally perennial areas with complex aquatic vegetation. Provide habitat, feeding and breeding, for all major aquatic groups, with considerable use by terrestrial vertebrate species	Probably area of desert rivers with highest abundance and diversity of all groups of aquatic and terrestrial species (e.g. African buffalo, cranes)

Lake	M, L	Freshwater and salt lakes that usually hold water for longer than floodplains and provide extensive feeding habitats for aquatic species, particularly waterbirds and fishes. Terrestrial species converge on lakes for drinking and bathing	Usually high biodiversity and abundance of waterbirds and fishes. Specialised habitat for some fish species (e.g. lungfish). Salt lakes may have specialised vertebrates (e.g. flamingos) as well as other waterbirds. Terrestrial species (e.g. waterbuck) often found around margins
Depression or claypan	M, L	Relatively small freshwater wetlands inundated by local rainfall; usually highly turbid. Provide habitat for frogs and waterbirds in particular, and drinking habitat for terrestrial vertebrates	Feeding and breeding habitat for frog species and feeding habitat for waterbirds
Estuary	L	The terminal part of desert rivers flowing out to sea. Highly productive ecosystem created between freshwater and marine processes	Includes many fish species of marine and freshwater origin. Waterbirds, particularly fish-eating and migratory wading birds, collect in large numbers on some estuaries of dryland rivers

desert regions (Table 7.1), despite most people's impressions. Desert rivers and local rainfall create extensive aquatic habitat with diverse and productive food webs for fishes (Bunn *et al.*, 2003; Chapters 4, 5 and 6, this volume). But it is the temporal and spatial dynamics of this habitat that strongly influence desert fish populations: the droughts and floods. Not surprisingly, our current knowledge of fish ecology from desert regions of the world is relatively poor, with most published studies coming from North America (Deacon & Minckley, 1974; Minckley & Deacon, 1991) and Australia (Merrick & Schmida, 1984; Wager & Unmack, 2000; Arthington *et al.*, 2005). Information on fishes from other desert regions is largely limited to taxonomic and distributional accounts (Table 7.2): Middle East (Bannister & Clarke, 1977; Al Kahem & Behnke, 1983; Krupp, 1983, 1988; Ross, 1985; Krupp & Schneider, 1989; Krupp *et al.*, 1990; Coad, 1991; Victor & Al-Mahrouqi, 1996; Goren & Ortal, 1999); Eurasia (Afghanistan (Coad, 1981)); and Africa (Beadle, 1974; Roberts, 1975; Dumont, 1982; Skelton, 1986: Lévêque, 1990).

Many extant fish species in deserts reflect effects of past climates and geomorphology because fish movement between isolated catchments is restricted (Minckley *et al.*, 1986; Unmack, 2001b). Progressive aridity has filtered fish species occurrence, as habitats recede and change from glacial periods (every 100 000–150 000 years) to long-term climate changes (more than one million years). For example, there are dozens of isolated basins throughout the Great Basin of North America with and without fish fauna (Hubbs & Miller, 1948; Hubbs *et al.*, 1974; Smith *et al.*, 2002). Ten thousand years ago, many of these basins contained large lakes filled from melting glaciers that spilled into successive basins. With increasing aridity, large lakes and streams disappeared, leaving a few small springs as fish habitat. Fish species unable to persist in these small springs disappeared. Similar processes shaped desert fish faunas in all deserts, including those of Australia (Wager & Unmack, 2000), Africa (Dumont, 1982; Lévêque, 1990) and the Middle East (Mirza, 1986; Krupp, 1983, 1987; Coad, 1987; Por, 1989).

The distribution of water and spatial and temporal variability of flooding and drying patterns dominate the ecology of fishes living in desert rivers, producing life-history traits that allow them to survive droughts and major floods while capitalising on the benefits derived from floods. Local geomorphology determines water permanence, which determines fish occurrence (Unmack, 2001a; Arthington *et al.*, 2005). In some desert rivers of Australia, fine clay sediments seal the

Table 7.2. *Total numbers of species (No.) and numbers of species in major fish groups, from ten desert regions of the world (annual rainfall < 500 mm)*

Compiled from Orange River, Africa (OR) (Skelton, 1986); Sahara Desert (SD), Africa (Dumont, 1982; Lévêque, 1990); Arabian Peninsula (AP) (Al Kahem & Behnke, 1983; Krupp, 1983); Jordan River/Israel (JR) (Krupp & Schneider, 1989; Goren & Ortal, 1999); Iran (I) (Coad, 1987); Tigris–Euphrates rivers (Coad, 1991); Afghanistan (AF) (Coad, 1981); Central Asia (CA) (Petr, 1999); North America (NA) (DFC, 2003); and Australia (AU) (Unmack, 2001b). Note that some fishes occur in the Namibian Desert but species were not recorded (Day, 1990) and that no information was available for South America. Species numbers for Central Asia probably substantially underestimate the number of fish species.

Common names	Taxonomic group	No.	OR	SD	AP	JR	I	TE	AF	CA	NA	AU
Lampreys	Petromyzontidae	1	–	–	–	–	1	–	–	–	–	–
Requiem sharks	Carcharhinidae	1	–	–	–	–	–	1	–	–	–	–
African lungfishes	Protopteridae	1	–	1	–	–	–	–	–	–	–	–
Sturgeons	Acipenseridae	10	–	–	–	–	4	–	3	2	1	–
Gars	Lepisosteidae	4	–	–	–	–	–	–	–	–	4	–
Eels	Angullidae	2	1	–	–	–	–	–	–	–	1	–
Herrings and shads	Clupeidae	13	–	–	–	–	9	–	–	–	3	1
Southern smelts	Retropinnidae	1	–	–	–	–	–	–	–	–	–	1
Salmons and trouts	Salmonidae	20	–	1	–	–	1	1	1	8	11	–
Pikes	Esocidae	2	–	–	–	–	1	–	–	1	–	–
Minnows and carps	Cyprinidae	509	10	17	14	14	70	37	67	200	80	–
Suckers	Catostomidae	37	–	–	–	–	–	–	–	–	37	–
Loaches	Cobitidae/ Neomacheilidae	134	–	–	–	5	20	11	28	70	–	–
River loaches	Balitoridae	4	–	–	–	–	–	–	–	4	–	–
	Psilorhynchidae	5	–	–	–	–	–	–	–	5	–	–
Sea catfishes	Ariidae	1	–	–	–	–	–	–	–	–	–	1
Bagrid catfishes	Bagridae	8	–	–	–	–	1	1	3	3	–	–
Airbreathing catfishes	Clariidae	4	1	2	–	1	–	–	–	–	–	–
Airsac catfishes	Heteropneustidae	1	–	–	–	–	1	–	–	–	–	–
Freshwater catfishes	Ictaluridae	7	–	–	–	–	–	–	–	–	7	–
Eeltail catfishes	Plotosidae	6	–	–	–	–	–	–	–	–	–	6

Table 7.2. (cont.)

Common names	Taxonomic group	No.	OR	SD	AP	JR	I	TE	AF	CA	NA	AU
Schilbid catfishes	Schilbeidae	4	–	–	–	–	–	–	1	3	–	–
Sheathfishes	Siluridae	16	–	–	–	–	2	2	5	7	–	–
Sisorid catfishes	Sisoridae	50	–	–	–	–	4	5	7	34	–	–
Clingfishes and singleslits	Gobiesocidae	1	–	–	–	–	–	–	–	–	1	–
New World silversides	Atherinopsidae	2	–	–	–	–	–	–	–	–	2	–
Old World silversides	Atherinidae	8	–	–	–	–	1	–	–	–	–	7
Rainbowfishes	Melanotaeniidae	3	–	–	–	–	–	–	–	–	–	3
Pupfishes	Cyprinodontidae	47	–	4	2	3	4	5	–	3	33	–
Topminnows and killifishes	Fundulidae	5	–	–	–	–	–	–	–	–	5	–
Poeciliids	Poeciliidae	32	–	–	–	–	–	–	–	–	32	–
Splitfins	Goodeidae	14	–	–	–	–	–	–	–	–	14	–
Sticklebacks	Gasterosteidae	2	–	–	–	–	1	–	1	–	1	–
Pipefishes and seahorses	Syngnathidae	1	–	–	–	–	1	–	–	–	–	–
Mullets	Mugilidae	5	–	–	–	2	1	1	–	–	3	–
Asiatic glassfishes	Ambassidae	2	–	–	–	–	–	–	–	–	–	2
Temperate basses	Percichthyidae	3	–	–	–	–	–	–	–	–	–	3
Sunfishes	Centrarchidae	5	–	–	–	–	–	–	–	–	5	–
Grunters and tigerperches	Terapontidae	6	–	–	–	–	–	–	–	–	–	6
Drums and croakers	Sciaenidae	1	–	–	–	–	–	–	–	–	1	–
Perches	Percidae	16	–	–	–	–	3	–	3	4	8	–
Cichlids	Cichlidae	22	2	5	–	7	1	–	–	–	8	–
Sleepers and gudgeons	Eleotridae	18	–	–	–	–	–	–	–	–	5	13
Gobies	Gobiidae	33	–	–	–	–	24	3	1	–	2	3
Combtooth blennies	Blenniidae	1	–	–	–	1	–	–	–	–	–	–
Sculpins	Cottidae	6	–	–	–	–	–	–	–	1	5	–
Spiny eels	Mastacembelidae	3	–	–	–	–	1	1	1	1	–	–
Totals		1076	14	29	16	33	151	68	121	346	269	46

Figure 7.3. This endangered desert fish, the humpback chub *Gila cypha*, lives in canyon-bound reaches of the Colorado River, where it depends on variable river flows (Photo P. J. Unmack).

bottom of waterbodies, preventing water loss by infiltration and creating permanent or semi-permanent aquatic habitats (Knighton & Nanson, 2000). In deserts with high stream gradients, deep alluvial sedimentary basins act as sinks that fill with water; when these are intersected by impervious geological strata (e.g. canyon walls), water is forced to the surface, providing permanent and flowing reaches of streams. This provides distinctive permanent habitat for desert fish species in parts of North America (e.g. Colorado River Basin) (Fig. 7.3), the Middle East and parts of central Australia (e.g. The MacDonnell Ranges) (Minckley, 1991; Unmack, 2001a).

Most desert fish species are habitat generalists with a broad diet. They opportunistically capitalise on times of plenty after flooding but also withstand dry times when there is less food. Desert fishes have no unique adaptations compared with other fishes (see Deacon & Minckley, 1974), but they have broad tolerances to high temperature, high suspended solids, low oxygen concentrations and high salinity (Beumer, 1979; Gehrke & Fielder, 1988; Deacon & Minckley, 1974; Wager & Unmack, 2000). Several desert fishes have small and deeply embedded scales that may reduce abrasion due to turbidity (e.g. most native minnows in the Colorado River Basin (Fig. 7.3), golden perch *Macquaria ambigua* from Australia). The woundfin *Plagopterus argentissimus* (Colorado River Basin) has specialised barbels around the mouth for feeding in turbid conditions (Minckley, 1973). A few species are long-lived (25–45 years) (Minckley & Deacon, 1991; Mallen-Cooper & Stuart, 2003), allowing them to survive many years without recruitment. These include several North American catostomid (e.g. razorback sucker *Xyrauchen texanus*, cui-ui *Chasmistes cujus*) and cyprinid fishes (e.g. Colorado pikeminnow *Ptychocheilus lucius*, humpback chub *Gila cypha* (Fig. 7.3)) and several Australian percichthyids (e.g. Murray cod *Maccullochella peelii*, golden perch). All long-lived species usually have high fecundity with several hundreds of thousands to a few million eggs (Minckley & Deacon, 1991; Merrick & Schmida, 1984).

Fishes living in deserts need to survive dry periods when floods recede. No true desert fishes survive without water as adults or eggs (aestivation), except African lungfishes *Protopterus* spp. from mesic and semi-arid regions, with biseasonal patterns of wet and dry seasons. They survive low oxygen concentrations and drying, buried in the mud, as they can breathe air. Eggs of some killifishes (Cyprinodontiformes) from Africa and South America, living in dry tropical seasonal areas, have to dry out each year before they can hatch (Helfman *et al.*, 1997). Other desert fishes survive short dry periods. The longfin dace *Agosia chrysogaster* (Colorado River Basin) survived for 14 days after evapotranspiration removed surface water until rainfall replenished the creek. They sheltered under algal mats during the day and fed at night when water was present (Minckley & Barbour, 1971). Similarly, spangled perch *Leiopotherapon unicolor* and catfish (Plotosidae) were alive in drying mud at the bottom of waterholes in Australia (Wager & Unmack, 2000). In Australia, desert fishes concentrate in permanent waterholes most of the time, with little available food, few opportunities for reproduction, declining oxygen concentrations, increasing salinity and high levels of predation. Here, there is little or no baseflow, sometimes lasting months or perhaps years (Walker *et al.*, 1995; Puckridge *et al.*, 2000). Eventually desert rivers dry up, trapping fishes with rising salinity causing massive mortality (Ruello, 1976; Kingsford & Porter, 1993; Arthington *et al.*, 2005).

Flooding is essential for desert rivers, often producing complex habitats for fishes in three ways, varying with stream channel type and sediment load. First, massive floods fill deep parts of a river while small floods scour the river bottom, removing sediment and forming large deep pools or waterholes (Deacon & Minckley, 1974; Graf, 1988;

Unmack, 2001a). Second, floods inundate floodplains and fill semi-permanent wetlands (Puckridge *et al.*, 2000; Sheldon *et al.*, 2002). For example, during major floods Cooper Creek in Australia changes from a string of isolated waterholes to a vast inundated floodplain (Puckridge *et al.*, 2000; Roshier *et al.* 2001a; Arthington *et al.*, 2005). Lastly, floods fill local groundwater basins that maintain surface water through major droughts by seepage (Minckley, 1991). Fish breeding is often related to flooding or rises in water levels, reflecting the temporal and spatial variability of flow regimes in desert rivers. Not all desert rivers are the same, with some having deeply incised channels and infrequently flooded floodplains. Many desert fishes reproduce or migrate in response to flooding, but only a few species require flooding or a rise in water level to stimulate reproduction. These include Welch's grunter *Bidyanus welchi*, silver perch *B. bidyanus* and golden perch from Australia (Merrick & Schmida, 1984). In dry years some species still have limited reproduction (Humphries *et al.*, 1999; Wager & Unmack, 2000) but massive breeding only occurs during years with large floods (Puckridge *et al.*, 2000; Wager & Unmack, 2000; Hocutt & Johnson, 2001). Within the Colorado River Basin, spawning for most species extends over many months because there is little variability in flow without rainfall (Minckley, 1973). However, higher rainfall years typically produce more native fishes (P. Marsh, personal communication). Some North American species (e.g. speckled dace *Rhinichthys osculus*) spawn immediately after a flood rearranges the substrate (Deacon & Minckley, 1974). In the Orange River in South Africa, most fish species spawn during spring and summer in response to higher flows (Skelton, 1986). In the variable flows of Cooper Creek in Australia, there was high variability among years in reproduction and recruitment of fishes (Puckridge *et al.*, 2000). In floodplain areas with large quantities of organic matter or toxic chemicals from plants (e.g. *Eucalyptus* spp.), floods are not always immediately beneficial to fish populations as they leach out tannins from vegetation, causing mortality (Gehrke, 1991; Gehrke *et al.*, 1993). In addition, small freshwater inflows may increase nutrient concentrations and stimulate algal growth that may deplete oxygen levels and result in mortality and lesions in native fish populations (e.g. Diamantina River and Cooper Creek waterholes in Australia) (V. Bailey, personal communication).

Creation of habitat with flooding results in significant movements in some species. Longfin dace and pupfishes *Cyprinodon* spp. from North America disperse to temporary habitats during flooding (Deacon

& Minckley, 1974) whereas suckers (*Catostomus* and *Pantosteus* spp.) migrate to spawning grounds (Weiss *et al.*, 1998; Douglas & Douglas, 2000). Colorado pikeminnow *Ptychocheilus lucius* migrated about 140 km along the Green River (Colorado River Basin) (Tyus, 1990) and golden perch moved over 1000 km in the Murray–Darling Basin of Australia (Reynolds, 1983). Floods produce other extraordinary movements in Australian fishes. Spangled perch *Leiopotherapon unicolor* swam along 16 km of wheel ruts in six hours (Shipway, 1947), a feat earning it the local reputation of appearing as rains of fishes (Wager & Unmack, 2000). Sometimes dispersal during flooding leads to desiccation and death. Small spates or large floods devastate larval fishes by removing them from suitable habitat to ephemeral downstream reaches, where they die (Deacon & Minckley, 1974).

Dams, diversions and floodplain developments have significantly reduced and affected native fish populations in desert rivers by altering flow regimes (Minckley, 1991; Contreras & Lozano, 1994; Thomas, 1995; Gehrke *et al.*, 1995; Lemly *et al.*, 2000; Chapter 8, this volume). Other contributory impacts include overfishing, pollution and grazing pressure (Robertson & Rowling, 2001). In many desert rivers, introduced species may dominate native species (Fig. 7.4). Overgrazing denudes the watershed, resulting in high sedimentation rates, as well as decreasing riparian vegetation. Native fish populations in North America have declined significantly (Minckley & Deacon, 1991), sometimes to extinction (Williams *et al.*, 1985). For example, the lower Colorado River from below Lake Havasu to the Gulf of California (*c.* 400 km) has no native fish species and is now populated by 30–40 introduced fish species (Minckley, 1991). Most habitats within the Colorado River Basin have two to three times more introduced fish species, which also dominate in abundance relative to native fishes. Native fish species have declined significantly in the Murray–Darling Basin, with some lowland rivers having only 20% native species compared with exotics (Gehrke & Harris, 2001). Similar shifts in composition have occurred in Iran (Coad, 1980; Coad & Abdoli, 1993), Israel (Goren & Ortal, 1999) and Saudi Arabia (Al Kahem & Behnke, 1983; Ross, 1985; Krupp *et al.*, 1990). River regulation (altered flows, migration barriers, floodplain alienation) has considerably affected native fish populations in Australia (Gehrke *et al.*, 1995), along with cold-water pollution, overfishing, exotic species and pesticides; exotic species may not be favoured by natural flooding patterns (Puckridge *et al.*, 2000; Bunn & Arthington, 2002).

Figure 7.4. Exotic species such as silver carp *Hypophthalmichthys molitrix* from the Massingir reservoir on the Elephants River, Mozambique, often establish with river regulation and the building of dams on desert rivers. They may be an important source of protein for local people but may also affect the ecology of native fish species (Photo M. Wishart).

Some exotic fish species were introduced into rivers whereas others were probably accidental introductions but have been favoured by river regulation, which has generally reduced flow variability (Gehrke *et al.*, 1995). Several examples from North America demonstrate that adult native fish species were unaffected by extreme flooding, but introduced species were extirpated or severely reduced in numbers (Deacon & Minckley, 1974; Meffe & Minckley, 1987; Minckley & Meffe,

1987). In the Guadiana River of Spain, exotic species numbers increased with river channelisation whereas native species numbers declined (Corbacho & Sánchez, 2001). Incidence of pathogens and fish disease could conceivably increase with regulation, increased nutrients and pesticides. Such global effects on desert fishes will continue to affect their survival because of the burgeoning human population in deserts and their adjacent regions and their expanding need for freshwater (Minckley & Deacon, 1991; Chapters 8 and 11, this volume).

AMPHIBIANS

Amphibians are the least likely of vertebrates to be regarded as denizens of the desert. Most experience high rates of evaporative loss, as their skin typically lacks structures to retard water loss. A boundary layer of water vapour around the skin of most amphibians resists evaporation (Spotila & Berman, 1976). Amphibian urine is usually dilute, seldom exceeding osmotic concentration of the body fluids (Adolph, 1927). Amphibian life histories are biphasal with tadpoles dependent on free-standing water for their development. So amphibians cannot achieve the degree of independence of free-standing water possible for other vertebrates, apart from fishes.

Despite these limiting attributes, many of the world's frog species (Table 7.3) successfully inhabit arid regions of the world, reliant on the flows of rivers and streams that fill wetlands. Australian frogs of the genera *Cyclorana* (12 species, 8 of which occur in arid regions) and *Neobatrachus* (9 species, all from dry or arid regions) are well known for their physiological and behavioural adaptation to desert existence (Main, 1968). The spadefoot toads of North America (2 genera, 7 species, all arid-adapted) occupy extremely arid regions including the Sonoran Desert (Mayhew, 1968); the deserts of Africa and South America each have a distinctive frog fauna (Poynton, 1964; Blair, 1976). Desert frogs live on floodplains and in permanent and ephemeral pools of desert rivers, usually not far from water (Predavec & Dickman, 1993). There are no truly desert forms of salamanders or urodeles (Heatwole, 1984).

Arid-adapted frogs cope with unpredictability in the availability of water with different morphological, physiological and behavioural adaptations, varying among species in relation to each one's own unique solution to the problems of desert living (van Beurden, 1982). Morphological, physiological and behavioural convergence across disparate

Table 7.3. *Diversity of habits for examples of reptile and amphibian species in Africa (AF), Australia (AU), North America (NA) and South America (SA) dependent or partly dependent on aquatic habitats created by desert rivers*

Class	Family	Species	Habits
Amphibia	Bufonidae (toads)	*Bufo alvarius* (NA)	Toads spend most of the year underground, in burrows they have dug themselves or abandoned rodent burrows. Active after rain, they breed in ephemeral pools
	Myobatrachidae (southern ground frogs)	*Crinia deserticola* (AU)	Inhabits creek beds, soaks and claypans associated with broad river channels throughout its range
		Limnodynastes spenceri (AU)	A burrowing frog found in sandy beds of ephemeral streams in arid regions
		Limnodynastes tasmaniensis (AU)	Found under logs and stones at the edge of permanent and semi-permanent waterholes
		Neobatrachus sp. (AU)	Species of burrowing frogs inhabiting arid regions where they are active above ground only after rain; many breed in flooded claypans
		Notaden nichollsi (AU)	
		Uperoleia micromeles (AU)	
	Hylidae (tree frogs)	*Cyclorana spp.* (AU)	A group of stout burrowing frogs from Australia showing remarkable convergence in body form to the burrowing ground frogs of the Myobatrachidae. Cocoons. Active only after rain
		Litoria rubella (AU)	Seeks refuge in trees and shrubs beside permanent and semi-permanent waterholes in central Australia
		Hyla cadavarina (NA)	Associated with oases in desert regions. Seeks refuge in crevasses close to water
		H. regilla (NA)	
		H. arenicolor (NA)	Restricted to riparian areas in rocky canyons in desert grassland regions of Colorado, where it is typically found along streams among medium to large boulders

Table 7.3. (cont.)

Class	Family	Species	Habits
		Pternohyla fodiens (NA)	A burrowing species that forms a cocoon from shed skin layers. Also uses the casque on its head to block the opening to its burrow.
		Phyllomedusa sauvagii (SA)	Adapted to its arid environment, this frog coats itself with a waxy secretion to keep in moisture. It also expels its urine in the form of semi-solid urates
	Hyperoliidae (African reed frogs)	*Hyperolius* spp. (AF), *Chiromantis* spp. (AF)	Extremely low rates of evaporative water loss allow them to persist in arid regions
	Pelobatidae (spadefoot toads)	*Scaphiopus* spp. (NA) *Spea* spp. (NA)	Burrowing frogs active only after rain. Breed in ephemeral ponds
Reptilia	Chelidae (southern side-necked turtles)	*Emydura macquarii emmotti* (AU)	Occupies permanent waterholes in the channel country of arid central Australia
		Chelodina steindachneri (AU)	Occupies permanent and ephemeral waters of arid western Australia, surviving the periods of no surface water by aestivating
	Emydidae (pond turtles)	*Clemmys marmorata* (NA)	Relict populations occupying man-made and natural ponds in the floodplain of the Mojave River, of the central Mojave Desert, California
	Kinosternidae (mud and musk turtles)	*Kinosternon sonoriense* (NA)	Occupies permanent and semi-permanent bodies of water in the Sonoran Desert, aestivating during periods when surface water is absent
	Crocodilidae (crocodiles and relatives)	*Crocodylus niloticus* (AF)	Occupies permanent and semi-permanent bodies of water in the Sahara Desert, aestivating during periods when surface water is absent. Also in the Nile River and waterbodies of its floodplain, in reaches that flow through desert

lineages are common, as for example in striking convergences of the myobatrachid and hylid desert frogs of Australia. Until relatively recently, the genus *Cyclorana* was thought to belong to the Myobatrachidae because its true affinity with the Hylidae was obscured by convergence in body form and external characteristics for burrowing (Tyler, 1978).

Some species, such as *Hyla cadaverina* of the Colorado Desert (McClanahan *et al.*, 1994), survive in permanent seeps or waterholes. With free-standing water readily available, they control body temperatures through evaporative cooling to overcome the oppressive heat of the desert. Arboreal forms in the genus *Phyllomedusa* depart greatly from the norm and live away from water by producing uric acid and by using their feet to smear waterproofing lipids produced by specialised glands over their bodies (Blaycock *et al.*, 1976; Shoemaker, 1988).

Despite these notable exceptions, most desert-adapted frogs are inactive when free water is unavailable. Some species survive the dry times in crevices, sealing the entrance with part of their bodies. The tree frog *Corythomantis greeningi* from semi-arid regions of Brazil has a casque-like head: it is flat and rough with dermal bony elements near the skull (Jared *et al.*, 1999). The frog blocks its retreat with its casque-like head protecting it from desiccation and predation (Seibert *et al.*, 1976).

Many other desert-adapted species escape desiccation by burrowing beneath the soil. These burrowing frogs commonly have a globose body form, short stout limbs, spade-like metatarsal tubercules and a urinary bladder of large capacity. The North American desert spadefoot *Scaphiopus couchii* spends up to 10 months of the year in a deep burrow, buffered from desiccation and high temperature (McClanahan, 1967), and only emerges after heavy rainfall (Dimmitt & Ruibal, 1980). The Australian water-holding frog *Cyclorana platycephalus* can be inactive for many years, conserving water by enveloping its body in an epidermal cocoon while buried. Three genera of Australian frog (Lee & Mercer, 1967), two genera of South American frog and one genus of North American frog (McClanahan, 1967) have cocoons. In Australian frogs, cocoons can reduce evaporative loss to as low as 6% of that of inactive animals without cocoons (Lee & Mercer, 1967).

Periods of activity for desert frogs are brief and unpredictable, usually after heavy rains that produce free-standing water for breeding when a flush of insect activity (Predavec & Dickman, 1993) and later plant growth are available for adults and young frogs. Some desert frogs may be able to predict rain, emerging eight hours before rain began in the Australian desert (Predavec & Dickman, 1993). Emerging frogs rapidly take up water with specialised seat patches, highly permeable regions where the skin is thin, highly vascularised (Roth, 1973) and responsive to hormonal activation (Baldwin, 1974). Desert frogs have short, opportunistic, explosive mating systems (Sullivan, 1989) and rapid tadpole development (taking as little as 8 days to complete in *Scaphiopus couchii* (Newman, 1989)). There is also flexible timing of metamorphosis when water levels drop or tadpole densities increase (Newman, 1994), and young can sometimes burrow and survive for limited periods. For example, adults of the water-holding frog *Cyclorana platycephalus* emerge after heavy summer rains and mate on the evening of emergence. Eggs are laid in claypans or temporary ponds. Development is extremely rapid and the emergent froglet is relatively large, well endowed with abdominal fat reserves and ready to burrow (van Beurden, 1982). Such activity produces high densities, reaching more than 2000 g ha^{-1} in some Australian desert frogs (Predavec & Dickman, 1993). These features are directly related to the unpredictability in the timing and duration of flooding and consequent availability of food for adult and tadpole survival.

Amphibians are surprisingly capable of inhabiting desert regions and many occupy the floodplains, and permanent and ephemeral pools of desert rivers. We found no studies that showed that the life history of a frog was driven by the dynamics of the desert river itself. Local rainfall events often produce widespread flooding (Kingsford *et al.*, 1999) that may coincide with flood events from upstream, so frog activity and flooding often coincide. However, their activities seem primarily stimulated by local rainfall, rather than flow patterns of desert rivers and widespread inundation. This may be due to failure of frogs to take advantage of the opportunities brought by episodic inundation of the desert landscape by flooded rivers or simply a knowledge gap for future research.

BIRDS

Of the more than 9000 species of bird in the world (Perrins, 1990), two main ecological groups depend on desert rivers and their habitats: waterbirds and 'terrestrial' bird species whose life cycles depend on aquatic systems (Table 7.4). Terrestrial bird species all require water to drink, even if only occasionally (Young, 1981), and many depend on other resources of aquatic habitats, such as food or nesting materials.

Table 7.4. *Major bird groups of the world dependent or partly dependent on aquatic habitats created by desert rivers, and their habits*

Group	Common names	Taxonomic group	No.[a]	Habits in desert river systems
Aquatic	Grebes	Podicipediformes	14	Almost completely aquatic, feeding on fishes and invertebrates
	Pelicans	Pelecaniformes	6	Almost completely aquatic, feeding on fishes by scooping fishes into their gular pouch. They breed on islands
	Cormorants and darters		11	Almost completely aquatic, feeding mostly on fishes and large crustaceans, capturing prey underwater. Breeding on islands and trees bordering wetlands
	Herons and bitterns	Ciconiiformes	61	Long-legged wading birds that feed in shallows around edge of wetlands
	Storks		18	Large long-legged and long-necked birds, generally with heavy bills. Feed on crustaceans, fishes and amphibians and nest in trees, cliffs or buildings
	Spoonbills and ibis		31	Feed on fishes, amphibia crustaceans, insects in and around edges of wetlands or terrestrial areas. They breed in large colonies on platforms made from vegetation

Table 7.4. (cont.)

Group	Common names	Taxonomic group	No.[a]	Habits in desert river systems
	Flamingos		4	Long-legged wading birds with webbed feet. They sieve microscopic plants and animals from alkaline lakes and lagoons
	Ducks, geese and swans	Anseriformes	147	Large group of waterbirds with wide variety of habits from herbivores to invertebrate- and fish-eaters that use the range of wetland habitats from the terrestrial (e.g. geese) to almost completely aquatic (e.g. diving ducks). Nesting is variable
	Screamers		3	Long-legged and with part webbed feet; mainly herbivorous birds that nest close to water
	Cranes	Gruiformes	15	Large, long-legged omnivorous birds that live around water but are capable of feeding in terrestrial habitats. They usually nest on aquatic vegetation in shallow water
	Rails		129	Not all are aquatic species. Small cryptic birds. Some have lobed feet (e.g. coots). They nest on aquatic vegetation
	Finfoots		3	Aquatic birds with long heads and necks, lobed toes and stiff tail

Table 7.4. (cont.)

Group	Common names	Taxonomic group	No.[a]	Habits in desert river systems
	Plovers, jacanas, avocets, stilts, sandpipers, terns, gulls and skimmers	Charadriformes	245	A diverse group of birds that includes small wading birds that usually feed along the edges of aquatic habitats, jacanas that seldom venture from the vegetation in the middle of wetlands, and gulls, terns and skimmers that usually forage from the air
Terrestrial	Kingfishers	Coraciiformes	87	Many species feed and breed near aquatic habitats, relying on aquatic animals (e.g. fishes) for food. They nest in holes in trees, banks of rivers or termite holes
	Eagles, kites and hawks	Falconiformes	286	A small number of these species (e.g. fish eagles, ospreys, harriers) prey on aquatic species and nest in trees close to water
	Bee-eaters	Coraciiformes	24	Some species feed near habitats of desert rivers and nest in large colonies in river banks
	Parrotbills and reed warblers	Passeriformes	49	Many occupy dense reedbeds around swamps, feeding on insects in the wetlands and building nests within the vegetation

[a] Data from Perrins (1990), but not all species found on desert rivers. Gaviiformes (divers and loons, four species) are excluded because their wetland habitats are not usually in desert regions.

Figure 7.5. Large concentrations of waterbirds collect on inland lakes in Australia, such as Lake Wyara. Their numbers fluctuate with regional and continental availability of wetland habitats, which are a response to floods and droughts of desert rivers (Photo R. T. Kingsford).

Among desert birds, waterbirds are often the most conspicuous and abundant (Fig. 7.5) and they tend to concentrate in the lower parts of desert aquatic habitats: the lakes, floodplains, swamps, river channels and estuaries (Turpie, 1995; Kingsford *et al.*, 1999) (Table 7.1, Fig. 7.1). Of all the taxonomic groups of waterbirds, only loons (Gaviiformes) are not found in desert regions. Specialised body forms and feeding structures allow waterbirds to occupy specific aquatic habitats (e.g. deep water areas, shallows, reed beds, floodplains) on a desert river, feeding on a wide range of different foods: invertebrates (ducks, flamingos), plants (geese and swans), fishes and frogs (storks, ibis, egrets). Bills vary enormously. Skimmers *Rynchops* spp. capture small fish and crustaceans while flying close to the water's surface by snapping their long flattened lower mandible, trailing in the water, shut against the short round upper mandible. Whale-headed storks *Balaeniceps rex* catch lungfish and other prey with their enormous bills (Table 7.4). Flamingos *Phoenicopterus* spp. use specialised bills to sieve microscopic biota. Ubiquitous pelicans are piscivores and have a large bill with a sack-like pouch to temporarily store prey. The specific habits of waterbirds also vary. Some species seldom venture onto land (e.g. grebes, diving ducks *Aythya* spp.) whereas others spend most of their time either around the edge of wetlands (egrets, herons) or often on land adjacent to waterbodies (geese, ibis).

'Terrestrial' bird species dependent on aquatic habitats comprise three taxonomic groups (Table 7.4). There are specialist birds of prey that are totally dependent on aquatic habitats (e.g. African fish eagle *Haliaeetus vocifer*, hammerkop *Scopus umbretta*, (Fig. 7.2) but some falcons and hawks also collect around wetlands in arid regions because of the abundance of waterbirds, fishes and carrion. During aerial surveys of Cooper Creek in Australia, concentrations of hundreds of wedge-tailed eagles *Aquila audax*, black kites *Milvus migrans* and whistling kites *Haliastur sphenurus* were recorded along the edges of the river and its freshwater wetlands (R. T. Kingsford, personal observations). Some species of bee-eater (e.g. carmine bee-eater *Merops nubicus*) nest in large colonies in the banks of desert rivers, and others forage near desert rivers. Parrotbills and reed warblers *Acrocephalus* spp. are passerines primarily dependent on vegetation along desert rivers, particularly wetland areas such as swamps and lakes (Table 7.4).

Birds are the best equipped of all vertebrates to exploit rapidly changing resources because they can fly to or from rivers during 'boom' and 'bust' periods (Schodde, 1982). Many aquatic and terrestrial bird species use the aquatic habitats of desert rivers only at certain times. In particular, migratory species (e.g. white stork *Ciconia ciconia*) breed during the northern hemisphere spring and then fly south during the winter months to feed in the habitats of desert rivers. Seasons may be predictable but temporal and spatial habitats on desert rivers are usually not (Roshier *et al.*, 2001a), resulting in nomadic movements (Roshier & Reid, 2003). Other species, particularly colonially breeding waterbird species, return to where they were hatched to breed. For example, flamingos breed in a few locations (e.g. Lake Baringo in eastern Africa), but feed at others (Vareschi, 1978). Although locations for breeding egrets, ibis and herons are predictable, they only nest following unpredictable widespread flooding of desert rivers in Australia (Kingsford & Johnson, 1998; Leslie, 2001). Most other Australian waterbirds also breed in response to flooding, particularly in desert regions (Kingsford & Norman, 2002). Terrestrial birds also have a rapid breeding response, following flooding and rainfall (Schodde, 1982; Davies, 1984; Lloyd, 1999).

Waterbirds in desert regions capitalise on 'boom' periods when rivers are in flood, creating considerable habitat (Kingsford *et al.*, 1999; Herremans, 1999). Temporary and saline habitats appear to be more productive in terms of density, numbers of species and abundance of waterbirds than those perennially flooded (Fig. 7.5) (Kingsford & Porter, 1993; Kingsford *et al.*, 2004). For waterbirds, movements extend beyond river catchments to continents and beyond (Haig *et al.*, 1998; Roshier *et al.*, 2001b, 2002; Roshier & Reid, 2003). During dry periods, waterbirds concentrate in remaining habitats (Kingsford, 1996; Schlatter *et al.*, 2002). Unpredictable wetland habitats (Roshier *et al.*, 2001a) produce variable movement patterns for desert waterbirds (Roshier *et al.*, 2002), considerably more than is found in movements of northern hemisphere species (Kingsford & Norman, 2002).

REPTILES

Many reptile species are aquatic, semi-aquatic or derive a substantial proportion of their foods from aquatic ecosystems (Table 7.3). They include the freshwater and marine turtles, the crocodilians, lizards of several families including the Agamidae, Iguanidae and Varanidae, and numerous freshwater and marine snake species. Relatively few aquatic species live in desert regions, and most of those that do survive as either relict populations from historically wet times, or with specific adaptations for desert life.

The western pond turtle *Clemmys marmorata* occupies constructed and natural ponds in the floodplain of the Mojave River of the Mojave Desert, California. Their life history differs little from populations in the moister coastal regions, with no identifiable adaptations to desert living. They are probably relict populations from the Pleistocene (Lovich & Meyer, 2002), with tenuous long-term prospects because of drawdown of their habitat for water resource development.

Populations of the Nile crocodile *Crocodylus niloticus* (Fig. 7.2) in Saharan Africa seem to share similar tenuous prospects (de Smet, 1998) but for different reasons. Middle Holocene remains and rock paintings indicate that the crocodiles used to occur across the Sahara. Some populations persist as relicts of this more widespread distribution. A few specimens survive in pools in a few river canyons of the Ennedi plateau of northern Chad. Another relict population, in the Tagant hills of Mauritania, appears to have become locally extinct as recently as 1996 (de Smet, 1998). Nile crocodiles also inhabit two types of wetland on the periphery of the Sahara (Mayell, 2002): lowland wetlands and a type of wetland known as a guelta. A guelta is formed when rain, or sometimes underground springs, forms a pool of water in a depression of a rocky plateau. Lowland wetlands in Mauritania are formed when rainwater collects in clay-lined depressions in otherwise dryland conditions. In both cases, no permanent water exists in the vicinity, and the crocodiles survive dry periods by aestivating in crevasses and burrows. Unlike the Western pond turtle, these crocodile populations appear to be showing behavioural adaptation to desert life.

The Sonoran mud turtle *Kinosternon sonoriense* occupies remote arid mountain ranges in southern Arizona, New Mexico (USA), and northern Sonora (Mexico). It lives in permanent bodies of water but can also survive dry periods by aestivating (Peterson & Stone, 2000). In Australia, the helmet-shell turtle *Chelodina steindachneri* lives almost entirely within desert regions in creeks and rivers that flow only after infrequent and sporadic heavy rain (Burbidge, 1967). During dry periods of months or even years when there is no water, this species also aestivates, with physiological adaptations for surviving these dry periods. These include low rates of evaporative water loss, tolerance of high temperatures, efficient conversion of ammonia to urates, and capacity to store water in its urinary bladder (Burbidge, 1967).

Our knowledge of reptile ecology in relation to disturbance regimes of desert rivers is relatively poor, but valuable insights are coming from turtle research on Cooper Creek in Australia. This desert river is an extensive system of anastomosing channels, discharging into the inland Lake Eyre (Kingsford *et al.*, 1999). Local rainfall and rain depressions in the catchment, driven by summer tropical monsoons and aseasonal cycles of the El Niño – Southern Oscillation (Puckridge *et al.*, 2000), produce widespread flooding of thousands of hectares (Kingsford *et al.*, 1999; Roshier *et al.*, 2001a). Most of the time, the system is dry (Puckridge *et al.*, 2000) but with many almost permanent waterholes in its mid sections (Knighton & Nanson, 1994). Turtles have three options in this unpredictable environment. They can live in temporary habitats until dry periods come and then move to remaining waters, always live in the permanent waterholes, or live underground in cavities and undercut banks until the next flood. The eastern snake-necked turtle *Chelodina longicollis* is a freshwater turtle that moves overland between ephemeral and permanent waters (Kennett & Georges, 1990, 1995). Although established in the headwaters of Cooper Creek, it has not extended its range into the drier parts of the river, presumably

Figure 7.6. The Cooper Creek turtle *Emydura macquarii emmotti* lives in semi-permanent and permanent waterholes of Cooper Creek, such as this one, and the Diamantina River in Australia, seldom venturing overland. During 'boom' flood periods they move onto the floodplain to feed but then retreat to the waterholes when dry 'bust' periods follow (Photo A. Emmott).

because it cannot move between habitats. Large spatial scales of hundreds of kilometres and long temporal scale of decades of inhospitable desert between temporary and permanent waters are too great a barrier for survival.

Instead, Cooper Creek and the nearby Diamantina River are occupied by another river turtle, the Cooper Creek turtle *Emydura macquarii emmotti* (Georges & Adams, 1996) (Fig. 7.6). It does not possess the specific adaptations to the vicissitudes of life in ephemeral habitats of *Chelodina* (Burbidge, 1967; Kennett & Georges, 1990, 1995; Kennett & Christian, 1994). It seldom moves overland in search of water or aestivates during dry periods. Instead this turtle is tied to permanent and semi-permanent waterholes of the channel country on the Diamantina River and Cooper Creek (Fig. 7.6). It capitalises on the 'boom–bust' cycle of resource availability (White, 2002). During infrequent flooding, the turtles move onto the floodplain to feed, retreating with the receding floodwaters to the discrete permanent waterbodies during 'bust' or dry periods. There is usually a lag in the 'boom' period for the turtles that feed on the great concentrations of dead and dying fishes in the waterholes for months after flooding. Extraordinarily high population densities of turtles live in these permanent waterholes, sustained by the 'boom' periods and persisting through the dry 'bust' periods by virtue of a low metabolic rate (compared with that of birds and mammals) and longevity (up to 80 years). Their ability to survive periods of low production in high population densities is an important ingredient in their success.

None of the waterholes of the channel country is truly permanent, although they may appear to be so on human timeframes. Drying occurs and causes catastrophic mortality (Cann, 1998) with local population extinction. Although the turtles show high site fidelity, regional persistence probably relies on a pattern of local extinction and reinvasion on a timescale of centuries, following the geomorphic scales of the river (Knighton & Nanson, 1994, 2000). The occupation of both 'permanent' and semi-permanent waterholes is also important to their long-term persistence in desert regions.

In summary, aquatic reptiles persist in desert regions through three primary strategies. First, they occupy permanent waterbodies (e.g. waterholes, lakes, rivers, swamps), often as relictual populations. Second, when extensive inundation of the desert landscape occurs sporadically, aquatic reptiles disperse to the newly created habitats. Third, some turtles can also aestivate for months or even years to survive 'bust' or dry periods.

MAMMALS

Relatively few mammals are entirely aquatic compared with other vertebrates (Table 7.5). Mammals that spend most of their lives in water (Table 7.5) are well equipped for such an existence. Apart from completely aquatic river dolphins, other aquatic mammals sometimes move on land (Table 7.5) as exemplified by the hippopotamus *Hippopotamus amphibius*, which feeds on land at night and spends the day in the water. Many aquatic mammals have partly or fully webbed feet (e.g. otters, water rat). In contrast the sitatunga, an antelope of terrestrial origin, is able to move over the unstable aquatic vegetation of African swamps, supported by splayed hooves. Most aquatic mammals also have sensory organs near the top of their heads (e.g. hippopotamus, capybara,

Table 7.5. *Habits of major mammal groups of the world dependent or partly dependent on aquatic habitats created by desert rivers*

Group	Common names	Taxonomic group	No.	Habits in desert river systems
Aquatic	Platypus	Monotremata	1	Almost completely aquatic, feeding on invertebrates and primarily confined to mesic parts of desert rivers
	River dolphins		5	Two species of river dolphin live in parts of desert rivers: the Indus dolphin *Platanista minor* and the Ganges dolphin *P. gangetica.*
	Rodents	Rodentia	<30	Many rodent species live primarily in water. They have nostrils, eyes and ears near the top of their heads and many have partially or fully webbed feet and flattened tails for an aquatic life. Species include the water rat *Hydromys chrysogaster*, South American water rat *Nectomys*, coypu *Mycogaster coypus*, capybara *Hydrochaeris hydrochaeris*, marsh rats (*Holochilus* spp.) and swamp rats (*Malocomys* spp.)
	Otters	Carnivora	13	These species are primarily aquatic with webbed feet, short fur, small ears and eyes on the top of their heads. They feed on fishes and crustaceans and insects in desert rivers. Species include the Cape clawless otter *Aonyx capensis*, smooth-coated otter *Lutrogale perspicillata* and spot-necked otter *Lutra maculicollis*

Table 7.5. (cont.)

Group	Common names	Taxonomic group	No.	Habits in desert river systems
	Hippopotamus	Artiodactyla	2	The hippopotamus *Hippopotamus amphibius* has large lungs and ears, eyes and nose on top of the head, muscles for closing eyes and ears under water, and broad muzzle. Daytime is spent in water; nocturnal feeding on floodplains
	Spiral-horned antelope		9	Some species live within swamps and floodplains and include sitatunga *Tragelaphus spekeii*, a species with specialist splayed hooves to support its weight in swamps, nyala *T. angasii* and species of reedbuck *Redunca arundinum, R. redunca*
	Grazing antelope		23	Some species live primarily on floodplains and include lechwe *Kobus leche, K. megaceros* and waterbuck *K. ellipsiprymnus*
	Insectivores	Insectivora	5	Specialist species (aquatic tenrec *Limnogale mergulus*, giant otter shrew *Potamogale velos*) are adapted for an aquatic existence with elongated body form and eyes and ears on top of a flattened head
Terrestrial	Bats	Chiroptera	900	Some species use riparian areas and aquatic area for feeding (e.g. large slit-faced bat *Nycteris grandis*)
	Lions	Carnivora	1	Primarily terrestrial but prey on herbivores around waterholes and wetlands

Table 7.5. (cont.)

Group	Common names	Taxonomic group	No.	Habits in desert river systems
	Humans	Primates	1	Many humans live close to water in desert regions, catching fish and using floodplain plants
	Elephants	Proboscidea	2	Elephants *Loxodonta africana* require considerable water and during the dry times may dig for water in dry river beds
	Tapir	Perissodactyla	4	Frequently found in rivers and lakes, where they take refuge from predators. Mostly in mesic areas of desert rivers
	Rhinoceros		5	Often found close to water; they use waterholes for wallowing and caking their skin with mud to deter parasites
	Buffalo	Artiodactyla	2	Some African buffalo *Syncerus caffer* spend much of their time on the floodplains and wetlands of desert rivers; water buffalo, as their name implies, live primarily in wetlands of desert rivers

long-haired rat *Rattus villosissimus*, thick-tailed opossum *Lutreolina crassicaudata*), allowing them to look out for prey or predators while mostly submerged. Many of them also swim well, with streamlined bodies propelled by flattened tails that also sometimes act as rudders.

Although few aquatic mammals depend entirely on desert rivers (Table 7.5), most mammals have to drink from a river or wetland. There are a few exceptions that can last for periods without water (oryx *Oryx gazella*, bilby *Macrotis lagotis*). Desert rivers and their dependent habitats are the focus for interactions between many mammal species and other vertebrates (Fig. 7.1). In the upper catchment of a desert river, the

Figure 7.7. Terrestrial species depend on rivers and wetlands in desert regions for their water, sometimes needing to travel up to 200 km to find water during dry periods. African elephants bathe and drink from the Ewaso Ngiro River, Kenya (Photo H. Grant).

riparian corridor (Malanson, 1993) and water attract mammals. Forest-dwelling mammals come to drink or feed in the riparian vegetation (Table 7.1). Much of our knowledge of interactions between mammals, other than humans, and rivers comes from research from northern latitude catchments and the importance of the riparian corridors (Nilsson & Dynesius, 1994). Otters, aquatic rodents and platypus, with specialist aquatic adaptations, are among the few aquatic mammal species, living in channels of desert rivers in the upper catchment (Table 7.1), where they feed on invertebrates and fishes.

In the mid-section of a river, increasing aridity brings many mammals to the river to drink, bathe or feed. For example, African elephants require 80–160 l of fluid a day (Macdonald, 2001) and either live near permanent water or travel up to 200 km to reach water during dry periods (Fig. 7.7) (Verlinden & Gavor, 1998). Large waterholes and a wide channel allow aquatic mammals such as hippopotamus to establish territories (Jacobsen & Kleynhans, 1993). Diversity and abundance of mammals is probably highest in the lowest part of a desert river: the floodplains, terminal lakes and estuaries (Table 7.1; Fig. 7.1). Here, extensive and variable habitats provide considerable opportunity for many species of mammals to survive and reproduce. Reduction of such areas with river regulation leads to declines in abundance of aquatic mammals (e.g. hippopotamus), which then concentrate in the reduced available habitat (Jacobsen & Kleynhans, 1993; Nilsson & Dynesius, 1994). Even the floodplain feeding habitat of the terrestrial chacma baboon *Papio ursinus* may be reduced (Attwell, 1970).

River channels and deltas provide habitat for aquatic species (e.g. river dolphins, hippopotamus, capybara, tapir) whereas floodplains, wetlands, waterholes and river channels are a magnet for terrestrial mammals that concentrate and interact, drawn by essential requirements of water and abundant food. The lechwe antelope *Kobus leche* lives in large herds on aquatic plants from floodplains or wetlands; the movement patterns and habitat use of African elephants were correlated with proximity to water and nutrient-rich areas during dry periods (Verlinden & Gavor, 1998). Within Australian deserts, the marsupial koala *Phascolarctos cinereus* tends to be found along the main riparian corridors where eucalypts grow largest. The need for mammalian herbivores to drink at waterholes, floodplains, wetlands or river banks produces high encounter rates between predators and prey. Lions *Panthera leo* and crocodiles ambush herbivores near waterholes; smaller mammalian carnivores (e.g. serval cat *Leptailurus serval*) may prey on birds coming to drink. Seasonally abundant vegetation also results in migrations of animals that make difficult river crossings (e.g. wildebeeste *Connochaetes* spp. in Africa), providing nutrients and prey to carnivores and scavengers. Although our knowledge of how disturbance patterns of desert rivers shape the ecology of mammals remains relatively poor, there is accumulating evidence that desert rivers drive the ecology of aquatic mammals and many terrestrial mammals in desert regions.

The availability of food and water meant that humans traditionally concentrated near desert rivers. Material for housing (e.g. reed beds) and food, particularly fishes (Thomas, 1995), are usually available from desert rivers and wetlands. Aboriginal communities in Australia used desert rivers as trade routes (Veth *et al.*, 1990). For the same reason, European colonists built towns on the banks of Australia's desert rivers, near dependable water. Desert rivers were not only utilitarian but also culturally important (e.g. dreamtime stories of Australian aborigines) (Goodall, 1999). Such dependency remains strong in many parts of the world but has been severely changed, with regulation of desert rivers (Lemly *et al.*, 2000; Chapter 8, this book).

Humans are not the only mammals that can change desert rivers. Herbivores exert considerable influence on vegetation patterns, affecting availability of food for other mammals. African elephants affect vegetation structure and composition on a floodplain and may affect the geomorphology by accessing river banks and wetlands and digging holes in rivers for drinking. Hippopotamus affect spatial flooding to wetlands and lakes in the Okavango Swamp by changing geomorphology and vegetation distribution (McCarthy *et al.*, 1998). The South American coypu similarly affects flooding patterns and has successfully colonised Africa and Europe, affecting flow patterns (MacDonald, 2001).

FLOOD 'BOOMS' AND 'BUSTS' AND EVERYTHING IN BETWEEN

The distribution and abundance of all organisms are dependent on access to sufficient food and water for survival and reproduction. This challenges vertebrates in desert regions because river flows are often so unpredictable (Puckridge *et al.*, 1998; Chapters 1 and 2, this volume). Such hydrological variation creates a range of inundation patterns or habitats from extensive floods, called 'boom' periods, to the bottlenecks or 'bust' periods (Kingsford *et al.*, 1999). These are the extremes, but the 'in-between' floods should not be discounted for their ecological importance. Clusters of floods create productive feeding habitats and recruitment opportunities for some vertebrates, allowing rapid recruitment responses during 'boom' periods (Puckridge *et al.*, 2000). They may be critical in refilling drying waterholes and wetlands that provide refuges for vertebrates (Chapter 4, this volume), priming the aquatic system and ensuring that the next flood creates more habitat (Puckridge *et al.*, 2000).

Floods in desert rivers stimulate tremendous productivity and biological activity. These are the 'boom' times when waterbirds arrive to colonise newly flooded habitats (Roshier *et al.*, 2002; Roshier & Reid, 2003) and plant growth occurs (Bacon *et al.*, 1993). Vegetation within wetlands and floodplains provides extensive areas for herbivorous mammals to forage. Fishes and waterbird populations breed rapidly (Kingsford & Johnson, 1998; Kingsford *et al.*, 1999; Puckridge *et al.*, 2000), providing abundant food for terrestrial vertebrates (e.g. birds of prey).

Inevitable 'bust' periods are more frequent and produce ecological bottlenecks for vertebrates living in the desert aquatic habitats (Table 7.1). As available habitat contracts, many waterbirds 'escape' to other wet habitats, or migrate, but even in birds, many die during 'bust' periods (Kingsford *et al.*, 1999). Some mammals move away from the area but many remain concentrated around remaining habitat patches (e.g. waterholes). Some desert frogs shut down most of their physiological processes and cocoon themselves underground. Fish populations become more concentrated. For a time, there is an abundance of prey for predators as prey are forced to concentrate in remaining areas. Fish-eating waterbirds collect in considerable numbers around drying lakes and waterholes where there is high density of prey. Similarly, carnivores find easy prey in other vertebrates, unable to stray far from water. The changing boundary between aquatic and terrestrial habitats creates opportunities for plants and animals to colonise. For example, small marsupials and rodents colonise the cracks created as lake beds on desert rivers dry up (Briggs, 1992; Briggs *et al.*, 2000). The Lake Eyre dragon *Ctenophorus maculosus* specifically forages on the dry lake-bed. Such species may be prone to drowning if they cannot escape quickly. In lowland sections of rivers, floods often move reasonably slowly.

Not all aquatic habitats on desert rivers disappear during dry periods. Some (e.g. lakes and waterholes) retain water for years, usually until another flood replenishes their water capacity. These more permanent habitats become island refuges that ensure desert vertebrates can recolonise wetland habitats in the next flood: the colonisation epicentres of desert rivers (Jacobsen & Kleynhans, 1993). Fishes and turtles retreat to these waterholes. For example, the oldest Cooper Creek turtles in central Australia occur only in the deepest waterholes that seldom dry up (Fig. 7.6) (White, 2002).

The two disturbance factors, flooding and drying, shape the quantity, frequency and extent of habitat created by desert rivers. They influence life histories of all aquatic dependent vertebrates in generally predictable ways but habits of desert vertebrates also vary considerably between and within groups. For example, fish-eating waterbirds generally breed later in a flooding regime, compared with other waterbird species, which wait until fish populations reproduce and increase (Kingsford *et al.*, 1999). Different vertebrates also use different parts of a desert river. For example, many desert frogs occupy habitats separated from the main channel of desert rivers, areas primarily reliant on local rainfall for filling of temporary wetlands. But as well as these aquatic vertebrates, we also argue that desert rivers influence the ecology of most terrestrial vertebrate species living in desert regions. These species need to drink mostly from rivers but reliance goes well beyond this physiological dependence. 'Boom' periods on desert rivers allow establishment of an abundant and diverse biota, supporting complex food webs, including those of terrestrial vertebrates. Stochastic rainfall is a strong determinant of the ecology of desert regions (Stafford Smith & Morton, 1990). Variable flows on desert rivers create and destroy network habitat patches, changing in space and time, for vertebrates. They not only create extensive foraging and recruitment habitats during 'boom' periods but they also concentrate resources during 'bust' periods.

CONCLUSIONS

Vertebrate ecology on desert rivers is considerably different from that on spatially and temporally more predictable rivers in mesic regions of the world. Behaviour of most vertebrates in desert regions is dependent on the counteracting disturbances of flooding and drying. Even most terrestrial vertebrates have their ecology primarily affected by flooding and drying of desert rivers when the habitats created by desert rivers provide abundant food. High species diversity and abundance of vertebrates on riverine habitats in desert regions provides good evidence for the importance of these areas. Only a few vertebrates (some reptiles and mammals) have cut their dependencies on the habitats of desert rivers. Strategies of survival and reproduction among vertebrate groups are remarkably similar. Frogs, fish, turtles and waterbirds all capitalise with extraordinary breeding events during flood periods when there is a relatively short time to complete recruitment. Similarly, during dry periods, the species are able to wait in refugia, whether this is waterholes, lakes or their own cocoons. We argue that the natural hydrological disturbance regimes of desert rivers remain the most important factor shaping the ecology of much of the vertebrate fauna that lives in these areas of the world.

We know relatively little about the interrelationships between vertebrates and desert river systems but there are tantalising pieces of information suggesting that desert rivers without vertebrates would be considerably poorer in diversity and abundance without the 'boom' and 'bust' periods. Some plant and invertebrate species may depend on transportation by birds to colonise different environments (Green *et al.*, 2002). Unlike most other biota (but see vegetation example (Tooth & Nanson, 2000)), vertebrates can change the disturbance patterns of rivers. Large mammals, such as hippopotamus, carve channels through large wetlands, changing flooding and drying patterns of dependent habitats; large herbivorous mammals may increase the patchiness and suitability of floodplains for other organisms by grazing.

Only humans can completely alter the disturbance patterns of desert rivers with dams and diversion of water, mainly for irrigation (see Lemly *et al.*, 2000; Chapter 8, this volume). Human damage to the ecology of vertebrates at this scale ranks among the most serious conservation problems affecting the world (Chapter 8, this volume).

River regulation and water resource development can remove much of the temporal and spatial complexity and extent of aquatic ecosystems (Chapter 8, this volume). Dry periods become longer and more frequent and flooded habitats shrink in size. What effects would this have on vertebrates in desert rivers and which groups would be most vulnerable? Clearly, such impacts are going to most affect vertebrates unable to withstand extended dry periods and with reduced opportunities to 'bounce back' after breeding events that would be inevitably smaller. If refugia, including lakes, dry up then turtles, waterbirds and fish would have significantly reduced populations. Examples exist where species have became locally extinct (Chapter 8, this volume). For regulated rivers that deliver flow for irrigation or human communities, low flows may also favour exotic species to the detriment of native species that are primarily capable of capitalising on large flood events. Even humans reliant on a subsistence existence are affected, particularly if dependent on fishing and other uses of desert rivers (see Chapter 8, this volume). If waterholes dry up during catastrophic droughts, vertebrates that use rivers primarily for drinking will be just as affected as aquatic dependent vertebrates.

Many questions about the interactions between vertebrates and rivers remain unanswered (Nilsson & Dynesius, 1994), but the reliance of their ecology on disturbance patterns seems clear. The 'boom' and 'bust' periods and everything in between allow many vertebrate species to persist in otherwise hostile environments. When we irrevocably change disturbance patterns, the essential ecology of these unique desert rivers collapses (see Chapter 8, this volume). Natural disturbance patterns of desert rivers are the essence of desert ecosystems and their vertebrates.

ACKNOWLEDGEMENTS

We thank Andrew Boulton for assisting with editing and comments. We also thank Susan Davis for Fig. 7.1.

REFERENCES

Adolph, E. F. (1927). The excretion of water by the kidneys of frogs. *American Journal of Physiology*, **81**, 315–24.

Al Kahem, H. R. and Behnke, R. J. (1983). Freshwater fishes of Saudi Arabia. *Fauna of Saudi Arabia*, **5**, 545–67.

Arthington, A. H., Balcombe, S. R., Wilson, G. A., Thoms, M. C. and Marshall, J. (2005). Spatial and temporal variation in fish assemblage structure in isolated waterholes during the 2001 dry season of an arid-zone river, Cooper Creek, Australia. *Marine and Freshwater Research*, **56**, 25–35.

Attwell, R. I. G. (1970). Some effects of Lake Kariba on the ecology of a floodplain of the Mid-Zambezi Valley of Rhodesia. *Biological Conservation*, **2**, 189–96.

Bacon, P. E., Stone, C., Binns, D. L., Leslie, D. J. and Edwards, D. W. (1993). Relationships between water availability and *Eucalyptus camaldulensis*: growth in a riparian forest. *Journal of Hydrology*, **150**, 541–61.

Baldwin, R. A. (1974). The water balance response of the pelvic "patch" of *Bufo punctatus* and *Bufo boreas*. *Comparative Biochemistry and Physiology*, **47A**, 1285–95.

Bannister, K. E. and Clarke, M. A. (1977). The freshwater fishes of the Arabian Peninsula. *Journal of Oman Studies, Special Report*, **1**, 111–54.

Beadle, L. C. (1974). *The Inland Waters of Tropical Africa: an Introduction to Tropical Limnology*. London: Longman Press.

Beumer, J. P. (1979). Temperature and salinity tolerance of the spangled perch, *Therapon unicolor* Gunther, 1859, and the east Queensland rainbowfish, *Nematocentris splendida* Peters, 1856. *Proceedings of the Royal Society of Queensland*, **90**, 85–91.

Blair, W. F. (1976). Adaptations of anurans to equivalent desert scrub of North and South America. In *Evolution of Desert Biota*, ed. D. W. Goodall, pp. 197–222. Austin, TX: University of Texas Press.

Blaycock, L. A., Ruibal, R. and Platt-Aloia, K. (1976). Skin structure and wiping behaviour of phyllomedusine frogs. *Copeia*, **1976**, 283–95.

Briggs, S. V. (1992). Wetlands as drylands: a conservation perspective. *Bulletin of the Ecological Society of Australia*, **22**, 109–10.

Briggs, S. V., Seddon, J. A. and Thornton, S. A. (2000). Wildlife in dry lake and associated habitats in western New South Wales. *Rangelands Journal*, **22**, 256–71.

Bunn, S. E. and Arthington, A. H. (2002). Basic principles and ecological consequences of altered flow regimes for aquatic biodiversity. *Environmental Management*, **30**, 492–507.

Bunn, S. E., Davies, P. M. and Winning, M. (2003). Sources of organic carbon supporting the food web of an arid zone floodplain river. *Freshwater Biology*, **48**, 619–35.

Burbidge, A. A. (1967). The biology of south-western Australian tortoises. Ph.D. thesis, University of Western Australia, Nedlands, Australia.

Cann, J. (1998). *Australian Freshwater Turtles*. Singapore: Beaumont Publishing.

Coad, B. W. (1980). Environmental change and its impact on the freshwater fishes of Iran. *Biological Conservation*, **19**, 51–80.

Coad, B. W. (1981). Fishes of Afghanistan, an annotated check-list. *National Museum of Natural Sciences Publications in Zoology*, **14**, 1–26.

Coad, B. W. (1987). Zoogeography of the freshwater fishes of Iran. In *Proceedings of the Symposium on the Fauna and Zoogeography of the Middle East, Mainz, 1985*, ed. F. Krupp, W. Schneider and R. Kinzelbach, pp. 213–28. Wiesbaden, Germany: Reichert.

Coad, B. W. (1991). Fishes of the Tigris-Euphrates Basin: a critical checklist. *Syllogeus*, **68**, 1–49.

Coad, B. W. and Abdoli, A. (1993). Exotic fish species in the fresh waters of Iran. *Zoology in the Middle East*, **9**, 65–80.

Contreras, B. S. and Lozano, V. M. L. (1994). Water, endangered fishes, and development perspectives in arid lands of Mexico. *Conservation Biology*, **8**, 379–87.

Corbacho, C. and Sánchez, J. M. (2001). Patterns of species richness and introduced species in native freshwater fish faunas of a mediterranean-type basin: the Guadiana River (southwest Iberian Peninsula). *Regulated Rivers: Research and Management*, **17**, 699–707.

Davies, S. J. J. F. (1984). Nomadism as a response to desert conditions in Australia. *Journal of Arid Environments*, **7**, 183–95.

Day, J. A. (1990). Environmental correlates of aquatic faunal distribution in the Namib Desert. In *Namib Ecology: 25 Years of Namib Research*, ed. M. K. Seely, pp. 99–107. (Transvaal Museum Monograph No. 7.) Pretoria: Transvaal Museum.

Deacon, J. E. and Minckley, W. L. (1974). Desert fishes. In *Desert Biology*, vol. 2, ed. G. W. Brown, Jr., pp. 385–488. New York, NY: Academic Press.

Desert Fishes Council (DFC) (2003). Desert Fishes Council index to fish images, maps and information. www.desertfishes.org/na/index.shtml.

Dimmitt, M. A. and Ruibal, R. (1980). Environmental correlates of emergence in spadefoot toads (*Scaphiopus*). *Journal of Herpetology*, **14**, 21–9.

Douglas, M. R. and Douglas, M. E. (2000). Late season reproduction by big-river Catostomidae in Grand Canyon (Arizona). *Copeia*, **2000**, 238–44.

Dumont, H. J. (1982). Relict distribution patterns of aquatic animals: another tool in evaluating Late Pleistocene climate changes in the Sahara and Sahel. *Palaeoecology of Africa*, **14**, 1–32.

Ellery, W. N. and McCarthy, T. S. (1994). Principles for the sustainable utilization of the Okavango Delta ecosystem, Botswana. *Biological Conservation*, **70**, 159–68.

Gehrke, P. C. (1991). Avoidance of inundated floodplain habitat by larvae of golden perch (*Macquaria ambigua* Richardson): influence of water quality or food distribution? *Australian Journal of Marine and Freshwater Research*, **42**, 707–19.

Gerhke, P. C. and Fielder, D. R. (1988). Effects of temperature and dissolved oxygen on heart rate, ventilation rate and oxygen consumption of spangled perch *Leiopotherapon unicolor* (Günther 1859) (Percoidei, Teraponidae). *Journal of Comparative Physiology*, **B157**, 771–82.

Gehrke, P. C. and Harris, J. H. (2001). Regional-scale effects of flow regulation on lowland riverine fish communities in New South Wales, Australia. *Regulated Rivers: Research and Management*, **17**, 369–91.

Gehrke, P. C., Revell, M. B. and Philby, A. W. (1993). Effects of river red gum, *Eucalyptus camaldulensis*, litter on golden perch, *Macquaria ambigua*. *Journal of Fish Biology*, **43**, 265–79.

Gehrke, P. C., Brown, P., Schiller, C. B., Moffatt, D. B. and Bruce, A. M. (1995). River regulation and fish communities in the Murray-Darling River system, Australia. *Regulated Rivers: Research and Management*, **11**, 363–75.

Georges, A. and Adams, M. (1996). Electrophoretic delineation of species boundaries within the short-necked chelid turtles of Australia. *Zoological Journal of the Linnean Society, London*, **118**, 241–60.

Goodall, H. (1999). Contesting changes on the Paroo and its sister rivers. In *A Free-flowing River: the Ecology of the Paroo River*, ed. R. T. Kingsford, pp. 23–50. Sydney: New South Wales National Parks & Wildlife Service.

Goren, M. and Ortal, R. (1999). Biogeography, diversity and conservation of the inland water fish communities in Israel. *Biological Conservation*, **89**, 1–9.

Graf, W. L. (1988). *Fluvial Processes in Dryland Rivers*. New York: Springer-Verlag.

Green, A. J., Figuerola, J. and Sánchez, M. I. (2002). Implications of waterbird ecology for the dispersal of aquatic organisms. *Acta Oecologica*, **23**, 177–89.

Haig, S. M., Mehlman, D. W. and Oring, L. W. (1998). Avian movements and wetland connectivity in landscape conservation. *Conservation Biology*, **12**, 749–58.

Heatwole, H. (1984). Adaptations of amphibians to aridity. In *Arid Australia*, ed. H. G. Cogger and E. E. Cameron, pp. 177–222. Chipping Norton, NSW: Surrey Beatty.

Helfman, G. S., Collette, B. B. and Facey, D. E. (1997). *The Diversity of Fishes*. Malden, MA: Blackwell Science.

Herremans, M. (1999). Waterbird diversity, densities, communities and seasonality in the Kalahari Basin, Botswana. *Journal of Arid Environments*, **43**, 319–50.

Hocutt, C. H. and Johnson, P. N. (2001). Fish response to the annual flooding regime in the Kavango River along the Angola/Namibia border. *South African Journal of Marine Science (Suid Afrikaanse Tydskrif vir Seewetenskap)*, **23**, 449–64.

Hubbs, C. L. and Miller, R. R. (1948). The zoological evidence: correlation between fish distribution and hydrographic history in the desert basins of western United States. In *The Great Basin with Emphasis on Glacial and Postglacial Times*, pp. 17–166. (Bulletin of the University of Utah 38, Biological Series 10.) Salt Lake City: The University of Utah.

Hubbs, C. L., Miller, R. R. and Hubbs, L. C. (1974). Hydrographic history and relict fishes of the north-central Great Basin. *Memoirs of the California Academy of Sciences*, **7**, 1–259.

Humphries, P., King, A. J. and Koehn, J. D. (1999). Fish, flows and flood plains: links between freshwater fishes and their environment in the Murray-Darling system, Australia. *Environmental Biology of Fishes*, **56**, 129–51.

Jacobsen, N. H. G. and Kleynhans, C. J. (1993). The importance of weirs as refugia for hippopotami and crocodiles in the Limpopo River, South Africa. *Water SA*, **19**, 301–6.

Jared, C., Antoniazzi, M. M., Katchburian, E., Toledo, R. C. and Freymuller, E. (1999). Some aspects of the natural history of the casque-headed tree frog *Corythomantis greeningi* Boulenger (Hylidae). *Annales des Sciences Naturelles*, **3**, 105–15.

Kear, J. (1990). *Man and Wildfowl*. London: T. & A. D. Poyser.

Kennett, R. and Christian, K. (1994). Metabolic depression in aestivating longnecked turtles (*Chelodina rugosa*). *Physiological Zoology*, **67**, 1087–102.

Kennett, R. and Georges, A. (1990). Habitat utilization and its relationship to growth and reproduction of the eastern long-necked turtle *Chelodina longicollis* (Testudinata: Chelidae). *Herpetologica*, **46**, 22–33.

Kennett, R. and Georges, A. (1995). The eastern long-necked turtle – dispersal is the key to survival. In *Jervis Bay: A Place of Cultural, Scientific and Educational value*, ed. G. Cho, A. Georges and R. Stoujesdijk, pp. 104–6. Canberra: Australian Nature Conservation Agency.

King, J., Cambray, J. A. and Impson, D. N. (1998). Linked effects of dam-released floods and water temperature on spawning of the Clanwilliam yellowfish *Barbus capensis*. *Hydrobiologia*, **384**, 245–65.

Kingsford, R. T. (1995). Occurrence of high concentrations of waterbirds in arid Australia. *Journal of Arid Environments*, **29**, 421–5.

Kingsford, R. T. (1996). Wildfowl (Anatidae) movements in arid Australia. *Gibier Faune Sauvage (Game and Wildlife)*, **13**, 141–55.

Kingsford, R. T. and Johnson, W. J. (1998). The impact of water diversions on colonially nesting waterbirds in the Macquarie Marshes in arid Australia. *Colonial Waterbirds*, **21**, 159–70.

Kingsford, R. T. and Norman, F. I. (2002). Australian waterbirds – products of the continent's ecology. *Emu*, **102**, 47–69.

Kingsford, R. T. and Porter, J. L. (1993). Waterbirds of Lake Eyre. *Biological Conservation*, **65**, 141–51.

Kingsford, R. T., Curtin, A. L. and Porter, J. (1999). Water flows on Cooper Creek in arid Australia determine 'boom' and 'bust' periods for waterbirds. *Biological Conservation*, **88**, 231–48.

Kingsford, R. T., Jenkins, K. M. and Porter, J. L. (2004). Imposed hydrological stability imposed on lakes in arid Australia and effect on waterbirds. *Ecology*, **85**, 2478–92.

Kingsford, R. T., Thomas, R. F. and Curtin, A. L. (2001). Conservation of wetlands in the Paroo and Warrego catchments in arid Australia. *Pacific Conservation Biology*, **7**, 21–33.

Knighton, A. D. and Nanson, G. C. (1994). Waterholes and their significance in the anastomosing channel system of Cooper Creek, Australia. *Geomorphology*, **9**, 311–24.

Knighton, A. D. and Nanson, G. C. (2000). Waterhole form and process in the anastomosing channel system of Cooper Creek, Australia. *Geomorphology*, **35**, 101–17.

Krupp, F. (1983). Freshwater fishes of Saudi Arabia and adjacent regions of the Arabian Peninsula. *Fauna of Saudi Arabia*, **5**, 568–636.

Krupp, F. (1987). Freshwater ichthyogeography of the Levant. In *Proceedings of the Symposium on the Fauna and Zoogeography of the Middle East, Mainz, 1985*, ed. F. Krupp, W. Schneider and R. Kinzelbach, pp. 229–37. Wiesbaden, Germany: Reichert.

Krupp, F. (1988). Freshwater fishes of the Wadi Baha drainage. *Journal of Oman Studies, Special Report*, **3**, 401–4.

Krupp, F. and Schneider, W. (1989). The fishes of the Jordan River drainage basin and Azraq Oasis. *Fauna of Saudi Arabia*, **10**, 347–416.

Krupp, F., Schneider, W., Nader, I. A. and Khushaim, O. (1990). Zoological survey in Saudi Arabia, Spring 1990. *Fauna of Saudi Arabia*, **11**, 3–9.

Lee, A. K. and Mercer, E. H. (1967). Cocoon surrounding desert-dwelling frogs. *Science*, **157**, 87–8.

Lemly, A. D., Kingsford, R. T. and Thompson, J. R. (2000). Irrigated agriculture and wildlife conservation: conflict on a global scale. *Environmental Management*, **25**, 485–512.

Leslie, D. J. (2001). Effect of river management on colonially-nesting waterbirds in the Barmah-Millewa forest, south-eastern Australia. *Regulated Rivers: Research and Management*, **17**, 21–36.

Lévêque, C. (1990). Relict tropical fish fauna in central Sahara. *Ichthyological Exploration of Freshwaters*, **1**, 39–48.

Lloyd, P. (1999). Rainfall as a breeding stimulus and clutch size determinant in South African arid-zone birds. *Ibis*, **141**, 637–43.

Lovich, J. and Meyer, K. (2002). The western pond turtle (*Clemmys marmorata*) in the Mojave River, California, USA: highly adapted survivor or tenuous relict? *Journal of Zoology, London*, **256**, 537–45.

Macdonald, D. (2001). *The New Encyclopedia of Mammals.* Oxford: Oxford University Press.

Main, A. R. (1968). Ecology, systematics and evolution of Australian frogs. In *Advances in Ecological Research*, ed. J. B. Craff, pp. 37–86. New York: Academic Press.

Malanson, G. P. (1993). *Riparian Landscapes.* Cambridge: Cambridge University Press.

Mallen-Cooper, M. and Stuart, I. G. (2003). Age, growth and non-flood recruitment of two potamodromous fishes in a large semi-arid/temperate river system. *River Research and Applications*, **19**, 697–719.

Matthews, W. J. (1998). *Patterns in Freshwater Fish Ecology.* New York: Chapman and Hall.

McCarthy, T. S., Ellery, W. N. and Bloem, A. (1998). Some observations on the geomorphological impact of hippopotamus (*Hippopotamus amphibius* L.) in the Okavango delta, Botswana. *African Journal of Ecology*, **36**, 44–56.

McClanahan, L. Jr. (1967). Adaptations of the spadefoot toad, *Scaphiopus couchi*, to desert environments. *Comparative Biochemistry and Physiology*, **12**, 73–99.

McClanahan, L. L., Ruibal, R. and Shoemaker, V. H. (1994). Frogs and toads in deserts. *Scientific American*, **270**, 64–70.

Mayell, H. (2002). Desert-Adapted Crocs Found in Africa. *National Geographic News.* June 18, 2002.

Mayhew, W. W. (1968). Biology of desert amphibians and reptiles. In *Desert Biology*, vol. 1, ed. G. W. Brown, pp. 195–356. New York: Academic Press.

Meffe, G. K. and Minckley, W. L. (1987). Persistence and stability of fishes and invertebrate assemblages in a repeatedly disturbed Sonoran Desert stream. *The American Midland Naturalist*, **117**, 177–91.

Merrick, J. R. and Schmida, G. E. (1984). *Australian Freshwater Fishes: Biology and Management.* Netley: J. R. Merrick.

Minckley, W. L. (1973). *Fishes of Arizona.* Phoenix, AZ: Arizona Game Fish Department.

Minckley, W. L. (1991). Native fishes of the Grand Canyon region: an obituary? In *Colorado River Ecology and Dam Management*, pp. 124–77. Washington, DC: National Academy of Sciences Press.

Minckley, W. L. and Barber, W. E. (1971). Some aspects of the biology of the longfin dace, a cyprinid fish characteristic of streams in the Sonoran Desert. *The Southwestern Naturalist*, **15**, 459–64.

Minckley, W. L. and Deacon, J. E. (1991). *Battle Against Extinction: Native Fish Management in the American West.* Tucson, AZ: University of Arizona Press.

Minckley, W. L. and Meffe, G. K. (1987). Differential selection for native fishes by flooding in streams of the arid American Southwest. In *Ecology and Evolution of North American Stream Fish Communities*, ed. W. J. Matthews and D. C. Heins, pp. 93–104 + literature. cited. Norman, OK: University of Oklahoma Press.

Minckley, W. L., Hendrickson, D. A. and Bond, C. E. (1986). Geography of western North American freshwater fishes: description and relationships to intracontinental tectonism. In *Zoogeography of North American Freshwater Fishes*, ed. C. H. Hocutt and E. O. Wiley, pp. 516–613 + literature cited. New York: John Wiley & Sons.

Mirza, M. R. (1986). Ichthyogeography of Afghanistan and adjoining areas. *Pakistan Journal of Zoology*, **18**, 331–9.

Newman, R. A. (1989). Developmental plasticity of *Scaphiopus couchii* tadpoles in an unpredictable environment. *Ecology* **42**, 763–73.

Newman, R. A. (1994). Effects of changing density and food level on metamorphosis of a desert amphibian, *Scaphiopus couchii*. *Ecology*, **75**, 1085–96.

Nilsson, C. and Dynesius, M. (1994). Ecological effects of river regulation on mammals and birds – a review. *Regulated Rivers: Research and Management*, **9**, 45–53.

Perrins, C. M. (1990). *The Illustrated Encyclopedia of Birds: the Definitive Guide to Birds of the World*. London: Headline.

Peterson, C. C. and Stone, P. A. (2000). Physiological capacity for estivation of the Sonoran mud turtle, *Kinosternon sonoriense*. *Copeia*, **2000**, 684–700.

Petr T. (1999). *Fish and Fisheries at Higher Altitudes: Asia*. (FAO Fisheries Technical Paper No. 385.) Rome: Food and Agriculture Organization.

Por, R. D. (1989). The legacy of Tethys: an aquatic biogeography of the Levant. *Monographiae Biologicae*, **63**, 1–192.

Poynton, J. C. (1964). The amphibia of southern Africa: A faunal study. *Annals of the Natal Museum*, **17**, 1–334.

Predavec, M. and Dickman, C. R. (1993). Ecology of desert frogs: a study from southwestern Queensland. In *Herpetology in Australia*, ed. D. Lunney and D. Ayres, pp. 159–69. Mosman: Royal Zoological Society of New South Wales.

Puckridge, J. T., Sheldon, F., Walker, K. F. and Boulton, A. J. (1998). Flow variability and the ecology of arid zone rivers. *Marine and Freshwater Research*, **49**, 55–72.

Puckridge, J. T., Walker, K. F. and Costelloe, J. F. (2000). Hydrological persistence and the ecology of dryland rivers. *Regulated Rivers: Research and Management*, **16**, 385–402.

Reynolds, L. F. (1983). Migration patterns of five fish species in the Murray-Darling River system. *Australian Journal of Freshwater and Marine Research*, **34**, 857–71.

Roberts, T. R. (1975). Geographical distribution of African freshwater fishes. *Zoological Journal of the Linnean Society*, **57**, 249–319.

Robertson, A. I. and Rowling, R. W. (2001). Effects of livestock on riparian zone vegetation in an Australian dryland river. *Regulated Rivers: Research and Management*, **16**, 527–41.

Roshier, D. A. and Reid, J. R. W. (2003). On animal distributions in dynamic landscapes. *Ecography*, **26**, 539–44.

Roshier, D. A., Robertson, A. I. and Kingsford, R. T. (2002). Responses of waterbirds to flooding in an arid region of Australia and implications for conservation. *Biological Conservation*, **106**, 399–411.

Roshier, D. A., Robertson, A. I., Kingsford, R. T. and Green, D. G. (2001b). Continental-scale interactions with temporary resources may explain the paradox of large populations of desert waterbirds in Australia. *Landscape Ecology*, **16**, 547–56.

Roshier, D. A., Whetton, P. H., Allan, R. J. and Robertson, A. I. (2001a). Distribution and persistence of temporary wetland habitats in arid Australia in relation to climate. *Austral Ecology*, **26**, 371–84.

Ross, W. (1985). Oasis fishes of eastern Saudi Arabia. *Fauna of Saudi Arabia*, **7**, 303–17.

Roth, J. J. (1973). Vascular supply to the ventral pelvic region of anurans as related to water balance. *Journal of Morphology*, **140**, 443–60.

Ruello, N. V. (1976). Observations on some fish kills in Lake Eyre. *Australian Journal of Freshwater and Marine Research*, **27**, 667–72.

Schlatter, R. P., Navaroo, R. A. and Corti, P. (2002). Effects of El Niño southern oscillation on numbers of black-necked swans at Río Cruces sanctuary, Chile. *Waterbirds*, **25**, 114–22.

Schodde, R. (1982). Origin, adaptation and evolution of birds in arid Australia. In *Evolution of the Flora and Fauna of Arid Australia*, ed. W. R. Barker and P. J. M. Greenslade, pp. 191–224. Frewville: Peacock Publications.

Seibert, E. A., Lillywhite, H. B. and Wassersug, R. J. (1976). Cranial co-ossification in frogs: relationship to rate of evaporative water loss. *Physiological Zoology*, **47**, 261–5.

Sheldon, F., Boulton, A. J. and Puckridge, J. T. (2002). Conservation value of variable connectivity: aquatic invertebrate assemblages of channel and floodplain habitats of a central Australian arid-zone river, Cooper Creek. *Biological Conservation*, **103**, 13–31.

Shipway, B. (1947). Rains of fishes? *Western Australian Naturalist*, **1**, 47–8.

Shoemaker, V. H. (1988). Physiological ecology of amphibians in arid regions. *Journal of Arid Environments*, **14**, 145–53.

Skelton, P. H. (1986). Fish of the Orange-Vaal system. *Monographiae Biologicae*, **60**, 143–61.

Smet, K. de (1998). Status of the Nile crocodile in the Sahara desert. *Hydrobiologia*, **391**, 81–6.

Smith, G. R., Dowling, T. E., Gobalet, K. W., *et al.* (2002). Biogeography and timing of evolutionary events among Great Basin fishes. *Smithsonian Contributions to the Earth Sciences*, **33**, 175–234.

Spotila, J. R. and Berman, E. N. (1976). Determination of skin resistance and the role of the skin in controlling water loss in amphibians and reptiles. *Comparative Biochemistry and Physiology*, **55A**, 407–11.

Stafford Smith, D. M. and Morton, S. R. (1990). A framework for the ecology of arid Australia. *Journal of Arid Environments*, **18**, 255–78.

Sullivan, B. K. (1989). Desert environments and the structure of anuran mating systems. *Journal of Arid Environments*, **17**, 175–83.

Thomas, D. H. L. (1995). Artisanal fishing and environmental change in a Nigerian floodplain wetland. *Environmental Conservation*, **22**, 117–42.

Timms, B. V. (1998). A study of Lake Wyara, an episodically filled saline lake in southwest Queensland, Australia. *International Journal of Salt Lake Research*, **7**, 113–32.

Timms, B. V. (2001). Large freshwater lakes in arid Australia: a review of their limnology and threats to their future. *Lakes and Reservoirs: Research and Management*, **6**, 183–96.

Timms, B. V. and Boulton, A. J. (2001). Typology of arid-zone floodplain wetlands of the Paroo River (inland Australia) and the influence of water regime, turbidity, and salinity on their aquatic invertebrate assemblages. *Archiv für Hydrobiologie*, **153**, 1–27.

Tooth, S. and Nanson, G. C. (2000). The role of vegetation in the formation of anabranching channels in an ephemeral river, northern plains, arid central Australia. *Hydrological Processes*, **14**, 3099–117.

Turpie, J. K. (1995). Prioritizing South African estuaries for conservation: a practical example using waterbirds. *Biological Conservation*, **74**, 175–85.

Tyler, M. J. (1978). *Amphibians of South Australia*. Adelaide, Australia: Government Printer.

Tyus, H. M. (1990). Potamodromy and reproduction of Colorado squawfish in the Green River Basin, Colorado and Utah. *Transactions of the American Fisheries Society*, **119**, 1035–47.

Unmack, P. J. (2001a). Fish persistence and fluvial geomorphology in central Australia. *Journal of Arid Environments*, **49**, 653–69.

Unmack, P. J. (2001b). Biogeography of Australian freshwater fishes. *Journal of Biogeography*, **28**, 1053–89.

van Beurden, E. (1982). Desert adaptations of *Cyclorana platycephalus*: an holistic approach to desert-adaptation in frogs. In *Evolution of the Flora and Fauna of Arid Australia*, ed. W. R. Barker and P. J. M. Greenslade, pp. 235–40. Adelaide: Peacock Publications.

Vareschi, E. (1978). The ecology of Lake Nakuru (Kenya). 1 Abundance and feeding of the lesser flamingo. *Oecologia*, **32**, 11–35.

Verlinden, A. and Gavor, I. K. N. (1998). Satellite tracking of elephants in northern Botswana. *African Journal of Ecology*, **36**, 105–16.

Veth, P., Hamm, G. and Lampert, R. J. (1990). The archaeological significance of the lower Cooper Creek. *Records of the South Australian Museum*, **24**, 43–66.

Victor, R. and Al-Mahrouqi, A. I. S. (1996). Physical, chemical and faunal characteristics of a perennial stream in arid northern Oman. *Journal of Arid Environments*, **34**, 465–76.

Wager, R. N. E. and Unmack, P. J. (2000). *Fishes of the Lake Eyre Catchment of Central Australia*. Brisbane: Queensland Department of Primary Industries.

Walker, K. F., Sheldon, F. and Puckridge, J. T. (1995). A perspective on dryland river ecosystems. *Regulated Rivers: Research and Management*, **11**, 85–104.

Weiss, S. J., Otis, E. O. and Maughan, O. E. (1998). Spawning ecology of flannelmouth sucker, *Catostomus latipinnis* (Catostomidae), in two small tributaries of the lower Colorado River. *Environmental Biology of Fishes*, **52**, 419–33.

White, M. (2002). The Cooper Creek turtle persisting under pressure: A study in arid Australia. Honours Thesis, Applied Ecology Research Group, University of Canberra, ACT, Australia.

Williams, J. D., Bowman, D. B., Brooks, J. E. *et al.* (1985). Endangered aquatic ecosystems in North American deserts with a list of vanishing fishes of the region. *Journal of the Arizona-Nevada Academy of Sciences*, **20**, 1–62.

Young, J. Z. (1981). *The Life of Vertebrates*. (3rd edn.) Oxford: Oxford University Press.

II   Human disturbance in desert river systems

8

Impacts of dams, river management and diversions on desert rivers

R. T. KINGSFORD, A. D. LEMLY AND J. R. THOMPSON

INTRODUCTION

The interaction of humans with the environment represents the greatest threat to the sustainability of biodiversity and many human communities. For much of history, our interactions with desert rivers were relatively benign, confined by the ecological dynamics of riverine systems that provided food, shelter and drinking water. Increasing human populations, consumption and improved quality of life have driven the imperative to wring more out of rivers over the past 50 years (Pimentel *et al.*, 1997; Postel, 2000), profoundly changing their ecology forever.

We control flow in many of the world's rivers (Dynesius & Nilsson, 1994; Vörösmarty *et al.*, 1997, 2003; Kingsford, 2000a), primarily for irrigation, industry, hydro-electricity generation, domestic water supply, navigation and flood control. Irrigation is the most prevalent reason for controlling desert rivers (Agnew & Anderson, 1992; Kingsford, 2000a; Lemly *et al.*, 2000) (Fig. 8.1). Humans have developed some desert rivers because of ambitions to, 'make the desert bloom' (Heffernan, 1990), accompanied by the widespread perception that flows in desert rivers are wasted in evaporation, groundwater recharge, floodplain wetlands (Kingsford, 2000a) or out at sea (Gillanders & Kingsford, 2002). This has driven societies around the world to build dams and levees to

Ecology of Desert Rivers, ed. R. T. Kingsford. Published by Cambridge University Press.
© Cambridge University Press 2006.

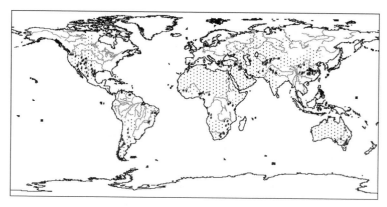

Figure 8.1. One hundred and ninety-two large dams, storage capacity ≥ 0.5km³ (filled circles) (data from Vörösmarty *et al.*, 1997, 2003) that affect rivers and dependent aquatic systems in desert regions of the world (annual rainfall < 500mm) (after Middleton & Thomas, 1997).

harness river flows for irrigation. Rarely was the ecological role of this water appreciated or realised, or the impacts on people living on rivers, particularly floodplain cultivators and fishers, evaluated (Lemly *et al.*, 2000).

The desert river is extremely complex, spatially and temporally, with ever-present disturbance (Chapters 2 and 3, this volume). Every part of it is linked from the upper catchment, riparian habitats, hyporheic zones and river channels to the terminal floodplains or estuaries. Many desert river systems are endorheic with extensive floodplain wetlands, including lakes and seas, among the most spectacular and biodiverse ecosystems in the world. For example, the Okavango Delta in Southern Africa (Fig. 8.2) (Davies *et al.*, 2000) and Lake Eyre in Australia (Kingsford & Porter, 1993) provide habitat for large concentrations of diverse assemblages of organisms. It is the dynamic nature of the desert river systems (Puckridge *et al.*, 1998, 2000; Chapter 2, this volume) and their dependent aquatic systems that produce areas rich in biodiversity. Complex patterns of spatial and temporal connectivity (Ward *et al.*, 1999; Sheldon *et al.*, 2002) create diverse arrays of food webs and habitats, supporting many specialised organisms (Power *et al.*, 1996; Chapters 4–7, this volume). The hydrological disturbance of floods (Sparks *et al.*, 1990; Walker *et al.*, 1995; Chapter 2 this volume) and subsequent

Figure 8.2. Aerial photograph of Okavango Delta, Botswana, one of the
world's most diverse wetlands that is threatened by diversion of water
upstream to Namibia and other water resource development projects
(Photo K. Rogers).

drying processes (Lake, 2000, 2003) are controlling factors, defined by
climate. Human disturbance, in the form of river regulation, imposes
simplicity on complexity, removing or dampening dynamic processes
that define desert rivers.

In this chapter, we detail how humans control flow in desert rivers and the subsequent effects on hydrological behaviour, dependent ecosystems and their biodiversity. In keeping with the theme of this book, we show how this regulation has simplified the disturbance that is the creator of hydrological and biological variability. Of 34 examples where river regulation has dramatically changed river ecology, we demonstrate the scale of destruction of aquatic ecosystems of high biodiversity and effects on people living on desert rivers, using six case studies.

RIVER REGULATORY STRUCTURES

The size, location and type of regulatory structure fundamentally affect the hydrology and ecology of desert river systems at different temporal and spatial scales. The pump is the simplest regulatory structure on a river and is the modern equivalent of water-lifting devices such as shadufs and norias, used for centuries (Agnew & Anderson, 1992). Pumps divert flows directly for agriculture, urban use or storage, where the resulting ecological impact depends on cumulative effects of withdrawals and their timing. Weirs (small dams) usually pool water, allowing water to be diverted down a canal or channel (Table 8.1). Large dams control much of a desert river's flow and geomorphological processes (see Vörösmarty *et al.*, 2003), usually from a position in the upper catchment where surface area is low relative to storage capacity and where much of a river's flow is captured before it reaches the floodplains (Table 8.1) (Kingsford, 2000a). Large storages also exist on lowland parts of floodplains. For example, the Condamine–Balonne river system in arid Australia has much of its flow diverted mostly into large private storages on the floodplain with a total capacity of $150 \times 10^7 \text{ m}^3$ (Kingsford, 2000b; Thoms, 2003; Queensland Department of Natural Resources and Mines, unpublished data), upstream of the major terminal wetland Narran Lakes (Fig. 8.3). There are at least 192 large dams (Fig. 8.4) of 633 worldwide (storage capacity $\geq 0.5 \text{ km}^3$) (Vörösmarty *et al.*, 1997, 2003) that influence desert rivers of the world (Fig. 8.1). Once stored, water is released from dams for urban use, hydroelectricity generation or irrigation. Sometimes rivers accept water harvested from another river basin, through an interbasin transfer (Table 8.1) (Davies *et al.*, 2000). Efficiency of water delivery in rivers can be 'improved' by straightening river channels, running the river at bankfull capacity, blocking off distributaries and removing obstructions

Figure 8.3. Large pumps (600mm diameter) on the Condamine–Balonne River in arid Australia capture floods and divert these into off-river storages, established on the floodplain, mainly to irrigate cotton. These deny the major wetlands of the Lower Balonne and Narran Lakes of their water (pictured), affecting their long-term sustainability. Within a period of about 10 years from the 1990s, most of about 1 500 000 Ml of off-river storage was built in the catchment, predominantly on the floodplains of the river (Photo R. T. Kingsford).

(e.g. debris). Where navigation is essential, locks in the river act as dams and stabilise water levels, allowing movement of vessels up and down the river.

On floodplains, different regulatory structures control river flows. Channels or drains remove water from floodplains or wetlands so the land can be used for agriculture, while levees protect crops and urban communities from flooding and can also channel water to storages for agricultural consumption (Table 8.1). Further control of river flows in lowland sections of a desert river occurs when floodplain lakes are turned into storages with regulators built across inflowing creeks (Kingsford, 2000a; Partow, 2001).

EFFECTS OF RIVER REGULATION ON RIVER PROCESSES

River regulation affects all river processes, with predictable effects on instream ecology and downstream wetlands (Tables 8.1 & 8.2). A large

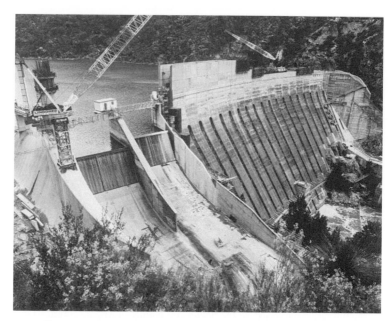

Figure 8.4. Burrinjuck Dam on the Murrumbidgee River was one of the first large dams built (1927) in southeastern Australia. With Blowering Dam, it remains the major dam that captures river flows for irrigation, upstream of the major floodplain, the Lower Murrumbidgee floodplain. Reduction in river flows combined with floodplain development have degraded or destroyed about 75% of the floodplain, significantly affecting waterbird populations and flood-dependent vegetation.

dam on a river changes hydrology and hydraulics of the river (Fig. 8.4 and 8.5), removing high variability of inflows by storing water and superimposing different variability during release periods (Maheshwari *et al.*, 1995; Kingsford & Thomas, 1995; McMahon & Finlayson, 2003). This changes the seasonality of the flows, usually from spring to summer flooding, when irrigation needs are highest (Table 8.1). Forceful channel-forming flows become less frequent with reduced magnitude, affecting downstream ecosystems (Benn & Erskine, 1994; Stevens *et al.*, 1995). Floodplain structures, such as levee banks and channels, confine flows and alienate or reduce connectivity of ecosystems from the main river channel. Total annual flows may remain the same upstream of diversion points on a river but fewer flows break the banks than naturally, because flow is controlled by release regimes designed to ensure the opposite.

Table 8.1. *River regulatory structures that affect river flows in desert river systems, ordered in terms of greatest impact, their purpose and major effects on hydrology, hydraulics and biota of the river*

Structures	Purpose	Effects on hydrology and hydraulics	Main effects on biota
Dams[a]	Irrigation; drinking water for humans and livestock; industry and generation of hydro-electricity	Stores flood flows, denying floodplains downstream; reduces channel-forming flows; reduces variability of flows at annual, monthly scales; increases variability at daily and hourly scales; changes seasonality of flows and decreases water temperatures	Reduces extent and complexity of habitat on floodplains and biodiversity; decreases survival, health and reproduction of biota and migration opportunities for riverine fauna; increases exotic species; reduces replenishment of groundwater ecosystems and freshwater flows for marine and estuarine processes; reduces access to water for riparian species and traps sediments and nutrients
Pumps	Irrigation; drinking water for humans and livestock; industry	Reduces flows downstream with effects dependent on quantities of flow diverted; affects temporal and spatial variability of flow regime	Reduces flows to floodplains and estuaries; decreases survival, health and reproduction of dependent biota
Levees or dykes	Protection of towns, crops and homesteads from flooding	Changes spatial distribution of flooding and prevents water reaching parts of the floodplain	Reduces habitat through alienation from flooding; reduces biodiversity; severs connectivity for sedentary organisms; decreases abundance and diversity of aquatic fauna and use of aquatic habitat by terrestrial fauna

Table 8.1. (cont.)

Structures	Purpose	Effects on hydrology and hydraulics	Main effects on biota
Interbasin transfers and pipelines	Irrigation and drinking	Reduces flows in the river where water is abstracted and increases flows in the river where water is transferred	Reduces aquatic and floodplain habitats on the river where water is abstracted; increases erosion in river reach where water is transferred; potential transfer of exotic organisms between river basins and establishment of organisms adapted to high flows in river
Locks	Navigation	Replacement of lentic conditions with lotic conditions; decreases variability in river heights and flows at monthly and yearly timescales; increases variability at hourly and daily timescales	Establishment of lotic biota; disappearance of lentic biota; trapping of sediment
Canals and Channels	Transfer water for human use; capture floodplain flows	Changed spatial distribution of flooding	Reduced aquatic floodplain habitat where flows captured

[a] Includes weirs, off-river storages (private dams on lowland parts of rivers where water during floods can be pumped into dams for storage) and lakes or wetlands turned into storages with banks or weirs over inlets.

Table 8.2. *Global examples of aquatic ecosystems with high biodiversity, supplied by rivers flowing through desert regions and the catchment countries, affected by river regulation. These include six case studies (CS), detailed in text. Regulatory structures (RS) include dams or reservoirs (R), levees or dykes (LV), locks (LO), interbasin transfers (IBT), pipelines (PL) and canals or channels (C). Primary purposes (P) of the structure are irrigation (I), drinking (D), hydro-electricity generation (H), flood control (F), navigation (N) and watering of livestock (L)*

Aquatic ecosystems	Rivers	Country	RS	P	Ecological and hydrological effects	References
Aibi Lake	Kuitan	China	R	I	Covered 1070 km² in 1958 and is now 570 km². Salinisation of springs and dust storms increasing. Dependent vegetation also dying.	Mengxiong (1995)
Aral Sea (CS)	Amar Dar'ya and Syr Dar'ya	Afghanistan, Iran, Kazakhstan, Kyrgyzstan, Turkmenistan, Tajikistan, Uzbekistan	R, LV, PL, C	I	Declining inflows reduced size of the wetland, causing extensive loss of biodiversity and increased salinity.	Micklin (1988); Aladin *et al.* 1998; Lemly *et al.* (2000)
Azraq Oasis	Wadi Rajil and springs	Jordan	R	I, D	Originally covered 12 000 ha, providing habitat to a million waterbirds during migration and 28 breeding species. A 98% reduction in flows almost completely dried habitat. Most of the dependent aquatic vegetation has died. Reduced flows stopped springs.	Scott (1995); Wishart *et al.* (2000)

Table 8.2. (cont.)

Aquatic ecosystems	Rivers	Country	RS	P	Ecological and hydrological effects	References
Burdur Lake	Bozçay	Turkey	R	I	About 50% reduction in inflows, contributing to declining waterbird populations	Green et al. (1996)
Lake Chad	Yobé, Chari, Logone	Cameroon, Chad, Niger, Nigeria	R	I, H, D	Decline in lake level with a resulting 95% reduction in area from 25 000 km² in early 1960s to 1350 km² today with perhaps as much as 50% due to irrigation. Increased mortality and extinction of fish populations. Reduced wet-season flooding and loss of grazing for wildlife including ungulates and elephant	Toro (1998); Coe & Foley (2001)

	River	Country		Description		Reference
Ejina Oasis, Juyan Lakes	Heihe	China	R	Westernmost Juyan Lake, largest terminal lake, was 213 km² in 1960 but remained dry since 1961. East Juyan Lake reduced in size from 36 km² in 1958 to drying up in 1997. In the 1980s salinity increased from 0.5 g l⁻¹ in 1962 to 0.98 mg l⁻¹ in 1987. Large areas of wetland vegetation have disappeared (12 × 10⁴ ha of swamp vegetation to 9.0 × 10⁴ ha today), seriously degraded grasslands cover 43 × 10⁴ ha. Sandstorms have increased. 35 × 10⁴ ha desertified and area of river dependent forest decreased by 5.76 × 10⁴ ha since the 1950s	I	Mengxiong (1995); Genxu & Guodong (1999a, b); Li *et al.* (2000)
Carson River floodplain	Carson	United States	R	84% of wetlands destroyed; most native fish species and many amphibians, reptiles and mammal species have declined or are locally extinct	I	Lemly *et al.* (2000)

Table 8.2. (cont.)

Aquatic ecosystems	Rivers	Country	RS	P	Ecological and hydrological effects	References
Colorado delta	Colorado	United States	R	I	Dramatic decrease in sedimentation, river flows and abundance of benthic invertebrates and migratory birds. Endemic fish totoaba *Totoaba maconaldii* once commercial but now endangered. Decreased shrimp catches. Increased salinity in the estuary	Plummer (1994); Lavín & Sanchez (1999); Kowalewski *et al.* (2000)
Dead Sea	Jordan	Israel, Jordan	R	I	Much of freshwater inflows diverted; water levels decreasing by 50–60 cm yr^{-1}; 1992–99 water level decreased by about 6 m, causing land subsidence, increasing salinity, reducing blooming of algae and bacteria	Watzman (1997); Baer *et al.* (2002)

Wetland	River	Countries			Description	References
Ganges delta	Ganges, Brahmaputra	Bangladesh, Bhutan, India, Nepal, China	R, C, IBT	I, D	Reduced flows so that water no longer reaches the delta (6×10^6 ha) for all or part of year, causing intrusion of salt into mangroves and reducing fish, river dolphin, turtle and crocodile populations	Postel (1996); Khan (1996); Milliman (1997); Gopal et al. (2000)
Floodplain wetlands of the Komadugu-Yobe Basin, including the Hadejia–Nguru wetlands (CS)	Hadejia, Jama'are, Yobe	Nigeria	R	I, H, D	Large reductions in wet season river flow and subsequent floodplain inundation (Fig 8.7). Reduction in wildlife habitat and degradation of riparian forests. Impacts to communities practising floodplain cultivation and fishing. Potential wider impacts due to reduced groundwater recharge. Lower flows promote the blockage of floodplain channels by vegetation	Hollis et al. (1993b); Thomas (1995); Thompson & Hollis (1995); Lemly et al. (2000); Thompson & Polet (2000)
Huleh Lake and Marshes	Jordan	Israel, Jordan, Lebanon, Syria	LV	I	Dry following water resource development. Lakes once covered about 1420 ha and papyrus swamps 3140 ha.	Karmon (1960); Agnew & Anderson (1992)

Table 8.2. (cont.)

Aquatic ecosystems	Rivers	Country	RS	P	Ecological and hydrological effects	References
Lac Iriki	wad Draa	Morocco	R	I	Lake of 8000 ha dried up in 1970s	Green et al. (2002)
Manas Lake	Manas	China	R	I	Covered 550 km^2 in 1968 but is now dry	Mengxiong (1995); Heng (2001)
Macquarie Marshes	Macquarie	Australia	R, LV	I	Loss of at least 40–50% of original wetland area from reduced flows, causing declines in abundance and diversity of waterbirds and breeding frequency and extent of colonial waterbirds	Kingsford & Thomas (1995); Kingsford & Johnson (1998)
Mesopotamian Marshlands (CS)	Tigris and Euphrates	Turkey, Iran, Iraq, Syria	R, C, LV	I, D, H	Destruction of most of Marshlands, resulting in devastating loss of biodiversity and homelands of Marsh Arabs	Wishart et al. (2000); Partow (2001)
Mono Lake	Owens	United States	R	D	Lake levels dropped by 14 m, causing breeding habitat islands to be exposed to predation. Salinity has doubled	Wiens et al. (1993)
Missouri River floodplain	Missouri	United States	R, LV	I, H	50% reduction in floodplain habitat turned into terrestrial habitat.	Gore & Shields (1995)

Lower Murrumbidgee floodplain (CS)	Murrumbidgee	Australia	R, LV, C, IBT	I, H	At least 60% reduction in annual flows and degradation of about 75% of wetlands, with a 80% decline in waterbird abundance across all functional groups during the period 1983–2000 (Fig. 8.4). Significant effects likely to affect other fauna and flora	Kingsford & Thomas (2002); Sheldon et al. (2000); Kingsford (2003)
Narmada estuary	Narmada	India	R	I, H	Reduced flows and trapping of sediments upstream. Dislocation of river communities	Black (2001), Foote et al. (1996), Gopal et al. (2000)
Narran Lakes (Fig 8.3) and Lower Balonne	Narran, Condamine, Balonne	Australia	R, LV	I	Reduction in magnitude, frequency and duration of flooding. Loss of floodplain areas and potential long-term impacts on waterbird populations as a result of 74% reduction in median flows. Annual reduction of 1293 tonnes in dissolved organic carbon supply to the floodplain	Kingsford (2000b); Thoms (2003)
Niger Delta	Niger	Benin, Guinea, Mali, Niger, Nigeria	R	H	Reduction in floods and 50% reduction in fish catch.	Lae (1995)

Table 8.2. (cont.)

Aquatic ecosystems	Rivers	Country	RS	P	Ecological and hydrological effects	References
Nile Delta	Nile	Egypt, Ethiopia, Sudan, Uganda	R, C	I, H, D	Nearly all freshwater flows diverted upstream. Reduced nutrients. Less than 10% of original sediment reaches the coast, leading to erosion. Fish populations have declined with 36% of commercial native fish species harvested following Aswan Dam; the annual sardine harvest declined by 83%. Increased pollution and eutrophication	Stanley & Warne (1993); Postel (1996); Stanley (1996); Milliman (1997); Stanley & Warne (1998)
Nyl River floodplain	Nyl	South Africa	R	D	Floodplain vegetation and waterbirds affected by reductions in flows	Higgins et al. (1996)
Okavango Delta (Fig. 8.2)	Okavango	Angola, Botswana, Namibia	PL	D	Proposed abstraction would reduce flows and affect one of the most diverse wetland ecosystems of the world	Ramberg (1997); Davies & Wishart (2000)

Qinghai Lake	Buh	China	R	I	Water levels dropped by 10 m (12 cm yr^{-1} for past 30 years), increasing salinity and reducing area (now 85% of original size in 1908). Supplied by 108 freshwater rivers in 1960s but 85% of these have since dried up, including Buh River in 2002. High biodiversity, particularly waterbirds including breeding species, vegetation and fish species probably affected	Comín & Williams (1994); Anon (2002)
San Joaquin River floodplain and delta (CS)	San Joaquin	United States	R, C, IV	I, H, D	Most flows diverted and serious decline in water quality (Fig 8.5). This has severely affected habitats and dependent wildlife	Lemly et al. (2000)
Senegal Delta and Lake Guiers	Senegal	Guinea, Mali, Mauritania, Senegal	R	I	Loss of wetland habitat, reduced fish populations and changes in vegetation. Reduced variability of lake levels, changing vegetation composition and abundance	Drijver & Marchand (1985); Vinke (1996); Cogels et al. (1997)

Table 8.2. (cont.)

Aquatic ecosystems	Rivers	Country	RS	P	Ecological and hydrological effects	References
South Saskatchewan River wetlands	South Saskatchewan River	Canada	R	I	Loss of over 40% of wetlands and shallow-water habitat	Livingstone & Campbell (1992); Gilbert & Ramey (1995)
Stillwater Marsh	Truckee, Humboldt, Carson Rivers	United States	R, C	I	Reduction in inflows of over 90% and water contamination, loss of 71% of wetlands	Hoffman et al. (1990)
Tana River and Delta	Tana	Kenya	R	H, I	Reduced flooding of riverine forests and increased fragmentation	Hughes (1984). Maingi & Marsh (2002)
Tarim floodplain wetlands, Lop Nur and Boston Lake (CS)	Tarim	China	R, C	I	81% reduction in flow, lower reaches dry, groundwater levels lowered by 8 m and increased in salinity, loss of 20 000 ha of grassland, poplar and red willow forests, increased desertification. Destruction of Lop Nur (3000 km^2) as an aquatic ecosystem	Mengxiong (1995); Hongfei et al. (2000); Xinguang et al. (2001)

			R, C, LV, IBT	F, H, I		
Yellow River delta	Yellow	China	R, C, LV, IBT	F, H, I	Reduced flows, particularly in lower parts (600 km) (dry for 70 days yr^{-1} in 1985–95, 122 days in 1995, 133 days in 1996 and 226 days in 1997). Deltaic land formed at 35 km^2 yr^{-1} naturally but now coast is eroding, endangered wetland habitats for migratory wading birds	Postel (1996); Kang & Yong (1998), Changming (1989); Li et al. (2000); Gillanders & Kingsford (2002)
Zambezi delta	Zambezi (Angola, Botswana, Mozambique, Namibia, Zambia, Zimbabwe)		R	I, D	Potentially affected by dams and water transfers, affecting endangered crane populations	Davies & Wishart (2000)

Figure 8.5. The Massingir Dam on the Elephants River on the border
between Mozambique and South Africa. This earthen dam stores water for
irrigation downstream, altering downstream flooding patterns (Photo
M. Wishart).

Downstream of major diversions for human uses, water quantity
decreases considerably compared with the natural flow volume, further
reducing flows to floodplains, wetlands, groundwater recharge areas
and estuaries or deltas (Table 8.2). Alternatively, interbasin transfers
can lead to increased flows (Davies *et al.*, 2000). They also connect desert
rivers that may have been isolated, creating colonisation opportunities
for exotic species (e.g. fish) and potentially affecting evolutionary processes. Dams also trap sediment and nutrients once deposited on floodplains and in the river (Stevens *et al.*, 1995). Concrete barriers (e.g. dam
walls, weirs) stop migratory species reaching breeding areas upstream;
levees stop species colonising floodplains.

Water quality of flows changes with regulation, with byproducts of
water use discharged into rivers. Leaching of trace elements from irrigated soils also pollutes rivers (Lemly *et al.*, 1993); human waste and
chemical byproducts seriously affect water quality in desert rivers in
many parts of the world (Foote *et al.*, 1996). In addition, dams often
discharge water at low temperatures (Sheldon *et al.*, 2000) which can take
hundreds of kilometres to equilibrate to natural temperatures. In 1991, a
cyanobacterial bloom extended for more than 1000 km of the Darling

River in central Australia, exacerbated by reduced flows (Bowling & Baker, 1996), affecting drinking water for people and livestock.

Ecological processes, governed by hydrology (Walker *et al.*, 1995), inevitably change with river regulation and water diversion (Table 8.1) owing to the simplification of highly variable temporal and spatial processes that affect organisms, habitats and people. At their most extreme, this results in destruction of aquatic habitats. For example, Hubei Province in China had 1052 lakes in the 1950s but now only 83 remain, covering only 200 km^2 (Heng, 2001). Ecological effects occur at three main locations: the dam site, between the dam and major diversion points, and the terminal floodplain, estuary or delta. Ecological impacts on the dam site, where terrestrial ecosystems are flooded and destroyed and local river people relocated, are the most obvious consequences (Fearnside, 1989). Downstream ecological effects are probably even more disruptive (Kingsford, 1999, 2000a; Gillanders & Kingsford, 2002; Lemly *et al.*, 2000). Only a fraction of terrestrial ecosystems are affected at the dam site (Fig. 8.5), compared with the entire dependent biota, habitats and ecological processes downstream.

Immediately downstream of a major dam, flows are modified and the dependent biota inevitably change. With structures acting as barriers for migration of native fauna, new flow regimes are often conducive to establishment of exotic fish species (e.g. European carp *Cyprinus carpio*) (Gehrke *et al.*, 1995) and plant species more typical of lentic systems (Blanch *et al.*, 2000). The frequency of low flows increases whereas the duration and frequency of dry periods declines (McMahon & Finlayson, 2003). The greatest effect of regulation on a desert river occurs at the end of the river: the floodplain, estuary, delta or groundwater recharge areas. High biodiversity congregates around these freshwater ecosystems (Boulton *et al.*, 2003; Chapters 4–7, this volume), reliant on variable flow regimes that create a range of different habitats. The removal of sustaining flows, particularly in endorheic river basins, destroys ecosystem processes and reduces biodiversity over extensive areas (Table 8.2). Ecosystems, once sustained by natural river flows, collapse or contract when the flow is diverted for humans (Table 8.2) (see case studies below). A cascade of change occurs as flood-dependent vegetation, established under a natural flow regime, dies on the margins and recolonises the river, reflecting reduced flooding (Bren, 1992). The considerable interaction between terrestrial and aquatic ecosystems decreases, reducing the transfer of organic matter from terrestrial systems (Robertson *et al.*, 1999). Such changes also affect terrestrial fauna that need to drink from wetlands, rivers and water-holes (Chapter 7, this volume). Compounding the problem for desert rivers, lag times for effects on ecology can be long (sometimes more than 50 years) before the full ecological damage can be detected, if at all (Thoms & Walker, 1993; Kingsford, 2000a; Sheldon *et al.*, 2002).

Effects of dams, river management and diversions on river processes are ubiquitous but some impacts are more pronounced in desert rivers. Large reservoirs trap most sediment (90%) in endorheic arid basins in Asia and Australia and almost all sediment on the desert rivers of the Nile and Colorado (Vörösmarty *et al.*, 2003). High variability of flows, relatively low flow volumes, high salt content, low relief and extensive floodplains are characteristics of desert rivers (Chapters 2,3 and 10, this volume). Reductions in temporal and spatial variability with river regulation have a large effect on processes in desert rivers. Removal of water from rivers particularly affects floodplain systems downstream, with reductions in frequency and extent of flooding (Table 8.2). Evaporative losses from storages and irrigation areas are high in desert regions (Fig. 8.5), increasing the need for extraction from desert rivers. Small levees, roads and even vegetation affect the hydraulics of inundation patterns across a highly complex floodplain where relief is low (Tooth & Nanson, 2000). Relatively low volumes of flows and evaporative losses in desert rivers accentuate effects of reduced water quality. Desert rivers flow through salt-laden soils; with decreasing volumes, salinity inevitably increases (Chapter 10, this volume). This particularly affects estuaries, where reduced freshwater inflows lead to hypersaline conditions and the destruction of habitats dependent on periodic dilution with river water (Duvail & Hamerlynck, 2003). The poorly understood replenishment role of rivers and groundwater systems becomes increasingly problematic with diversion of water upstream, leading to serious reductions in flows from springs (Contreras & Lozano, 1994). Reductions affect flows to groundwater-dependent ecosystems and baseflows and subsurface flows for riparian vegetation.

CASE STUDIES

Table 8.2 summarises 34 examples of aquatic ecosystems supplied by desert rivers, around the world, destroyed or seriously affected by river regulation. Seven of these are from Asia, three from Australia, ten from

Africa, one from Europe, six from the Middle East and seven from North America. We detail six of these examples in our case studies.

Aral Sea

The Aral Sea (historically 68 000 km^2) (Fig. 8.6a) is the world's worst example of the impacts of river regulation on desert rivers and associated aquatic ecosystems (Lemly et al., 2000). It used to receive about half the flows from Amu Dar'ya and Syr Dar'ya Rivers from snowmelt in the Pamir and Tyan-Shan mountain ranges, which also supplied deltaic floodplains, marshes and tugay forests or evaporated (Ferrari et al., 1999). More than 80 dams and canals (700 000 km of irrigation channels, (Williams, 2003)) control river flows (Fig. 8.6a), allowing 102.4 km^3 of flow to be diverted each year (Ferrari et al., 1999) to water seven million hectares of cotton and rice (Li et al., 2000). The unfinished Karakum Canal (1370 km long) (Fig. 8.6) diverts water into the Turkmenistan Desert (Williams, 2003).

The Aral Sea has declined by 46% in area and 65% in volume (originally 1000 km^3); water levels have dropped by 15–20 m (Micklin, 1988, 1996; Kotlyakov, 1991). Salinity had increased from about 10 g l^{-1} in 1961 to 40 g l^{-1} by 1994 (Aladin et al., 1998). Large dust storms mobilise the 27 000 km^2 of salt exposed on the dry lake-bed, blowing 43 million tonnes annually across 200 000 km^2 (Lemly et al., 2000). Most biodiversity has disappeared, including much of the deltaic floodplain (Novikova, 1996). By the 1980s, only 20% of the 173 large animal species once living in the Aral wetlands remained (Micklin, 1988). This included a 57% decline in mammal species and the loss of half of the nesting bird species (Kotlyakov, 1991). Sixteen of 20 native fish species are extinct (Micklin, 1988). Wetland plant communities, particularly extensive reed beds and tugay forests were destroyed or degraded (Novikova, 1996). Of 64 macroinvertebrate species found in the 1980s, 78% were not present in the 1990s, with all freshwater species extinct (Aladin et al., 1998). Originally, total biomass of all aquatic plants was 8.1–10.2 million tonnes with a substantial freshwater component but, by the 1980s, species diversity had declined, salt-tolerant plants became dominant and all aquatic macrophyte species were extinct (Aladin et al., 1998). Pollution by pesticides (Ivanov et al., 1996) and, more recently, oil from Uzbekistan into the Syr D'arya River, further affected the biota.

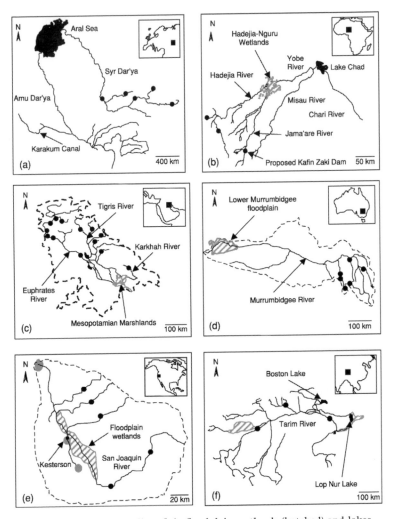

Figure 8.6. Case studies of six floodplain wetlands (hatched) and lakes (filled), supplied by desert river systems, regulated by major dams (filled circles): (a) Aral Sea Basin (Eastern Europe), (b) Hadejia–Nguru wetlands and Lake Chad Basin (West Africa), (c) Mesopotamian Marshlands (Middle East); (d) Lower Murrumbidgee floodplain (Australia); (e) San Joaquin River, floodplains and delta (North America); (f) Tarim River floodplains and Lop Nur and Boston Lakes (China). Dashed lines indicate catchment boundaries.

The fishing industry, employing 60 000 people, collapsed from 48 000 tonnes in the 1950s to almost nothing (Micklin, 1988; Postel, 1996; Williams, 2003), with fishing villages stranded on the former shoreline of the lake now up to 75–80 km from the water. The deltaic marshes of the rivers used to be areas for spawning commercial fish, feeding livestock, hunting and trapping, and harvesting of reeds. Salt deposits blown from the dry lakebed contribute to salinisation of irrigation areas. Increasing levels of anaemia, thyroid disease, viral hepatitis and oesophageal, stomach and liver cancer in local communities have decreased life expectancy by 20%, exacerbated by increasing infant mortality (Perera, 1993; Pearce, 1995), now among the highest in the world (Williams, 2003). Inevitably, despite international efforts at rehabilitation (Aladin et al., 1998; Williams, 2003), only a fragment of the ecology of this magnificent system will be restored without the water and the removal of structures that regulate river flows.

Floodplain wetlands of the Komadugu–Yobe River Basin

The Hadejia and the Jama'are Rivers of the Komadugu–Yobe River Basin (84 140 km^2) in north-eastern Nigeria (Hollis et al., 1993b) flow northeast to the Yobe River, which joins the Misau River to flow into Lake Chad (Fig. 8.6b). Annual rainfall exceeds 1300 mm in the south but progressively declines downstream to less than 500 mm (Polet & Thompson, 1996). There are distinct wet (May–September) and dry seasons (October–April), producing highly seasonal river flows (Thompson & Polet, 2000); about 80% of the annual runoff occurs in August and September (Hollis et al., 1993b).

Large floodplain wetlands, including the most extensive Hadejia–Nguru Wetlands (see, for example, Hollis et al., 1993a; Polet & Thompson, 1996; Figs. 8.6b, 8.7), flood in July–September and then slowly dry. Declining rainfall and diversions over the last four decades of the twentieth century have reduced flood extent. Floods extended over more than 2000 km^2 in the 1960s, 1000 – 2000 km^2 in the 1970s, only 300 km^2 in 1984 (Hollis & Thompson, 1993a; Thompson & Hollis, 1995) and 1000 – 2000 km^2 in the 1990s (e.g. 1994), with wet years. Lake Chad (Fig. 8.6b) experienced similar drying patterns (Coe & Foley 2001).

Upstream dams and irrigation diversions prolonged drought, reducing the natural pattern of rising and falling floods since the mid-1970s (Fig. 8.7) (Hollis & Thompson, 1993b; Thompson & Hollis, 1995). Large-scale irrigation schemes across northern Nigeria (Adams et al.,

Figure 8.7. Floodplain wetlands such as Nguru Lake (Photo J. R. Thompson) rely on flows from the Hadejia and Jama'are rivers in north-eastern Nigeria, Africa. Dams constructed in the upstream part of the Hadejia Basin, primarily for irrigation, have reduced flooding, affecting wildlife, subsistence farming and fisheries and the availability of grazing resources.

1993) are supported by more than 20 dams within the Komadugu–Yobe Basin, mostly in the upper catchment of the Hadejia River. Tiga Dam (1400 × 10^6 m^3) built in 1971–4 on the Kano River, is the largest dam, supplying the Kano River Irrigation Project of over 14 000 ha, with another 53 000 ha planned. Water from Challawa Gorge Dam supplies the second major irrigation scheme, the Hadejia-Valley Project (around 8000 ha, 12 500 ha planned), immediately upstream of the Hadejia–Nguru Wetlands (Fig. 8.6b).

When Tiga Dam filled in 1976, droughts became more severe compared with those of the early 1970s (Stock, 1978), reducing flood-plain cultivation of rice and destroying recession farms along the Hadejia River (Wallace, 1980). In 1977, local people complained that 'the land is dying' (Kulatunga *et al.*, 1997, cited in Olofin, 1996). Tiga Dam and irrigation reduced peak flood extent by an average of 11%; the effect was more pronounced in dry years (Thompson, 1995). Challawa Gorge Dam dramatically decreased downstream inundation in 1992 (Lemly *et al.*, 2000) and, with other dams and diversions for irrigation, further reduced peak wet-season flood extents by an average of 17% on the Hadejia River (Thompson, 1995), affecting agricultural and fishing communities. With reduced flows, vegetation (*Typha domingensis*) has blocked floodplain channels (Goes, 2002), further affecting flood patterns. Declining groundwater recharge has caused water levels in wells to drop, with potentially enormous social and economic consequences.

Some parts of the floodplain are particularly affected. The Keffin Hausa distributary rarely receives flows from the Hadejia River (Thompson & Polet, 2000), despite regular channel dredging (Adams *et al.*, 1993). Flood rice cultivation has nearly disappeared in this part of the wetlands, and the Baturia Wetland Reserve is degraded (Adams, 1993). Flows within the Yobe River now rely upon the Jama'are River (Goes, 2002) and so are considerably reduced. Nigeria's first Ramsar convention site, within the Hadejia–Nguru Wetlands, is an important habitat for waterbirds that are declining in abundance and diversity in years of poor flooding (Garba Boyi & Polet, 1996). Waterbirds congregate on the few remaining regularly flooded areas (Olofin, 1996). Declines in floodplain fisheries from reductions in flood extent (Thomas *et al.*, 1993) compound the upstream impacts of dams and irrigation schemes. Calls for artificial floods (Aminu-Kano *et al.*, 1993; Barbier & Thompson, 1998) have been answered with some experimental releases. However, plans for irrigation expansion remain, as does a scheme to build the largest reservoir in the basin (Kafin Zaki Dam) on the largely unregulated Jama'are River (Fig. 8.6b). Further ecological effects are inevitable.

Mesopotamian Marshlands

Much of the Mesopotamian Marshlands (originally 15 000 – 20 000 km^2), supplied by the Tigris (1900 km) and Euphrates (3000 km) rivers, has been destroyed (Fig. 8.6c). Most of the runoff for the Marshlands comes from spring snowmelt in Turkey and Syria, in the upper catchment of the Tigris–Euphrates River Basin (950 876 km^2), producing highly variable seasonal flows (Partow, 2001). The Karun and Karkheh rivers originate in Iran and flow into the Marshlands at the confluence of the two major rivers. The Mesopotamian Marshlands, the largest wetland system in the Middle East, were a complex of seasonal and permanent freshwater marshes and lakes, temporary wetlands and brackish lagoons (Partow, 2001).

The first major river regulatory structures were the Ramadi and Samarra barrages, built in the 1950s across the Tigris and Euphrates rivers (Beaumont, 1978) to divert water into seasonal lakes or depressions (Partow, 2001). Dams in the upper catchment have a greater impact on flows. There are now 32 large dams in the basin (eight under construction, 13 planned) (Fig. 8.6), potentially storing 137% and 92% of the respective flows in the Euphrates and Tigris Rivers at the Turkish–Syrian border (Beaumont, 1978; Partow, 2001). These trap water and sediment (107 million tonnes yr^{-1}) that historically supplied the Marshlands (Partow, 2001). Most of the water stored is diverted for irrigation and drinking supply to major cities. Combined with massive drainage works in the Marshlands, river regulation destroyed 85–90% of this outstanding wetland between 1973 and 2000, replacing it with desert and salt-encrusted areas (Munro & Touron, 1997; Partow, 2001). About 97% of Central Marsh and 94% of Al Hammar Marsh had disappeared by 2000 (Partow, 2001).

Drainage works affecting the marshlands began in 1953 with the large Third River Canal (90 m wide, 565 km long) to deal with soil salinisation. It was completed in the 1990s to reclaim the marshes and capture flows for irrigation, perceived as wasted in the Marshlands (Pearce, 1993; Partow, 2001). A series of other large canals effectively stopped water from reaching most of the Marshes (e.g. Prosperity River, 2 km wide, 50 km long). Other embankments, dikes and polders completed the destruction, leaving only a small northern fringe of the Al-Hawizeh Marsh on the border between Iran and Iraq (Munro & Touron, 1997; Partow, 2001). By 1993, two thirds of the water to the Marshes had been stopped, with all flows in the Euphrates diverted into the Third River Canal while flows in the Tigris and its distributaries were diverted into a canal 90 km west of the river (Pearce, 1993).

Destruction of marshland had devastating effects on the wildlife. Much of the reed beds (reed *Phragmites communis* and reed mace *Typha augustata*) was destroyed (Partow, 2001). Fourteen bird species, three mammal species and a dragonfly species are listed as globally threatened (Groombridge, 1993). The smooth-coated otter *Lutra perspicillata maxwelli* and soft-shell turtle *Trionyx euphraticus* are already probably extinct; the bandicoot rat *Erythronesokia bunnii* and barbel *Barbus sharpeyi* could also be extinct (Partow, 2001). The Marshlands were a key habitat for several million waterbirds, up to two thirds of West Asia's wintering waterbirds, with internationally significant numbers of 66 bird species at risk (Scott, 1995; Partow, 2001). Two species, the African darter

Anhinga rufa and sacred ibis *Threskiornis aethiopicus*, are probably region-
ally extinct; the pygmy cormorant *Phalacrocorax pygmaeus* and Goliath
heron *Ardea goliath* have probably disappeared from Iraq. About 40
waterbird species are potentially affected (Partow, 2001), including
pelicans, flamingos, herons and ducks. Pesticide residues (DDT) are high
in the Marshlands, potentially affecting birds of prey. Many native fish
species (commercially about 17 300 tonnes in 1991) are severely affected
by wetland loss, poisoning (rotenone and cyanide), increased salinity as
a result of returning saline waters, and polluted water (munitions,
industrial and domestic waste) (Partow, 2001). There is the potential
for exotic species to dominate remaining fish populations, compounding large-scale effects on coastal fisheries because the Marshlands
formed an important recruitment area for coastal fish and shrimp
species, including 40% of Kuwait's shrimp catch (Partow, 2001). Nesting
turtle populations are also probably affected. With reduced flows into
the Gulf, salt water intrusion affects deltaic areas of the river basin.

About 350 000–500 000 Marsh Arabs, the Ma'dan, lived on the
wetlands for 5000 years (Partow, 2001). They raised water buffalo, cultivated rice and millet, fished, hunted waterbirds and exported reed mats
from the Marshlands (Scott, 1995). They constructed elaborate dwellings
on floating islands, made beds, fences, mats and baskets from reeds cut
from the Marshes, and moved around in canoes. These peoples were
displaced by the Iraqi Government, with 80 000–120 000 fleeing to Iran
and about 200 000 moving away to other parts of the country (Fell, 1996).

The future for this incredibly biodiverse area is bleak because
its water has become an essential part of upstream development of
irrigated agriculture and drinking water supply. A rehabilitation effort
(the Eden project), focussed initially on the Hawizah Marsh, holds
some promise (Furlow, 2003) but upstream the Ilisu Dam proposal
on the Tigris in Turkey will further reduce flows to the Marshlands
(Pearce, 2001). Even a proposed integrated river basin approach (Partow,
2001) will be extremely difficult, given nationalist priorities for water.
Unless water is released from storages and agricultural areas, the
Marshlands will remain buried by the desert and another reminder of
the devastating ecological effects of water resource development.

Lower Murrumbidgee floodplain

The Murrumbidgee River flows west from the Great Dividing Range in
southeastern Australia for 1690 km (Crabb, 1997) until it joins the

Murray River through a catchment of 84 000 km^2 (Fig. 8.6d). As in many desert river systems, large flows inundated an extensive floodplain in the lowest part of the river, the Lower Murrumbidgee floodplain, that covered more than 300 000 ha in the 1900s (Fig. 8.6d). The floodplain was regularly supplied with river flows during winter and spring, after snowmelt. The Murrumbidgee was one of the first rivers in Australia to be developed for irrigation and accounts for about 22% of all annual flows diverted from the Murray–Darling Basin (MDBMC, 1995) and 50% of Australia's irrigated rice crop (Kingsford, 2003).

River management began in the 1850s when the bed elevation of one of the river's major distributaries was reduced to divert more water away from the main channel. Even at this time, there were concerns about the impacts on the Lower Murrumbidgee floodplain (Kingsford, 2003). Now 26 major dams regulate river flows in the catchment (Fig. 8.4) (Kingsford, 2003) so that water can be diverted to some of the larger irrigation areas in Australia. In the 1980s and 1990s, a large part of the floodplain of the Lower Murrumbidgee was developed for irrigation, primarily because it no longer received the flows of the past and river regulatory structures existed to control water (Kingsford, 2003; Kingsford and Thomas, 2004).

These changes have destroyed or degraded about threequarters of the wetland, the central part of which is criss-crossed by channels and banks to spread water on 200 ha irrigation bays (Kingsford & Thomas, 2002, 2004; Kingsford, 2003). The edges of the floodplain no longer receive flooding; the remaining flood-dependent vegetation has died or is in poor condition. One part of the floodplain, the Redbank system, remains in reasonable condition but even this is affected by reduced flooding. Waterbird populations across all foraging groups (piscivores, herbivores, large wading birds, small wading birds) declined by 90% in the period 1983–2001 (Kingsford & Thomas, 2004). Other organisms that form the basis of the food web for these species are also probably affected.

San Joaquin River floodplains, wetlands and delta

The San Joaquin River (310 km long) in the south-western USA has a catchment of 25 000 km^2 (Fig. 8.6e). It has a long history of human disturbance with extensive irrigation projects beginning in 1870, covering 100 000 ha by 1880, 400 000 ha by 1915 and one million hectares by the late 1980s. Twenty-seven major reservoirs were built between 1910 and 1979 to supply irrigation water, electricity and flood protection. The Central Valley Project, the largest irrigation project, has 12 large dams and more than 20 000 km of diversion canals for irrigation (USBR, 1981). About 4 km³ of water (about 90% of total natural flows) are now impounded or diverted from the San Joaquin River each year, mostly (95%) for irrigated agriculture (SJVDP, 1990). On average about 780 m³ s⁻¹ were naturally discharged at the mouth of the river each year but this now averages less than 70 m³ s⁻¹, falling to almost zero in dry years. There is often little measurable flow in the upper 100 km of the river in summer, with the remaining 200 km of river regaining little water from major downstream tributaries, the Merced, Tuolomne and Stanislas rivers.

Landowners drained and leveed swamps and floodplains, including one million hectares of floodplain bottomlands. Riparian and wetland habitat was converted into irrigated areas (1850s–1920s) and riparian woodlands were cut down for fuel and fencing. These changes profoundly affected the river's ecology. Most of the floodplains, remaining riparian areas and wetlands were destroyed by 1985 (Fig. 8.6e), leaving less than 10% of the original areas (Frayer *et al.*, 1989). Twenty-four large animal species became rare or endangered with receding habitat (CSWRCB, 1987). Migratory waterfowl and shorebirds were forced onto remaining wetlands, where avian cholera and botulism frequently killed hundreds to tens of thousands of birds in the 1960s and 1970s (CDFG, 1980, 1983). Before Friant Dam was built (1941), 30 000 – 60 000 chinook salmon *Oncorhynchus tshawytscha* spawned each spring, but their runs were extirpated by 1950, with the decreased flows below the dam changing the river from perennial to temporary (SJVDP, 1990). Anadromous salmon no longer migrate along mid-elevation (500–2000 m) river habitat (CDFG, 1987).

Urban runoff, irrigation spillage, controlled storage releases and subsurface irrigation drainage have degraded water quality, exacerbated by reduced natural flows. Salinity has increased in the upper part of the river (Chapter 10, this volume). Some 125 km of the mid-reach of the river is subsurface drainage (SJVDP, 1990) because impermeable clay 2–6 m thick on the west side of the river prevents infiltration. Artificial drainage removes subsurface water with low amounts of herbicides and pesticides but elevated concentrations of salts and trace elements. Boron concentrations, which are normally below 30 μg l⁻¹, can exceed 2 900 μg l⁻¹; selenium concentrations, normally less than 1 μg l⁻¹, can exceed 6 µg l^{-1}. Only 2 µg l^{-1} of selenium kills aquatic life, poisoning fish, birds, insects, reptiles, amphibians and mammals in wetlands (Lemly *et al.*, 1993; Lemly, 1994), particularly the Kesterson Wetlands (Fig. 8.6e, Fig. 8.8). More than 700 ha of marshes were contaminated in the mid-1980s. Tens of thousands of young and adult waterfowl and shorebirds had severe congenital malformations (missing eyes and feet, protruding brains and grossly deformed beaks, legs and wings) and chronic selenosis (lesions in the liver, heart, spleen and ovaries). Thousand of fish died, followed by a high frequency (30%) of stillbirths in the remaining fish species, the mosquitofish *Gambusia affinis* (Lemly *et al.*, 1993).

North American Indians (Yokuts) hunted and gathered hundreds of different plants and animals from the wetlands of the San Joaquin River Basin but this culture disappeared with agriculture. Waterfowl hunters, now the main users of wetlands, contributed more than US$30 million in 1990 through equipment, ammunition, lodging and guided hunts (SJVDP, 1990). They strongly supported water management reforms. Fish and wildlife have never been a major factor in battles about water over the past 100 years (Reisner, 1986), but the Kesterson

Figure 8.8. Water quality sampling on Kesterson wetlands, where tens of thousands of young and adult waterfowl and shorebirds had severe congenital malformations and thousands of fish died in the mid-1980s from selenium poisoning (Photo United States Fish and Wildlife Service).

debacle forced examination of river flow policies (Zahm, 1986). Agribusiness fought cutbacks to their water, but legal challenges to policies changed this in 1990, providing more environmental flows for the river (Lemly, 1994).

The history of human influences on the San Joaquin River Basin has one main theme: a struggle to balance the water for irrigation and the ecological needs of a desert river and its biota, accentuated by trace element pollution. For sustainable populations of fish (including anadromous salmonids), waterbirds, and other wetland wildlife, 70% of historical flows would need to be restored in the San Joaquin River (SJVDP, 1990). A slow and difficult rehabilitation process has begun, with marginal ecological gains, until river regulation is wound back considerably.

Tarim River Basin wetlands, including Boston and Lop Nur Lakes

The Tarim River Basin (1.02×10^6 km^2) is the largest desert basin in China, with 144 rivers. These flow seasonally, following the melting of glaciers, from the mountains that cover 58% of the basin and include the Tianshan Mountains and Pamir highlands of western China, Kyrgyzstan and Tadjikistan and the Kunlun Mountains of western China and Pakistan (Hongfei et al., 2000). About 74% of runoff volume is between June and September (Feng et al., 2001). The rivers converge into three main tributaries of the Tarim River, the Aksu, Yarkant and Hotan rivers (Hongfei et al., 2000) (Fig. 8.6f). Annual precipitation in the mountains is 200–800 mm but in the lower reaches is only 50–107 mm (Feng et al., 2001).

There is 5269 km^2 of irrigation in the basin (Feng et al., 2001), requiring 108×10^8 m^3 of water each year, stored in 11 reservoirs with a capacity of 10.7×10^8 m^3 (Feng et al., 2001). Flows are distributed down 192 000 km of irrigation canals and 71 000 km of drainage canals. About 35%–50% of water diverted from the rivers is lost before it reaches irrigation areas (Hongfei et al., 2000; Feng et al., 2001). A quarter to a third of farmland has a salinity problem (Mengxiong, 1995; Hongfei et al., 2000), partly caused by high water tables, which are within two metres of the surface, across 60% of the irrigated areas in the lower region (Xinguang et al., 2001). Salinity of the Tarim River has increased from a maximum of 1.28 g l^{-1} in 1960 to 7.8 g l^{-1} in desert areas (Feng et al., 2001; Hongfei et al., 2000). Pollution is a major problem, with concentrations of many elements up to 20 times higher than recommended for human health (Feng *et al.*, 2001).

Before river regulation and development of water resources, most of the water in the rivers reached floodplains and large terminal lake systems, such as Lop Nur Lake (Fig. 8.6f). Water overflowed from Boston Lake before flowing into the Kongque River, which then filled Lop Nur Lake (Fig. 8.6f). Now most of this water is regulated and withdrawn upstream. The freshwater Boston Lake has become increasingly saline and mineralised (Xinguang *et al.*, 2001). Pesticides, fertilisers and human waste increasingly pollute Boston Lake, raising nutrient levels (Xinguang *et al.*, 2001). The water level of Boston Lake has decreased by about three metres, reducing the area of reed swamps by more than 25% (Xinguang *et al.*, 2001). Some endemic fish species are almost extinct, and waterbirds are decreasing in numbers (Xinguang *et al.*, 2001).

Downstream, Lop Nur Lake no longer exists; its dry lakebed is used for nuclear testing. It was the focus for the Loulan culture that thrived for 800 years up to about 200 BC, centred on the Loulan oasis, which has also dried up (Xinguang *et al.*, 2001). Lop Nur Lake once covered 3000 km^2 (Feng *et al.*, 2001), shrinking to 1900 km^2 in 1943 and 530 km^2 in 1962 (Mengxiong, 1995). It was two metres deep in 1959 and dried up some time before 1972 (Xinguang *et al.*, 2001). Little accessible information exists on the biodiversity of this system but, given that of wetlands of similar size around the world, it was undoubtedly high. The annual flow in the Tarim River is now reduced by about 81%, drying up 300 km of lower reaches of the river and causing groundwater levels to fall by eight metres (Mengxiong, 1995; Feng *et al.*, 2001). Hundreds of kilometres of lakes and rivers in the lower part of the basin have dried up (Xinguang *et al.*, 2001), killing floodplain and wetland vegetation (Mengxiong, 1995; Hongfei *et al.*, 2000). Between the 1950s and 1990s, the area and biomass of *Populus euphtatica* decreased by two thirds and half, respectively; the main forest (*P. euphtatica* and *Elaeagnus augustifolia*) declined by 3820 km^2; and meadow area declined by 200 km^2 (Feng *et al.*, 2001). In 1958, 24 600 km^2 was desertified, but this increased by 30% by 1987 (Feng *et al.*, 2001), increasing the frequency of dust storms (Xinguang *et al.*, 2001). Rehabilitation proposals may flood a small part of this once magnificent aquatic ecosystem but dependence of people on irrigation water remains the dominant factor affecting the aquatic ecosystems of the Tarim River Basin.

CONCLUSION

Desert rivers and their dependent ecosystems exhibit considerable temporal and spatial variability, reflecting the variation in flows of the rivers that supply them (Puckridge *et al.*, 1998; Chapter 2 this book). This provides ever-changing habitat for many organisms (Chapters 4–7, this volume). The building of dams, diversion of water and river management simplifies such complexity, reducing variability of disturbance patterns and extent of aquatic habitats.

Although much of our understanding about river regulation comes from mesic areas in the Northern Hemisphere (Ward *et al.*, 2001), desert regions provide the starkest examples of how reduced variability, volume and frequency of river flows have devastated dependent biodiversity and ecosystems (see Table 8.2 and case studies). Such effects will continue because development pressure on more water resources for human use is inexorable, particularly in desert regions (Chapter 11, this volume), with strong political imperatives for more costly expenditure on dams, barrages and canals (Pearce, 2003a). Water shortage for human communities is often the catalyst to develop desert rivers. The 2002–3 drought in Australia stimulated a proposal to divert waters of major coastal rivers into desert areas, at a cost of US$25 billion (Pearce, 2003d). Recent droughts in India resulted in a proposal for 1000 km of canals and 300 reservoirs to pump 1500 m^3 s^{-1} of water from 14 tributary rivers of the Ganges and Brahmaputra Rivers to 17 southern rivers, costing US$70–US$200 billion (Pearce 2003c, d). For other countries, the need to provide water for burgeoning human populations inevitably means development of desert rivers. More water for the Chinese cities of Tianjin and Beijing will come from 50 km^3 of water diverted from the Yangtze River, through canals and tunnels to the Yellow River. Water for irrigation, tourist resorts and golf courses in the arid south of Spain was to be delivered by a planned 850 km canal taking three quarters of the flow of the Ebro River, until the recent change in government. This would have had inevitable ecological consequences for the Ebro delta, one of the more important reserves in southern Europe (Pearce 2003d). Grand water plans are often cyclic.

Graphic examples of the long-term ecological costs (Table 8.2; case studies) are building strong counter-arguments against water resource development, but much still needs to be done at a global scale

(Chapter 11, this volume). For example, there was little planning for water delivery to poor people or assessment of the ecological consequences at the 2003 World Water Forum (Pearce, 2003b). Occasionally, water debates occur before developments, sometimes based on ecology (Kingsford *et al.*, 1998; Kingsford, 2000b). This contrasts with traditional and institutionalised approaches to river management (Lee, 1993), which emphasised development and hydrology (Feng *et al.*, 2001) and generally ignored externalities. There is also an increasing emphasis on adaptive management of environmental flows (Bunn & Arthington, 2003; Arthington & Pusey, 2003; Poff *et al.*, 2003), recognising the ecological needs of rivers and their dependent ecosystems, as well as the social needs of people (Chapter 11, this volume). Identification of environmental flows and good management can usually restore some degraded aquatic habitat but cannot deal with all problems of river regulation. Sediment retention by large dams deprives the river and floodplain downstream of organic and inorganic material (Vörösmarty *et al.*, 2003).

We have shown that simplification of disturbance patterns in desert rivers has devastated many of the world's most spectacular ecosystems (Table 8.2). The building of dams for hydro-electricity, irrigation and drinking water supply (Table 8.2) inevitably changes downstream disturbance regimes for desert rivers, with predictable consequences. Large wetlands and terminal lake systems, once denied their water, become terrestrial ecosystems. Dependent aquatic organisms either disappear or contract in their distribution and abundance (Table 8.2). Sometimes exotic species are favoured by such changes. The extent to which this happens in a desert river depends on the extent of alteration of flows that supply dependent aquatic ecosystems. This includes the quantity, timing and variability, the essential elements of river flows that sustain complex ecosystem processes and the range of organisms. Unfortunately, further development of desert rivers will continue until human needs for water are adequately controlled. An ecological focus that manages and protects the quantity, timing, variability and quality of discharges of desert river flows remains the only hope of halting current degradation of desert rivers. Variable disturbance patterns produce a diversity and abundance of biota and ecological processes that we are only just beginning to understand. We have much to learn from our treatment of desert rivers and its repercussions, for the management of all rivers.

ACKNOWLEDGEMENTS

We particularly thank Andrew Boulton for editing this chapter and coordinating its refereeing. We thank Kate Brandis for her help with research and the figures for this chapter. Michelle Stevens reviewed the Mesopotamian Marshlands case study. We are also indebted to the Water Systems Analysis Group, University of New Hampshire, for providing access to locations of large reservoirs influencing flows in desert regions.

REFERENCES

Adams, W. M. (1993). The wetlands and conservation. In *The Hadejia-Nguru Wetlands: Environment, Economy and Sustainable Development of a Sahelian Floodplain Wetland*, ed. G. E. Hollis, W. M. Adams and M. Aminu-Kano, pp. 211–14. Gland: IUCN.

Adams, W. M., Hollis, G. E. and Hadejia, I. A. (1993). Management of the river basin and irrigation. In *The Hadejia-Nguru Wetlands: Environment, Economy and Sustainable Development of a Sahelian Floodplain Wetland*, ed. G. E. Hollis, W. M. Adams and M. Aminu-Kano, pp. 119–48. Gland: IUCN.

Agnew, C. and Anderson, E. (1992). *Water Resources in the Arid Realm*. London: Routledge.

Aladin, N. V., Filippov, A. A., Plotnikov, I. S., Orlova, M. I. and Williams, W. D. (1998). Changes in the structure and function of biological communities in the Aral Sea, with particular reference to the northern part (Small Aral Sea), 1985–1994: a review. *International Journal of Salt Lake Research*, **7**, 301–43.

Aminu-Kano, M., Thompson, H. and Hollis, G. E. (1993). Land use, water-management and conservation in the Komadugu-Yobe Basin: conclusions. In *The Hadejia-Nguru Wetlands: Environment, Economy and Sustainable Development of a Sahelian Floodplain Wetland*, ed. G. E. Hollis, W. M. Adams and M. Aminu-Kano, pp. 215–29. Gland: IUCN.

Anon (2002). Farming, dry weather shrink China's largest lake. *China Daily* (7/11/02).

Arthington, A. H. and Pusey, B. J. (2003). Flow restoration and protection in Australian rivers. *River Research and Applications*, **19**, 377–95.

Baer, G., Schattner, U., Wachs, D., *et al.* (2002). The lowest place on earth is subsiding: an InSAR (interferometric synthetic aperture radar) perspective. *GSA Bulletin*, **114**, 12–23.

Barbier, E. B. and Thompson, J. R. (1998). The value of water: floodplain versus large-scale irrigation benefits in northern Nigeria. *Ambio*, **27**, 434–40.

Beaumont, P. (1978). The Euphrates River–an international problem of water resource development. *Environmental Conservation*, **5**, 35–43.

Benn, P. C. and Erskine, W. D. (1994). Complex channel response to flow regulation: Cudgeong River below Windamere Dam, Australia. *Applied Geography*, **14**, 153–68.

Black, M. (2001). The day of judgment; They only 'hold pen'; A temple too far; 'Narmada - The facts; Not even a fig-leaf of legality; Resettled by registered mail; At the end of the line. *New Internationalist*, **336**, 9, 12, 15, 18, 20, 22, 26.

Blanch, S. J., Walker, K. F. and Ganf, G. G. (2000). Water regimes and littoral plants in four weir pools of the River Murray, Australia. *Regulated Rivers: Research and Management*, **16**, 445–56.

Boulton, A. J., Humphreys, W. F. and Eberhard, S. M. (2003). Imperilled subsurface waters in Australia: biodiversity, threatening processes and conservation. *Aquatic Ecosystem Health and Management*, **6**, 41–54.

Bowling, L. C. and Baker, P. D. (1996). Major cyanobacterial bloom in the Barwon-Darling River, Australia, in 1991, and underlying limnological conditions. *Marine and Freshwater Research*, **47**, 643–57.

Bren, L. J. (1992). Tree invasion of an intermittent wetland in relation to changes in the flooding frequency of the River Murray, Australia. *Australian Journal of Ecology*, **17**, 395–408.

Bunn, S. E. and Arthington, A. H. (2002). Basic principles and ecological consequences of altered flow regimes for aquatic biodiversity. *Environmental Management*, **30**, 492–507.

CDFG (California Department of Fish and Game) (1980). *At the crossroads: a report on the status of California's endangered and rare fish and wildlife. Technical Report*. Sacramento, CA; California Department of Fish and Game.

CDFG (California Department of Fish and Game) (1983). *A plan for protecting, enhancing, and increasing California's wetlands for waterfowl. Technical Report*. Sacramento, CA: California Department of Fish and Game.

CDFG (California Department of Fish and Game) (1987). *The status of San Joaquin drainage chinook salmon stocks, habitat conditions, and natural production factors. Technical Report*. Sacramento, CA: California Department of Fish and Game.

Changming, L. (1989). Problems in management of the Yellow River. *Regulated Rivers: Research and Management*, **3**, 361–9.

Coe, M. T. and Foley, J. A. (2001). Human and natural impacts on the water resources of the Lake Chad basin. *Journal of Geophysical Research*, **106**, 3349–56.

Cogels, F. X., Coly, A. and Niang, A. (1997). Impact of dam construction on the hydrological regime and quality of a Sahelian Lake in the River Senegal basin. *Regulated Rivers: Research and Management*, **13**, 27–41.

Comín, F. A. and Williams, W. D. (1994). Parched continents: our common future? In *Limnology Now: a Paradigm of Planetary Problems*, ed. R. Margalef, pp. 473–527. Amsterdam: Elsevier Science.

Contreras, S. and Lozano, M. L. (1994). Water, endangered fishes, and development perspectives in arid lands of Mexico. *Conservation Biology*, **8**, 379–87.

Crabb, P. (1997). *Murray-Darling Basin Resources*. Canberra: Murray-Darling Basin Commission.

CSWRCB (California State Water Resources Control Board) (1987). *Regulation of agricultural drainage to the San Joaquin River. Technical Report No. W. Q. 85–1*. Sacramento, CA: California State Water Resources Control Board.

Davies, B. R. and Wishart, M. J. (2000). River conservation in the countries of the Southern African Development Community (SADC). In *Global Perspectives on River Conservation: Science, Policy and Practice*, ed. P. J. Boon, B. R. Davies and G. E. Petts, pp. 179–204. Chichester: John Wiley & Sons.

Davies, B. R., Snaddon, C. D., Wishart, M. J., Thoms, M. C. and Meador, M. (2000). A biogeographical approach to interbasin water transfers: implications for river conservation. In *Global Perspectives on River Conservation: Science, Policy and Practice*, ed. P. J. Boon, B. R. Davies and G. E. Petts, pp. 431–44. Chichester: John Wiley & Sons.

Drijver, C. A. and Marchand, M. (1985). *Taming the floods: Environmental Aspects of Floodplain Development in Africa*. Leiden, The Netherlands: Centre for Environmental Studies, University of Leiden.

Duvail, S. and Hamerlynck, O. (2003). Mitigation of negative ecological and socio-economic impacts of the Diama dam on the Senegal River Delta wetland (Mauritania), using a model based decision support system. *Hydrology and Earth System Sciences*, **7**, 133–6.

Dynesius, M. and Nilsson, C. (1994). Fragmentation and flow regulation of river systems in the northern third of the world. *Science*, **266**, 753–62.

Fearnside, P. M. (1989). Brazil's Balbina Dam: environment versus the legacy of the pharaohs in Amazonia. *Environmental Management*, **13**, 401–23.

Fell, N. (1996). Outcasts from Eden. *New Scientist*, **151** (2045), **24**.

Feng, Q., Endo, K. N. and Cheng, G. D. (2001). Towards sustainable development of the environmentally degraded arid rivers of China – a case study from Tarim River. *Environmental Geology*, **37**, 218–22.

Ferrari, M. R., Miller, J. R. and Russell, G. L. (1999). Modeling the effect of wetlands, flooding, and irrigation on river flow: application to the Aral Sea. *Water Resources Research*, **35**, 1869–76.

Foote, A. L., Pandey, S. and Krogman, N. T. (1996). Processes of wetland loss in India. *Environmental Conservation*, **23**, 45–54.

Frayer, W. E., Peters, D. D. and Pywell, H. R. (1989). *Wetlands of the California Central Valley: Status and Trends, 1939 to mid-1980's. Technical Report*. Portland, OR: United States Fish and Wildlife Service.

Furlow, B. (2003). A struggle for Eden. *New Scientist*, **178** (2392), 14–15.

Garba Boyi, M. and Polet, G. (1996). Birdlife under stress: The case of the Hadejia-Nguru Wetlands, northern Nigeria. In *Proceedings of the African Crane and Wetland Training Workshop*, Maun, Botswana, 8–15 August 1993, ed. R. D. Belilfuss, W. R. Tarboton and N. N. Gichuki, pp. 147–52. Baraboo, WI: International Crane Foundation.

Gehrke, P. C., Brown, P., Schiller, C. B., Moffatt, D. B. and Bruce, A. M. (1995). River regulation and fish communities in the Murray-Darling River system, Australia. *Regulated Rivers: Research and Management*, **11**, 363–75.

Genxu, W. and Guodong, C. (1999a). The ecological features and significance of hydrology within arid inland river basins of China. *Environmental Geology*, **37**, 218–22.

Genxu, W. and Guodong, C. (1999b). Water resource development and its influence on the environment in arid areas of China – the case of the Hei River basin. *Journal of Arid Environments*, **43**, 121–31.

Gilbert, C. L. and Ramey, J. B. (1995). *Riverine Wetlands of the Alberta Prairie*. Edmonton, Alberta: Alberta Parks and Wildlife.

Gillanders, B. M. and Kingsford, M. J. (2002). Impact of changes in flow of freshwater on estuarine and open coastal habitats and associated organisms. *Oceanography and Marine Biology: an Annual Review*, **40**, 233–309.

Goes, B. J. M. (2002). Effects of river regulation on aquatic microphyte growth and floods in the Hadejia-Nguru Wetlands and flow in the Yobe River, Northern Nigeria: implications for future water management. *River Research and Applications*, **18**, 81–95.

Gopal, B., Bose, B. and Goswami, A. B. (2000). River conservation in the Indian sub-continent. In *Global Perspectives on River Conservation: Science, Policy and Practice*, ed P. J. Boon, B. R. Davies and G. E. Petts, pp. 233–61. Chichester: John Wiley & Sons.

Gore, J. A. and Shields, F. D. (1995). Can large rivers be restored? *BioScience*, **45**, 142–52.

Green, A. J., Fox, A. D., Hilton, G. *et al.* (1996). Threats to Burdur Lake ecosystem, Turkey and its waterbirds, particularly the white-headed duck *Oxyura leucocephala*. *Biological Conservation*, **76**, 241–52.

Green, A. J., El Hamzaoui, M., El Agbani, M. A. and Franchimont, J. (2002). The conservation status of Moroccan wetlands with particular reference to waterbirds and to changes since 1978. *Biological Conservation*, **104**, 71–82.

Groombridge, B. (1993). *IUCN Red List of Threatened Animals*. Gland, Switzerland: IUCN.

Heffernan, M. J. (1990). Bringing the desert to bloom: French ambitions in the Sahara desert during the late nineteenth century - the strange case of '*la mer intérieure*'. In *Water, Engineering and Landscape*, ed. D. Cosgove and G. Petts, pp. 94–114. London: Belhaven Press.

Heng, L. (2001). 20 Natural Lakes/ Year dried up in China. *People's Daily*, 20-1-01.

Higgins, S. I., Coetzee, M. A. S., Marneweck, G. C. and Rogers, K. H. (1996). The Nyl River floodplain, South Africa, as a functional unit of the landscape: a review of current information. *African Journal of Ecology*, **34**, 131–45.

Hoffman, R. J., Hallock, R. J., Rowe, T. G., Lico, M. S., Burge, H. L. and Thompson, S. P. (1990). *Reconnaissance investigation of water quality, bottom sediment, and biota associated with irrigation drainage in and near Stillwater Wildlife Management Area, Churchill County, Nevada, 1986–87.* (Water-Resources Investigations Report 89–4105.) Carson City, NE: US Geological Survey.

Hollis, G. E. and Thompson, J. R. (1993a). Hydrological model of the floodplain. In *The Hadejia-Nguru Wetlands: Environment, Economy and Sustainable Development of a Sahelian Floodplain Wetland*, ed. G. E. Hollis, W. M. Adams and M. Aminu-Kano, pp. 69–79. Gland, Switzerland: IUCN.

Hollis, G. E. and Thompson, J. R. (1993b). Water resource developments and their hydrological impacts. In *The Hadejia-Nguru Wetlands: Environment, Economy and Sustainable Development of a Sahelian Floodplain Wetland*, ed. G. E. Hollis, W. M. Adams and M. Aminu-Kano, pp. 149–90. Gland, Switzerland: IUCN.

Hollis, G. E., Adams, W. M. and Aminu Kano, M. (eds) (1993a). *The Hadejia-Nguru Wetlands: Environment, Economy and Sustainable Development of a Sahelian Floodplain Wetland*. Gland, Switzerland: IUCN.

Hollis, G. E., Penson, S. J., Thompson, J. R. and Sule, A. R. (1993b). Hydrology of the river basin. In *The Hadejia-Nguru Wetlands: Environment, Economy and Sustainable Development of a Sahelian Floodplain Wetland*, ed. G. E. Hollis, W. M. Adams and M. Aminu-Kano, pp. 19–67. Gland, Switzerland: IUCN.

Hongfei, Z., Yudong, S. and Shunjun, H. (2000). Irrigated agriculture and sustainable water management strategies in the Tarim Basin. In New Approaches to water management in Central Asia *Proceedings of the Workshop held in Aleppo, Syria, 6–11 Nov. 2000*, pp.127–38. www.unu.edu/env/workshops/ Aleppo.

Hughes, F. M. R. (1984). A comment on the impact of development schemes on the floodplain forests of the Tana River of Kenya. *Geographical Journal*, **150**, 230–44.

Ivanov, Y. N., Chub, V. E., Subbotina, O. I., Tolkacheva, G. A. and Toryannikova, R. V. (1996). Review of the scientific and environmental issues of the Aral Sea Basin. In *The Aral Sea Basin*, ed. P. P. Micklin and W. D. Williams, pp. 9–21. (NATO ASI Series Partnership, Sub-Series 2. Environment - Vol 12.) Berlin, Heidelberg: Springer-Verlag.

Kang, O. and Yong, L. (1998). Management and utilisation of water resources in the People's Republic of China. In *Water Resource Management: a Comparative Perspective*, ed. D. K. Vajpeyi, pp. 33–50. Westport: Praeger Publishers.

Karmon, Y. (1960). The drainage of the Huleh Swamps. *Geographical Review*, **50**, 169–93.

Khan, T. A. (1996). Management and sharing of the Ganges. *Natural Resources Journal*, **36**, 455–79.

Kingsford, R. T. (1999). Managing the water of the Border Rivers in Australia: irrigation, Government and the wetland environment. *Wetlands Ecology and Management*, **7**, 25–35.

Kingsford, R. T. (2000a). Ecological impacts of dams, water diversions and river management on floodplain wetlands in Australia. *Austral Ecology*, **25**, 109–27.

Kingsford, R. T. (2000b). Protecting or pumping rivers in arid regions of the world? *Hydrobiologia*, **427**, 1–11.

Kingsford, R. T. (2003). Social, institutional and economic drivers for water resource development - case study of the Murrumbidgee River, Australia. *Aquatic Ecosystem Health and Management*, **6**, 69–79.

Kingsford, R. T. and Johnson, W. J. (1998). The impact of water diversions on colonially nesting waterbirds in the Macquarie Marshes in arid Australia. *Colonial Waterbirds*, **21**, 159–70.

Kingsford, R. T. and Porter, J. L. (1993). Waterbirds of Lake Eyre. *Biological Conservation*, **65**, 141–51.

Kingsford, R. T. and Thomas, R. F. (1995). The Macquarie Marshes in arid Australia and their waterbirds: a 50 year history of decline. *Environmental Management*, **19**, 867–78.

Kingsford, R. T. and Thomas, R. F. (2002). Use of satellite image analysis to track wetland loss on the Murrumbidgee River floodplain in arid Australia, 1975–1998. *Water Science and Technology*, **45**, 45–53.

Kingsford, R. T. and Thomas, R. F. (2004). Destruction of wetlands and waterbird populations by dams and irrigation on the Murrumbidgee River in arid Australia. *Environmental Management*, **34**, 383–96.

Kingsford, R. T., Boulton, A. J. and Puckridge, J. M. (1998). Challenges in managing dryland rivers crossing political boundaries: Lessons from Cooper Creek and the Paroo River, central Australia. *Aquatic Conservation: Marine and Freshwater Ecosystems*, **8**, 361–78.

Kotlyakov, V. M. (1991). The Aral Sea Basin: a critical environmental zone. *Environment (January/February)*, **33**, 4–38.

Kowalewski, M., Serrano, G. E. A., Flessa, K. W. and Goodfriend, G. A. (2000). Dead delta's former productivity: two trillion shells at the mouth of the Colorado River. *Geology*, **28**, 1059–62.

Lae, R. (1995). Climatic and anthropogenic effects on fish diversity and fish yields in the Central Delta of the Niger River. *Aquatic Living Resources*, **8**, 43–58.

Lake, P. S. (2000). Disturbance, patchiness, and diversity in streams. *Journal of the North American Benthological Society*, **19**, 573–92.

Lake, P. S. (2003). Ecological effects of perturbation by drought in flowing waters. *Freshwater Biology*, **48**, 1161–72.

Lavín, M. F. and Sanchez, S. (1999). On how the Colorado River affected the hydrography of the upper Gulf of California. *Continental Shelf Research*, **19**, 1545–60.

Lee, K. N. (1993). Greed, scale mismatch, and learning. *Ecological Applications*, **3**, 560–4.

Lemly, A. D. (1994). Irrigated agriculture and freshwater wetlands: a struggle for coexistence in the western United States. *Wetlands Ecology and Management*, **3**, 3–15.

Lemly, A. D., Finger, S. E. and Nelson, M. K. (1993). Sources and impacts of irrigation drainwater contaminants in arid wetlands. *Environmental Toxicology and Chemistry*, **12**, 2265-79.

Lemly, A. D., Kingsford, R. T. and Thompson, J. R. (2000). Irrigated agriculture and wildlife conservation: conflict on a global scale. *Environmental Management*, **25**, 485-512.

Li, L., Liu, C. and Mou, H. (2000). River conservation in central and eastern Asia. In *Global Perspectives on River Conservation: Science, Policy and Practice*, ed. P. J. Boon, B. R. Davies and G. E. Petts, pp. 263-79. Chichester: John Wiley & Sons.

Livingstone, A. J. and Campbell, I. A. (1992). Water supply and urban growth in southern Alberta: constraint or catalyst? *Journal of Arid Environments*, **23**, 335-49.

Maheshwari, B. L., Walker, K. F. and McMahon, T. A. (1995). Effects of regulation on the flow regime of the River Murray, Australia. *Regulated Rivers: Research and Management*, **10**, 15-38.

Maingi, J. K. and Marsh, S. E. (2002). Quantifying hydrologic impacts following dam construction along the Tana River, Kenya. *Journal of Arid Environments*, **50**, 53-79.

McMahon, T. A. and Finlayson, B. L. (2003). Droughts and anti-droughts: the low hydrology of Australian rivers. *Freshwater Biology*, **48**, 1147-60.

MDBMC (Murray-Darling Basin Ministerial Council) (1995). *An Audit of Water Use in the Murray-Darling Basin*. Canberra: Murray-Darling Ministerial Council.

Mengxiong, C. (1995). Impacts of human activities on the hydrological regime and ecosystems in an arid area of northwest China. In *Man's Influence on Freshwater Ecosystems and Water Use. Proceedings of a Boulder Symposium, IAHS Publ. No. 230*, ed. G. Petts, pp. 131-9. Wallingford, Oxon: IAHS Press, Institute of Hydrology.

Micklin, P. P. (1988). Desiccation of the Aral Sea: a water management disaster in the Soviet Union. *Science*, **241**, 1170-6.

Micklin, P. P. (1996). Introductory remarks on the Aral issue. In *The Aral Sea Basin*, ed. P. P. Micklin and W. D. Williams, pp. 3-8. (NATO ASI Series Partnership, Sub-Series 2. Environment - Vol 12.) Berlin, Heidelberg: Springer-Verlag.

Middleton, N. J. and Thomas, D. S. G. (1997). *World Atlas of Desertification (2nd Edn)*. London: United Nations Environment Programme, Edward Arnold.

Milliman, J. D. (1997). Oceanography - Blessed dams or damned dams? *Nature*, **386**, 325.

Munro, D. C. and Touron, H. (1997). The estimation of marshland degradation in southern Iraq using multitemporal Landsat TM images. *International Journal of Remote Sensing*, **18**, 1597-606.

Novikova, N. (1996). Current changes in the vegetation of the Amu Dar'ya delta. In *The Aral Sea Basin*, ed. P. P. Micklin and W. D. Williams, pp. 69-78. (NATO ASI Series Partnership, Sub-Series 2. Environment - Vol 12.) Berlin, Heidelberg: Springer-Verlag.

Olofin, E. A. (1996). Dam-induced drying-out of the Hadejia-Nguru Wetlands, northern Nigeria, and its implications for the fauna. In *Proceedings of the African Crane and Wetland Training Workshop*, Maun, Botswana, 8-15 August 1993, ed. R. D. Belilfuss, W. R. Tarboton and N. N. Gichuki, pp. 141-5. Baraboo, WI: International Crane Foundation.

Partow, H. (2001). *The Mesopotamian Marshlands: Demise of an Ecosystem. Early Warning and Assessment Technical Report, UNEP/DEWA/TR.01-3 Rev. 1*. Nairobi, Kenya: Division of Early Warning and Assessment, United Nations Environment Programme (UNEP).

Pearce, F. (1993). Draining life from Iraq's marshes: Saddam Hussein is using an old idea to force the Marsh Arabs from their home. *New Scientist*, **138** (1869), 11.

Pearce, F. (1995). Poisoned waters. *New Scientist*, **148** (2000), 29–33.

Pearce, F. (2001). Iraqi wetlands face total destruction. *New Scientist*, **170** (2291), 4–5.

Pearce, F. (2003a). Dismay over call to build more dams. *New Scientist*, **177** (2387), 11.

Pearce, F. (2003b). Safe water remains a mirage. *New Scientist*, **177** (2388), 8–9.

Pearce, F. (2003c). Conflict looms over India's river plan. *New Scientist*, **177** (2384), 4–5.

Pearce, F. (2003d). Replumbing the planet. *New Scientist*, **178** (2398), 30–4.

Perera, J. (1993). A sea turns to dust. *New Scientist*, **140** (1896), 24–7.

Pimentel, D., Houser, J., Preiss, E. *et al.* (1997). Water resources: agriculture, the environment and society. *BioScience*, **47**, 97–106.

Plummer, J. L. (1994). Western water resources: The desert is blooming, but will it continue? *Water Resources Bulletin*, **30**, 595–603.

Poff, N. L. Allan, J. D., Palmer, M. A. *et al.* (2003). River flows and water wars: emerging science for environmental decision making. *Frontiers in Ecology and Environment*, **1**, 298–306.

Polet, G. and Thompson, J. R. (1996). Maintaining the floods: hydrological and institutional aspects of managing the Komadugu-Yobe River Basin and its floodplain wetlands. In *Water management and Wetlands in Sub-Saharan Africa*, ed. M. C. Acreman and G. E. Hollis, pp. 73–100. Gland, Switzerland: IUCN.

Postel, S. (1996). Dividing the waters: food security, ecosystem health, and the new politics of scarcity. Worldwatch Paper no. 132. Washington, DC: Worldwatch Institute.

Postel, S. L. (2000) Entering an era of water scarcity: the challenges ahead. *Ecological Applications*, **10**, 941–8.

Power, M. E., Dietrich, W. E. and Finlay, J. C. (1996). Dams and downstream aquatic biodiversity: potential food web consequences of hydrologic and geomorphic change. *Environmental Management*, **20**, 887–95.

Puckridge, J. T., Sheldon, F., Walker, K. F. and Boulton, A. J. (1998). Flow variability and the ecology of arid zone rivers. *Marine and Freshwater Research*, **49**, 55–72.

Puckridge, J. T., Walker, K. F. and Costelloe, J. F. (2000). Hydrological persistence and the ecology of dryland rivers. *Regulated Rivers: Research and Management*, **16**, 385–402.

Ramberg, L. (1997). A pipeline from the Okavango River? *Ambio*, **26**, 129.

Reisner, M. (1986). *Cadillac Desert: the American West and its Disappearing Water*. New York, USA: Viking Penguin.

Robertson, A. I., Bunn, S. E., Boon, P. I. and Walker, K. F. (1999). Sources, sinks and transformation of organic carbon in Australian floodplain rivers. *Marine and Freshwater Research*, **50**, 813–29.

Scott, D. A. (1995). *A Directory of Wetlands in the Middle East*. Gland and Slimbridge: World Conservation Union and International Waterfowl and Wetlands Research Bureau.

Sheldon, F., Boulton, A. J. and Puckridge, J. T. (2002). Conservation value of variable connectivity: aquatic invertebrate assemblages of channel and floodplain habitats of a central Australian arid-zone river, Cooper Creek. *Biological Conservation*, **103**, 13–31.

Sheldon, F., Thoms, M. C., Berry, O. and Puckridge, J. T. (2000). Using disaster to prevent catastrophe: referencing the impacts of flow changes in large dryland rivers. *Regulated Rivers: Research and Management*, **16**, 403–20.

SJVDP (San Joaquin Valley Drainage Program) (1990). *Fish and Wildlife Resources and Agricultural Drainage in the San Joaquin Valley, California. Technical Report.* Sacramento, CA: San Joaquin Valley Drainage Program.

Sparks, R. E., Bayley, P. B., Kohler, S. L. and Osborne, L. L. (1990). Disturbance and recovery of large floodplain rivers. *Environmental Management,* **14,** 699–709.

Stanley, D. J. (1996). Nile delta: extreme case of sediment entrapment on a delta plain and consequent coastal land loss. *Marine Geology,* **129,** 189–95.

Stanley, D. J. and Warne, A. G. (1993). Nile Delta: recent geological evolution and human impact. *Science,* **260,** 628–34.

Stanley, D. J. and Warne, A. G. (1998). Nile Delta in its destruction phase. *Journal of Coastal Research,* **14,** 794–825.

Stevens, L. E., Schmidt, J. C., Ayers, T. J. and Brown, B. T. (1995). Flow regulation, geomorphology, and Colorado river marsh development in the Grand Canyon, Arizona. *Ecological Applications,* **5,** 1025–39.

Stock, R. E. (1978). The impact of the decline of the Hadejia River in Hadejia Emirate. In *The Aftermath of the 1972–74 Drought in Nigeria,* ed. G. J. van Apeldoorn, pp. 141–6. Zaria, Nigeria: Centre for Social and Economic Research, Ahmadu Bello University.

Thomas, D. H. L. (1995). Artisanal fishing and environmental change in a Nigerian floodplain wetland. *Environmental Conservation,* **22,** 117–42.

Thomas, D. H. L., Jimoh, M. A. and Matthes, H. (1993). Fishing. In *The Hadejia-Nguru Wetlands: Environment, Economy and Sustainable Development of a Sahelian Floodplain Wetland,* ed. G. E. Hollis, W. M. Adams and M. Aminu-Kano, pp. 97–115. Gland, Switzerland: IUCN.

Thompson, J. R. (1995). Hydrology, water management and wetlands of the Hadejia-Jama'are Basin, Northern Nigeria. Ph.D. thesis, University of London, UK.

Thompson, J. R. and Hollis, G. E. (1995). Hydrological modelling and the sustainable development of the Hadejia-Nguru Wetlands, Nigeria. *Hydrological Sciences Journal,* **40,** 97–116.

Thompson, J. R. and Polet, G. (2000). Hydrology and land use in a Sahelian floodplain wetland. *Wetlands,* **20,** 639–59.

Thoms, M. C. (2003). Floodplain-river ecosystems: lateral connections and the implications of human interference. *Geomorphology,* **56,** 335–49.

Thoms, M. C. and Walker, K. F. (1993). Channel changes associated with 2 adjacent weirs on a regulated lowland alluvial river. *Regulated Rivers: Research and Management,* **8,** 271–84.

Tooth, S. and Nanson, G. C. (2000). The role of vegetation in the formation of anabranching channels in an ephemeral river, Northern plains, arid central Australia. *Hydrological Processes,* **14,** 3099–117.

Toro, S. M. (1998). Strategies towards sustainable development in Nigeria's semi-arid regions. *Journal of the Chartered Institute of Water and Environmental Management,* 1998, 212–15.

USBR (United States Bureau of Reclamation) (1981). *Central Valley Project: its Historical Background and Economic Impacts.* Sacramento, CA: United States Bureau of Reclamation, US Department of the Interior.

Vinke, P. P. (1996). The integrated development programme for the left bank of the Senegal River Valley. In *Water Management and Wetlands in Sub-Saharan Africa,* ed. M. C. Acreman, and G. E. Hollis, pp. 145–53. Gland, Switzerland: IUCN.

Vörösmarty, C. J., Meybeck, M., Fekete, B., *et al.* (2003). Anthropogenic sediment retention: Major global-scale impact from the population of registered impoundments. *Global and Planetary Change,* **39,** 169–90.

Vörösmarty, C. J., Sharma, K. P., Fekete, B. M. *et al.* (1997). The storage and aging of continental runoff in large reservoir systems of the world. *Ambio*, **26**, 210–19.

Walker, K. F., Sheldon, F. and Puckridge, J. T. (1995). A perspective on dryland river ecosystems. *Regulated Rivers: Research and Management*, **11**, 85–104.

Wallace, T. (1980). Agricultural projects and land in northern Nigeria. *Review of African Political Economy*, **17**, 59–70.

Ward, J. V., Tockner, K. and Schiemer, F. (1999). Biodiversity of floodplain river ecosystems: ecotones and connectivity. *Regulated Rivers: Research and Management*, **15**, 125–39.

Ward, J. V., Tockner, K., Uehlinger, U. and Malard, F. (2001). Understanding natural patterns and processes in river corridors as the basis for effective river restoration. *Regulated Rivers: Research and Management*, **17**, 311–23.

Watzman, H. (1997). Left for dead. *New Scientist*, **153** (2068), 37.

Wiens, J. A., Patten, D. T. and Botkin, D. B. (1993). Assessing ecological impact assessment: lessons from Mono Lake, California. *Ecological Applications*, **3**, 595–609.

Williams, C. (2003). Long time no sea. *New Scientist*, **177** (2376), 34–7.

Wishart, M. J., Davies, B. R., Boon, P. J. and Pringle, C. M. (2000). Global disparities in river conservation: 'First World' values and 'Third World' realities. In *Global Perspectives on River Conservation: Science, Policy and Practice*, ed. P. J. Boon, B. R. Davies and G. E. Petts, pp. 353–69. Chichester: John Wiley & Sons.

Xinguang, D., Tao, J. and Huifang, J. (2001). Study on the pattern of water resources utilization and environmental conservation of Yanqi Basin of Xinjiang. In *XXIX IAHR Congress, 21st Century: New Era for Hydraulic Research and its application, Sept. 16–21, 2001, Beijing*, ed. G. Li, pp. 332–41. Beijing: Tsinghua University Press.

Zahm, G. R. (1986). Kesterson Reservoir and Kesterson National Wildlife Refuge: history, current problems, and management alternatives. *Transactions of the North American Wildlife and Natural Resource Conference*, **51**, 324–29.

9

Serial weirs, cumulative effects: the Lower River Murray, Australia

K. F. WALKER

Few 'desert rivers' flow entirely within xeric landscapes. Although many rivers do traverse arid or semi-arid regions, their discharge is often greater than could be sustained by regional rainfall. They rise in nearby mesic areas and flow for long distances over dry land, with only minor contributions from tributaries unless those too have exotic origins.

The hallmark of all such rivers is variability, at multiple scales (see, for example, Puckridge *et al.*, 1998, 2000; Chapter 2, this volume). Where rainfall occurs sporadically in time and space, and conversion to runoff is moderated by landforms, soils and vegetation (e.g. McMahon *et al.*, 1992), rivers have a distinctive, erratic hydrographic signature. One consequence is that the native flora and fauna are likely to include species with wide tolerance to environmental changes, opportunistic life cycles and a capacity for rapid dispersal (e.g. Walker, 1992; Chapters 4–7, this volume).

Another consequence of high flow variability is to amplify human demands. The dependence of irrigated agriculture on secure water supplies, for example, has encouraged massive resource developments in many of the world's big desert river systems (Chapter 8, this volume). One example is the Murray–Darling Basin in south-eastern Australia,

Ecology of Desert Rivers, ed. R. T. Kingsford. Published by Cambridge University Press.
© Cambridge University Press 2006.

where about two thirds of the mean annual discharge (4915 GL (MDBMC, 1995)) is diverted, mainly for irrigation.

Dams and reservoirs are conventional means to regulate river discharge. In the Murray–Darling Basin there are 200 storages with capacity of more than 1 GL, including 84 of over 10 Gl and 30 of more than 100 GL, nearly all on tributary rivers and above 150 m altitude. The 30 largest dams store 94% of the total capacity in the system (34 727 GL); indeed, storage volumes in the basin far exceed the mean annual discharge (MDBMC, 1995). There are few dams in the lowland areas of the system, but many weirs, levees, barrages, causeways, channelled diversions and off-stream storages. Individually these may have minor effects, but their effects are magnified where numbers occur in close proximity. Their cumulative effects may even eclipse those of dams.

There are no dams on the 830 km tract of the River Murray below the Murray–Darling junction, but flows to this region (the 'Lower Murray') are controlled by upstream dams and weirs (Fig. 9.1). On the Lower Murray there are ten weirs (Fig. 9.1), plus river-mouth barrages,

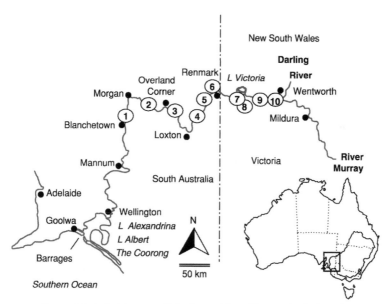

Figure 9.1. A sketch map of the Lower River Murray, including the locations of Locks 1–10, cities and towns (filled circles) and lakes (italicised) and location within the Murray–Darling Basin (identified in inset) in south-eastern Australia.

floodplain levees and many small regulating structures on riparian wetlands. Collectively, these structures intensify the control of flows and river levels, compounding the effects of upstream reservoirs.

This chapter is a case study of the effects of serial weirs on river and floodplain environments of the Lower Murray. It confirms that the effects of regulation are more subtle and extensive than those revealed by casual inspection of a hydrograph, and shows that even comparatively small structures can have profound effects on the physical and biological environment. It also suggests that some of the Murray weirs may need to be modified or removed to restore and maintain the river's natural character.

MURRAY–DARLING BASIN

The Murray–Darling Basin extends over more than one million square kilometres and occupies most of south-eastern Australia (Fig. 9.1). It extends over 13 degrees of latitude and longitude (24–37°S, 138–151°E), and is contained by desert to the north and west and more humid highlands to the south and east. The northern basin is subtropical, the east is cool and humid, the south is temperate and the intervening area, inland from the mountains, is dry and hot. Mean annual air temperatures range from 13.7 °C in the east to 20.2 °C in the west, and water temperatures in the Murray vary seasonally from 6–28 °C near the headwaters to 10–30 °C in the lower reaches (Mackay et al., 1988).

The basin is home to nearly two million people, mainly in rural areas, and is a vital resource for the more densely populated areas of coastal eastern Australia. It supports one quarter of the nation's cattle and dairy farms, half the sheep and half the cropland, and yields more than A$10 billion in annual production (Crabb, 1997). Irrigation accounts for 90%–95% of annual water consumption.

The river system is among the longest in the world (River Murray 2530 km, Darling River 2740 km). The Murray lies near the basin's southern perimeter. It rises at 1430 m altitude in the Snowy Mountains and flows north-west as the border between New South Wales and Victoria, meeting with the Ovens, Goulburn, Murrumbidgee and Darling Rivers. It continues westward into South Australia until near Morgan, where it turns south toward Lake Alexandrina and the sea, near Goolwa. The Darling rises in south-eastern Queensland and flows south-west across New South Wales, receiving the Macintyre, Gwydir, Namoi, Castlereagh, Macquarie, Bogan, Culgoa and Warrego Rivers as

Figure 9.2. The junction of the Murray and Darling Rivers at Wentworth, New South Wales, Australia, immediately upstream of Lock 10 on the Murray. The Darling is typically highly turbid owing to colloidal clays (Photo K. F. Walker).

it progresses toward the Murray–Darling confluence at Wentworth (Fig. 9.2), 830 km from the Murray mouth. The Murray generally has a broad floodplain, with many wetlands and woodlands, but the Darling's channel is incised 10 m below the land surface and flows are too low and variable to sustain permanent natural wetlands along its length. Water in the Darling is rendered milky by suspended colloidal clays.

As most of the basin is arid or semi-arid, mean annual runoff (14 mm) is only about 3% of the rainfall (Crabb, 1997). Annual rainfall varies from over 1400 mm in the east to less than 300 mm in the north-west, and generally is less than annual evaporation, except in highland areas. Peak rainfall is in winter–spring in the south and summer in the north, but rainfall and stream flows in inland areas are highly variable (Chapter 2, this volume). Most of the system's discharge originates from the headwater catchments of the Murray and its tributaries, where there is rain and snow in winter and spring. The Darling drains about half of the basin but contributes only about 10% of the long-term mean discharge (Maheshwari et al., 1995), as its headwaters are governed by unreliable summer monsoons and part of its flow is diverted by floodplain distributaries.

LOWER MURRAY

Regional features

The Lower River Murray is ecologically different from its parent rivers, and is a distinctive 'environmental unit' for research and management (cf. Walker, 1992). Much of the region lies within the state of South Australia (Fig. 9.1). The regional climate is semi-arid, with annual rainfall 200–500 mm and evaporation 1500–2400 mm, and the soils are formed from calcareous marine sediments. Irrigated agriculture and grazing are the main forms of land use, and most towns have populations of less than 10 000.

The distinctiveness of the region is reflected in the physical nature of the river. The Lower Murray has no significant tributaries, and its hydrologic behaviour is usually determined by flows from the middle and upper Murray rather than from the Darling. The geomorphic character of the Murray immediately below the Darling junction is like that of the middle reaches, including a 10–20 km floodplain with extensive wetlands and woodlands (Gill, 1973) and regional landforms shaped by the prevailing wind. Below Overland Corner, the river enters a 30m limestone gorge (Fig. 9.3) where marine, tectonic and fluvial factors have dominated at different times (Twidale et al., 1978; Walker & Thoms, 1993). The floodplain is constrained to 2–3 km and the typical riparian wetlands are 'channel margin swales' rather than oxbows (Pressey, 1986).

Hydrology

Flows to the Lower Murray are governed by climate and upstream storages and diversions, including small but operationally significant storages at Lake Victoria on the Lower Murray (680 GL, constructed 1928) and Menindee Lakes on the Darling (1682 GL, 1968). The flow regime is inherently variable (Fig. 9.4), reflecting the climate of the small region near the source of the Murray (this is less than 5% of the basin area, but contributes about half of the annual discharge). The variability is compounded by evaporation and other processes implicated in the conversion of rainfall to runoff, and by the El Niño Southern Oscillation (ENSO) and other atmospheric phenomena (McMahon et al., 1992; Simpson et al., 1993; Chapter 2, this volume). Drought- or flood-dominated

Figure 9.3. Limestone cliffs along part of the Lower Murray are a legacy of marine incursions over the past 20 million years (Photo K. F. Walker).

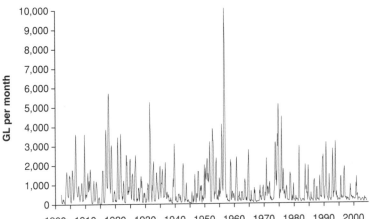

Figure 9.4. Hydrograph of monthly flows (1 GL = 10^3 ML = 10^6 m^3) in the River Murray entering South Australia, 1902–2004 (data courtesy A. Close, Murray-Darling Basin Commission).

regimes may persist for several years, demonstrating hydrological 'persistence' (Greenwood, 2000; Puckridge et al., 2000; Whiting et al., 2003).

The behaviour of the Murray changed markedly with the advent of regulation, beginning with minor diversions in the late nineteenth century and dam, weir, barrage and levee construction thereafter. Regulation intensified sharply with the expansion of irrigated agriculture after 1950 (Close, 1990; Maheshwari et al., 1995). The Darling too has come under pressure, especially in the last decade, from diversions associated with the expansion of cotton irrigation (Walker et al., 1997; Kingsford, 2000). In dry periods, as in the ENSO-dominated regime of the late 1980s, Darling water may be transferred to Lake Victoria and stored for release during the irrigation season (Mackay et al., 1988; Walker et al., 1992), and the Lower Murray may then receive increased proportions of turbid water. Operating rules for the system, administered by the Murray–Darling Basin Commission, guarantee a minimum annual 'entitlement flow' of 1850 GL to South Australia (see, for example, Jacobs, 1990). This is designed for consumers, particularly irrigators, and bears no relation to environmental management.

Under natural conditions, about half of the basin's mean annual runoff (24 300 GL) would have reached the sea, given substantial losses to floodplains, seepage and evaporation. From 1894 to 1993, the regulated annual discharge of the river system varied from 1626 GL to 54 168 GL, with mean 10 090 GL and median 8489 GL (Maheshwari et al., 1995) (cf. Fig. 9.4). Averages are misleading, however, because annual diversions from the basin (excluding Queensland) increased from 3000 to 11 000 GL in 1930–91 and were 'capped' by the Murray–Darling Basin Commission in 1995, nominally at levels that prevailed in 1993–94 (see, for example, Whittington et al., 2000). At the level of regulation in 1995, the long-term mean annual discharge would be 4915 GL (median 2539 GL) (MDBMC, 1995). About 36% of the natural mean annual discharge, or 27% of the natural median annual discharge, now reaches the sea (Fig. 9.5).

The depletion of volume is striking, but the temporal pattern of flows is also affected (see, for example, Walker & Thoms, 1993; Maheshwari et al., 1995). Regulation has reduced the variability of mid-range annual flows in the Lower Murray, leaving a regime dominated by low flows and occasional high flows. Low flows (less than 5000 GL) occur 66% of the time under regulation, but would have occurred 7% of the time under natural conditions. This pattern may be a common effect of river regulation in dryland rivers (cf. McMahon & Finlayson, 2003). Ninety-five

Figure 9.5. The mouth of the River Murray in South Australia is a shallow, sandy channel that has become increasingly prone to blockages caused by accumulated sand. It is now maintained by dredging (Photo K. F. Walker).

percent of regulated annual flows in the Lower Murray are in the range 0–15 000 GL, compared with 2500–20 000 GL for natural flows. Big floods (recurrence 20+ years) are little affected.

The seasonal extremes of monthly flows have changed immediately downstream of the big dams, but are less affected in the Lower Murray where the pattern still tends to a natural summer–autumn minimum and winter–spring maximum (Maheshwari *et al.*, 1995). The magnitude of the seasonal peak, however, has decreased, limiting the frequency and extent of floodplain inundation.

Geomorphology

The Murray–Darling Basin originated by subsidence in the Tertiary. The sea occupied the basin at various times until two million years ago, when the Murray–Darling confluence was inundated by a large freshwater lake (Bungunnia) that left extensive clay deposits (Gill, 1973; Stephenson, 1986; Walker, 1986). When the barrier impounding the lake was breached, less than a million years ago, the lake drained and the Murray began to incise a gorge through marine limestone, reaching the sea at the edge of the continental shelf (Twidale *et al.*, 1978). Since then the sea-level has risen and the land has been exposed to weathering, as the climate has oscillated between wet and dry conditions.

The age of the Lower Murray is reflected in its slight bed slope (mean 5.5 cm km^{-1}) and low stream power. There are four main geomorphic sections (Pressey, 1986; Walker & Thoms, 1993) (Fig. 9.1), as follows

1. *Valley Section*: from the Murray–Darling confluence to Overland Corner. The river meanders over a wide, terraced floodplain with many anabranches, billabongs (oxbows), deflation basins and Lake Victoria. The Chowilla wetland complex, near Renmark, is renowned for waterbirds and is 'a wetland of international importance' under the international Ramsar Convention. Locks 3–10 are in this section.

2. *Gorge Section*: from Overland Corner, where the river enters its limestone gorge, to Mannum. The channel includes long, straight reaches aligned by geological faults, but retains some inclination to meander. The diversity of wetlands is much less than in the Valley Section. Locks 1–2 are in this section.

3. *Swamplands Section*: from Mannum to Wellington. The river flows through a section flanked by reclaimed swamplands used for pasture and forage crops. Earthen levees protect the reclaimed areas. There are no weirs, but river-mouth barrages maintain a 450–600 mm increase in the river level below Lock 1.

4. *Lakes Section*: from Wellington, where the river enters Lake Alexandrina (2015 GL), to the sea. The lake was formerly estuarine, but is now isolated from the sea by a series of five barrages. Sedimentary diatom records show that the lake became fresh after barrage construction in 1939–40, with a shift towards epiphytic species associated with littoral plants (Fluin, 2002). The river enters the sea near the town of Goolwa, although the mouth is subject to closures caused by flow depletion and accumulated sand (MDBC, 2002). A small lake, Albert, is fed from Lake Alexandrina, and a hypersaline lagoon 100 km long, the Coorong, adjoins the river mouth. This area is a haven for migratory waterbirds and is another 'wetland of international importance' under the Ramsar Convention.

The principal geomorphic division is that between the Valley and Gorge sections. In the Valley, the bank sediments are clays mixed with easily eroded sand, whereas the sediments in the Gorge are typically heavy, cohesive clays.

Terminology

Weirs on the Murray are commonly referred to as numbered 'locks' (thus, Lock 1 at Blanchetown, Lock 10 at Wentworth: Fig. 9.1), although each weir and adjacent lock chamber are separate entities. There is no standard reference for the weir pools, but the form used here is like that used for weirs ('navigation dams') on the Mississippi River (cf. Walker *et al.*, 1994) (Fig. 9.6). The 3–5 km reach below each weir is the *tailwater*, leading successively to *upper*, *middle* and *lower* pools. Each pool is identified by the number of the impounding weir. For example, Pool 3 is impounded by Lock 3; the water immediately above Lock 3 is the lower part of Pool 3, and the Lock 3 tailwater is the upper part of Pool 2.

Location and configuration

The ten Lower Murray weirs are at 29–88 km intervals between Blanchetown and Wentworth, respectively 274 km and 830 km from the river mouth (Fig. 9.1). They were built in 1922–37, originally to promote riverboat transport, but are now used mainly to preserve stable levels

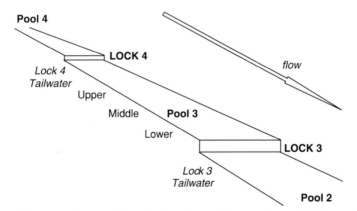

Figure 9.6. Suggested terminology for weir impoundments (Walker *et al.*, 1994).

for irrigation pumps. The lowermost reaches are further controlled by barrages along the seaward margins of Lake Alexandrina, designed to prevent the influx of sea water. Thus, the Lower Murray is virtually a series of cascading pools. Indeed, weir pools occupy 52% of the length of the Murray from the beginnings of its lowland tracts (Lake Hume, about 350 km from the source) to the river mouth (Pressey, 1986).

The Murray weirs are a Boulé design. They are 3 m high and 84–169 m long, excluding the lock chambers, with pools 29–88 km long and capacities of 13–74 GL (Table 9.1). Each incorporates a *navigable pass*, with $1m^2$ panels supported by needle beams and trestles, and a *sluice section*, with bays of 300–400 mm prismatic pre-stressed concrete 'stop logs' (Jacobs, 1990). During high flows, the panels and stop logs are removed by using a mobile crane and the trestles may be folded against the base of the weir. At other times, the weir keeper manipulates the panels and stop logs to keep the water level near the target 'pool level', nominally to within ±50 mm, mainly to service irrigation offtake pumps. Control increases downstream towards Lock 1, with variations damped by successive weirs.

Hydraulic effects

Despite variable inflows from the Murray and Darling rivers, the Lower Murray weirs have much less effect on flow than on river stage (water level). Since the 1920s, SA Water (the principal water management authority in South Australia) has maintained daily stage records for the upper and lower lock chambers at each weir, to estimate discharge. Records for Locks 1–6 have been analysed (Maheshwari *et al.*, 1995; Blanch *et al.*, 1999c); some features of the changed regime are outlined below (see further Walker *et al.*, 1992, 1995).

Weir operations maintain steady pool levels except when flows exceed storage capacity, but tailwater levels depend largely on incoming discharge (Lee *et al.*, 1998). When discharge is rising or falling, weir operations maintain a steady upstream pool level, but cause the amplitude of stage fluctuations in the tailwater to be nearly four times that in the pool immediately above the weir. In the tailwater of Lock 3 in 1980–89, there were frequent rises and falls in daily levels of ±200 mm, and changes of more than ±500 mm occurred about once annually (Fig. 9.7).

The effects of the weirs in damping stage variations are illustrated by data for several stations along Pool 3 in 1983–86 (Fig. 9.8). Clearly, fluctuations below Lock 4 are transferred down river towards Lock 3,

Table 9.1. *Features of the ten weirs on the River Murray below the Murray–Darling junction*

AHD: Australian Height Datum, equivalent to sea level.

Lock	Name	Distance to mouth (km)	Pool level (m AHD)	Pool length (km)	Weir length (m)	Capacity (GL)	Removed at flow (ML d^{-1})	Year completed
1	Blanchetown	274	3.30	–	169	63.7	68 000	1922
2	Waikerie	362	6.10	88	138	44.3	54 000	1928
3	Overland Cnr	431	9.80	69	123	74.4	63 000	1925
4	Bookpurnong	516	13.20	85	125	45.3	63 000	1929
5	Renmark	562	16.30	46	125	62.8	67 000	1927
6	Murtho	620	19.25	58	87	43.4	60 000	1930
7	Rufus River	697	22.10	77	84	12.9	29 000	1934
8	Wangumma	726	24.60	29	119	21.4	45 000	1935
9	Kulnine	765	27.40	39	94	42.5	53 000	1926
10	Wentworth	825	30.80	60	117	44.3	–	1929

Source: Data from Murray–Darling Basin Commission.

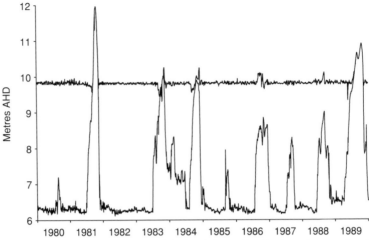

Figure 9.7. River stage or height levels immediately above (topmost) and below (lowermost) Lock 3 in 1980–89 (AHD, Australian Height Datum, equivalent to sea level).

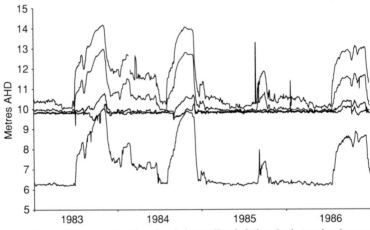

Figure 9.8. From top, river levels immediately below Lock 4 and at Loxton, Cobdogla, and above and below Lock 3, 1983–86 (AHD, Australian Height Datum). See Fig. 9.1 for locations of locks.

where the sequence is reset to begin a new progression towards Lock 2. Thus, there are gradients in the magnitude of water-level changes between the weirs. The changes at sites along the pool may be estimated by simple interpolation (Blanch *et al.*, 1999c).

Lock 3 was commissioned in 1925, and water levels in the river below the weir subsequently showed increased short-term (daily) variation (Fig. 9.9). Essentially, rises and falls were greater in magnitude and more sustained before regulation than afterward (Maheshwari *et al.*, 1993, 1995). Data for Lock 6 (Fig. 9.10) also illustrate the effects of regulation on the amplitudes of changes in river stage (Fig. 9.11). The graphs compare three flood pulses, each represented by actual stage records and modelled natural flows estimated by the MDBC *Monthly Simulation Model* (Walker *et al.*, 1995). During a big, unregulated flood (Fig. 9.11a), the river moves continually over the banks and the floodplain. A moderate pulse (Fig. 9.11b) shows more prolonged stable levels; for smaller, in-channel pulses (Fig. 9.11c), regulation has eliminated any water-level response.

The weirs also slow the passage of water and increase the flushing time in pools and backwaters, favouring algal blooms. Blooms are promoted by increased transparency associated with settling of particulate matter and flocculation caused by increased salinity, partly offset by high turbidity and seasonally low nutrients (see, for example, Bormans *et al.*, 1997; Grace *et al.*, 1997; Webster *et al.*, 1997; Bormans & Webster, 1999; Maier *et al.*, 1998, 2001; Baker *et al.*, 2000).

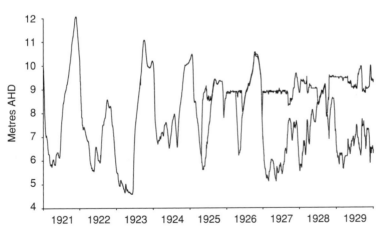

Figure 9.9. River height levels at Lock 3 in 1920–29. Completion of the weir in 1925 is signalled by the separation of upper and lower pools (AHD, Australian Height Datum).

Figure 9.10. An aerial view of Lock 6 on the Lower Murray, with a maintenance boat moored on the upstream side of the weir (Photo K. F. Walker).

Erosion and sedimentation

The internal dimensions and slope of the Murray channel are still adjusting to the presence of the weirs, even after 80 years (Thoms & Walker, 1992a,b, 1993). The process has been prolonged by the river's low energy, cohesive bank material and limited sediment supply. The sequential position of each weir, the time of its construction, and local factors (e.g. width of floodplain, composition of bank material) have produced redistributions of sediment rather than changes in the overall sediment budget. The channel is developing a stepped gradient, with deposition above each weir and erosion downstream. Remnant internal benches of the unregulated river, formed by prolonged low flows and rare high-discharge events, have been eroded in the upper pools, are covered to depths of more than 2 m in the lower pools, but remain intact as gently shelving banks in the middle reaches.

The progression of changes differs in the Valley and Gorge Sections (Fig. 9.1), reflecting constraints imposed by the bank material and limestone matrix (Thoms & Walker, 1992a). In Pool 3, in the Valley Section, erosion has increased the channel width : depth ratio by 32%

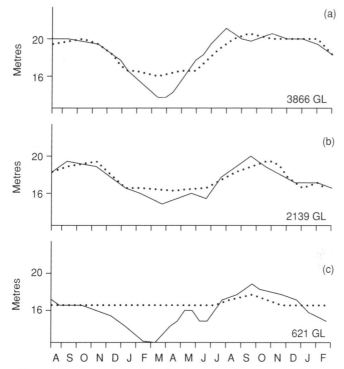

Figure 9.11. Comparison of flood pulses at Lock 6 (Walker *et al.*, 1995). Each panel shows simulated (unregulated) river stages (solid lines) superimposed on actual records (broken lines) for floods in (a) 1952, (b) 1936 and (c) 1945. Peak monthly flows (1 GL = 10^3 ML = 10^6 m^3) are inset.

but in Pool 2, in the Gorge Section, deposition has decreased the ratio by 22%. Although flow variations will ensure that the shape of the channel is never completely stable, the channel slope may reach a quasi-equilibrium in about 40 years. Weir pools in the Valley Section are likely to stabilise before those in the Gorge Section.

Bank erosion is a conspicuous problem in some areas, particularly in the weir tailwaters. In 1988–89, an estimated 1.8 million tonnes of bank material was lost over the 153 km reach occupied by Pools 2–3 (Thoms & Walker, 1989). Erosion is promoted by rapid drawdown following high flows, changes in water level associated with weir operations (Lee *et al.*, 1998) and, in some areas, backwash from boats (Munday *et al.*, 1998).

Superficial sediments in the weir pools are well mixed, with little vertical structure, but cores from the mid-channel show an abrupt transition from unregulated to regulated conditions. The predominant sediments prior to regulation were coarse sand. These are preserved below about 70 cm depth, but are overlain by fine silts and clays eroded from the river banks (Thoms & Walker, 1992a,b). The weirs are effective sediment traps, so that much of the fine material is being accumulated (cf. Vörösmarty et al., 2003).

In wetlands associated with the weir pools, sedimentary diatom records indicate distinct changes in response to flow regulation and persistent high water levels, including a shift toward planktonic species at the expense of littoral and plant-associated taxa (Gell et al., 2002, 2005; Reid et al., 2002). This suggests that there has been a decline in the abundance of aquatic plants. At Tareena Billabong, on the floodplain at Chowilla, these effects were amplified by installation of a regulator that raised the water level after floods in 1956–58 (Bulpin et al., 2005). At Loch Luna, above Lock 3, the sedimentary diatom record shows a near-complete turnover of species between the periods before 1925, when the wetland was flooded intermittently, and after construction of the weir, when it became permanently inundated (Baldwin, 2003).

The diminished flows caused by upstream diversions (cf. Maheshwari et al., 1995) have also caused sediment to accumulate near the river mouth. There have been long periods when no flow has entered the sea (Bourman & Barnett, 1995; MDBC, 2002), and the most recent closure (2003–04) has necessitated dredging. Mouth closures, and the deterioration of the adjacent Coorong lagoon as a bird sanctuary and a fishery for cockles Donax deltoides and the mulloway Sciaena antarctica and other fish, have led to reappraisals of options for regulating flows to the lower lakes. Prolonged closures also mean no flushing of salt from the river and floodplain, intensifying the effects of salinisation (Chapter 10, this volume).

Salinity

Groundwater and surface water in the Murray Valley are rendered saline by underlying marine sediments (Evans & Kellett, 1989), inappropriate irrigation methods, excessive vegetation clearance and river regulation (Allison et al., 1991; Jolly, 1996; Walker et al., 1996; Crabb, 1997; MDBC, 1997; MDBMC, 1999). In the Murray, typical salinities range from less than 30 mg l^{-1} in the headwaters to >450 mg l^{-1} near the mouth (Mackay et al., 1988; Close, 1990). The effects of salt are a major issue in river management because water more saline than 300–500 mg l^{-1} is unsuitable for most irrigated crops, and salinised land is of little use for agriculture. About one quarter of the floodplain in South Australia (100 000 ha) is salinised, and a further 12 500 ha of waterlogged riparian land is vulnerable (MDBC, 1997).

Salinity is the main determinant of diatom species composition in the Lower Murray, outweighing the effects of flow velocity, pH and nutrients (Tibby & Reid, 2004). As diatoms are dominant phytoplankters in the Murray (Hötzel & Croome, 1996; Baker et al., 2000), salinity changes may prove to have indirect effects on grazing invertebrates, particularly zooplankton (cf. Shiel et al., 1982).

Salt may not have greatly affected other riverine plants and animals because the adults of many species are tolerant (see, for example, Hart et al., 1990), although juveniles are less so (cf. Chapter 10, this volume). A riverine focus, however, obscures considerably higher salinities that prevail in floodplain areas, including wetlands (see, for example, Suter et al., 1993; Skinner et al., 2000). Floodplain trees, especially river red gum Eucalyptus camaldulensis and black box E. largiflorens, are salt-affected (Margules et al., 1990; O'Malley & Sheldon, 1990; Jolly et al., 1996a,b; Slavich et al., 1999), and some areas have been invaded by halophilic plants (e.g. the chenopod Sarcocornia quinqueflora).

The problem of salt accession is exacerbated by rapid recessions following floods, excessive clearance of vegetation, waterlogging in poorly drained soils, inefficient irrigation, poor land and water management, and the hydraulic effects of weirs. Weirs may be responsible for about one quarter of the salt load entering the Murray in South Australia (EWS, 1978). Saline groundwater is pushed downwards by the hydraulic head associated with the impounded water, but forced upwards in the downstream reach, where it enters the river. The prospects for lowering or removing one or more weirs have not seriously been investigated because of concerns that flushing the salt accumulated in the floodplain soil could take decades and incur short-term increases in river salinity (see, for example, NEC, 1988). It should be possible, however, to lower a weir pool slowly to control the rate of flushing of stored salt. This should be a priority for research and management.

ECOSYSTEM PROCESSES

Floodplain environments

Surface water 'connectivity' maintains floodplain–river ecosystems, but is prejudiced by flow regulation (see, for example, Ward & Stanford, 1995). Wetland organisms depend on the river for replenishment, and the riverine biota depend on wetlands for food, breeding sites and refugia. Regulation, particularly through weir operations, isolates the river and floodplain for longer than natural periods, and the ecological effects may be devastating (see, for example, Chapter 8, this volume). At an ecosystem level, for example, there may be radical changes in the patterns of storage and flux of carbon (Robertson *et al.*, 1999). Regulation may also extend the area of permanently flooded wetlands. In the Lower Murray, about 70% of wetlands (backwaters, 'side-arms', ana-branches, lakes, billabongs) are now connected to the river at pool level (Pressey, 1986). Many of these wetlands formerly had more frequent water-level changes, and some would have dried periodically.

A consequence of prolonged inundation is that affected wetlands may no longer exhibit the *pulse* of plant and animal growth associated with variable flooding and drying (cf. Kingsford *et al.*, 2004). Similarly, wetlands subjected to drying for too long may not produce a pulse of high productivity when floods occur. Thus, disruption of the natural drying and wetting cycle affects the capacity of the river–floodplain ecosystem to respond to floods. This idea (see, for example, Briggs & Maher, 1985) has not been rigorously tested but enjoys wide acceptance, and reinstatement of wetting–drying cycles is a priority in the restoration of Lower Murray wetlands (see, for example, Jensen, 1998).

The drying–wetting sequence is as significant for the river channel as for the wider floodplain. Indeed, the term 'flood' may be usefully applied to all increases in river level rather than merely to over-bank flows (Puckridge *et al.*, 1998; Chapter 3, this volume). Rises and falls in water level within the channel stimulate responses in the growth of biofilms (algae, bacteria and fungi growing on sediments, rocks and wood) that provide food for some fish, and for snails and other grazing invertebrates. Instream water-level changes caused by regulation may have promoted the growth of algae at the expense of bacteria, and reduced the nutritional value of the biofilms for grazing animals (Sheldon & Walker, 1993, 1997; Burns & Walker, 2000) (see further below).

The degradation of floodplain woodlands (Margules *et al.*, 1990) is linked to regulation, as the vigour of river red gums, black box and other plants is correlated with changes in the pattern of flooding (e.g. Craig *et al.*, 1991; Thorburn *et al.*, 1994; Jolly *et al.*, 1996a,b; Jolly, 1996; Jolly & Walker, 1996; Walker *et al.*, 1996; Akeroyd *et al.*, 1998; Chong & Walker, 2005).

Riverine littoral zone

The open river channel has strong currents, unstable sediments and a shallow photic zone, and is a harsh environment for many plants and animals. The littoral zone, however, supports a narrow band of emergent and submerged plants (e.g. *Cyperus gymnocaulos*, *Phragmites australis*, *Potamogeton crispus*, *Typha* spp., *Vallisneria americana*) that are a refuge for many terrestrial and aquatic animals. These plants are distributed broadly in relation to the frequency of flooding and exposure (Walker *et al.*, 1992, 1994; Blanch & Walker, 1998; Blanch *et al.*, 1998, 1999a,b, 2000) and there may be secondary zonation in the distributions of some invertebrates. Turbidity may have complementary effects. In the dry years of the late 1980s, turbid Darling River water was stored in Lake Victoria and released to the Lower Murray in summer and autumn. As a result, turbidity in the Lower Murray was three-fold higher than in the Murray above the Darling confluence, and plant growth was reduced. The effect was compounded because levels in the weir tailwaters fluctuated daily through a range comparable to the extent of the photic zone. With higher flows in 1990, diversions from the Darling were reduced and there were remarkable increases in the abundance of riverine plants and animals (see, for example, Walker *et al.*, 1994).

The stands of littoral plants are an artefact of weir construction, as historical photographs show that the banks of the unregulated river were largely bare (Walker *et al.*, 1992, 1994; Blanch *et al.*, 2000). The weir impoundments and seasonally more stable flow regime have allowed numerous wetland species to invade the channel. The present littoral community warrants special consideration in conservation and management, however, as it would be adversely affected if stage variations were restored as part of management. Initiatives to restore instream habitats would need to be developed in parallel with restoration of wetlands and restoration of connections between river and floodplain.

Willows (Salicaceae), widely used for ornament and bank stabilisation, were introduced to the Lower Murray in the nineteenth century. Weeping willow *Salix babylonica* and crack willow *S. fragilis* were planted mainly to stabilise levees protecting reclaimed riparian swamplands, and in some areas the former species now rivals the native river red gum as a riparian tree (Kennedy *et al.*, 2003). Attempts to remove or control the willows have met with strong community opposition, but there is also opposition to leaving them unchecked. The presumed adverse effect of willows on littoral invertebrate assemblages is unconfirmed (Schulze & Walker, 1997), but willows may consume considerably more water than the native eucalypts and this may prove significant as water-accounting becomes more intense in future management. *S. fragilis* occurs mainly downstream of Murray Bridge, below Lock 1 (Fig. 9.1), but *S. babylonica* predominates up to Blanchetown and thereafter is commonest in the lower weir pools, where water levels are stable. The latter trees are exclusively females and therefore dependent on vegetative reproduction, but there is concern that hybrid willows (e.g. *S. matsudana* × *alba*) can fertilise *S. babylonica* and may initiate seeding (cf. Cremer *et al.*, 1995).

Fish

The regulated river is a stressful environment for some native fish, particularly those with opportunistic, flow-dependent breeding cycles and patterns of movement requiring free access to instream and off-stream habitats (see, for example, Gehrke *et al.*, 1995; Thorncraft & Harris, 2000). Most native fish species have declined in range and abundance (see, for example, Harris & Gehrke, 1997) in favour of introduced species, for example the common carp *Cyprinus carpio* (Koehn *et al.*, 2000), which account for about one-quarter of species in the present Murray–Darling fish fauna. The decline is arguably most advanced in the Lower Murray (Lloyd & Walker, 1986). One of few native fish to have increased in numbers in this region is the bony herring *Nematalosa erebi*, which thrives in the weir pools (Puckridge & Walker, 1990) and rivals carp in numbers and biomass.

    Not all native fish have breeding cycles directly linked to flooding, but it is likely that recruitment in most, if not all, species is enhanced by high flows (cf. Harris & Gehrke, 1994; Gehrke *et al.*, 1995; Humphries *et al.*, 1999; Chapters 4 and 7, this volume). This may be mediated by access to and from floodplain wetlands, or by increased abundance of food for young fish, exported from the floodplain. Conceivably, the

Figure 9.12. The Murray cod *Maccullochella peelii peelii* is an iconic
species of the Murray–Darling river system. This captive specimen
weighs over 50 kg (Photo K. F. Walker).

reduced frequency of smaller floods limits low-level 'bridging' recruitment in some longer-lived species, e.g. Murray cod *Maccullochella peelii
peelii*, so that stocks are depleted and unable to respond to big floods
(Fig. 9.12).

Weirs and other regulating structures are significant obstacles to
fish migrations throughout the Murray–Darling Basin; removal of barriers or installation of fishways are a prerequisite for restoration and
conservation of native species (Thorncraft & Harris, 2000). Although
Lock 6 has a fishway, investigations into a structurally similar fishway
at Lock 15 (Euston, New South Wales) suggest that the design is inappropriate for most native fish (Mallen-Cooper, 1996). The Murray–Darling
Basin Commission recently began to 'retro-fit' fishways to all weirs on
the Murray and to the river-mouth barrages.

Invertebrates

A striking feature of the Lower Murray is the scarcity of freshwater
snails (Fig. 9.13). About 18 species were present before weir construction, but there have been sporadic records of only six over the past

Figure 9.13. The viviparid river snail *Notopala sublineata hanleyi*, once common in the the Lower River Murray, now survives only in an irrigation pipeline fed from the river. These two shells, shown with a single operculum, are about 15 mm in diameter (Photo K. F. Walker).

decade and now only the ancylid *Ferrissia* sp. and the introduced physid *Physa acuta* are common in field collections (see, for example, Sheldon & Walker, 1993, 1997). The latter is easily confused with the native planorbid *Glyptophysa gibbosa*, and the regional status of these two species needs clarification.

Irrigation pipelines fed from the river are a refuge for some gastropods, notably the caenogastropod ('prosobranch') species *Thiara balonnensis* (Thiaridae) and *Notopala sublineata hanleyi* (Viviparidae). In some years, *Notopala* occurs in very large numbers and becomes a major nuisance for irrigators, and efforts have been made to poison the population. This species is presently known in small numbers from only one place, apart from a few regional wetlands where small numbers have been translocated (B. Weir, personal communication), and warrants recognition as an 'endangered' species (cf. Sheldon & Walker, 1993; Walker, 1996).

Three factors may be implicated in the decline of the snails (Sheldon & Walker, 1997; Burns & Walker, 2000). One is the alienation of the river and its floodplain wetlands, exacerbated by weir operations. Second is the likelihood of predation by introduced carp, which occur in high densities, particularly in wetlands. Third is that water-level changes associated with weir operations may be responsible for changes in the composition of biofilms that provide food to snails and other grazing animals. The prevalence of frequent, rapid short-term variations in water level encourages the growth of algae at the expense of bacteria, and thereby reduces the nutritional value of the biofilms. The inner walls of pipelines support bacterial biofilms, as there is no light to sustain algal growth.

Freshwater mussels (Hyriidae) are common throughout the Murray (Walker, 1981; Walker et al., 2001). The obligate riverine species Alathyria jacksoni is intolerant of low oxygen concentrations and dehydration and so does not live in wetlands; Velesunio ambiguus, a species typical of floodplain wetlands, is a relatively weak burrower, and is thereby excluded from fast-flowing sections of the river (Sheldon & Walker, 1989). The range of A. jacksoni has contracted since construction of weirs, whereas V. ambiguus has invaded the weir pools and the sheltered margins of the main channel. A similar redistribution has occurred among the crayfish (Parastacidae) (Walker, 1986). The yabbie Cherax destructor is typical of still-water floodplain habitats, but now is also common in weir pools and along the river margins. The Murray crayfish Euastacus armatus is a riverine species and virtually extinct in the lower river. Its decline is circumstantially related to the construction of weirs, and it is possible that populations could be increased by re-stocking (see, for example, Geddes et al., 1993).

CONCLUSION

The Lower Murray region is draining slowly after repeated incursions by the sea over millions of years. The landforms, soils and groundwater all demonstrate that marine legacy. In hindsight, the imposition of irrigated agriculture, vegetation clearance and intensive flow regulation was a prescription for salinisation. The outcomes are reinforced by the presence of multiple large dams, weirs, river-mouth barrages, wetland regulators, floodplain levees and local diversions. The weirs are merely part of a larger problem.

Weir pools extend over two thirds (556 km) of the Murray's course below the Darling confluence, and the remainder is affected by the river-mouth barrages. River levels vary with distance along the pools (that is, with proximity to upstream and downstream weirs), during periods when flows are confined to the channel. The sediment regime is dominated by a continuing process of channel adjustment, as the river-bed gradient progressively assumes a stepped profile to match the hydraulic gradient imposed by the weirs. The pools and river margins are havens for plants and animals typical of floodplain wetlands. The rising and falling phases of the flood pulse have been disrupted, and periods of stasis have been reinforced. Regulation has reduced the frequency, amplitude and duration of floods and increased the frequency of short-term water-level fluctuations, disrupting connections between the river and its floodplain. It has stabilised the river level at a seasonal scale, by maintaining near-bankfull capacities, but also introduced gradients between weirs in the magnitude of water-level changes, and daily fluctuations in the weir tailwaters. The effects have been to weaken the *longitudinal* linkages within the channel and the *lateral* linkages between the river and its floodplain. These linkages need to be restored and maintained if the integrity of the river–floodplain ecosystem is to be maintained. This is a prerequisite also for conservation of many species of native flora and fauna.

Over the twentieth century, the Lower Murray environment has been physically and biologically transformed. Although the weirs are not solely responsible, they are an important part of the solution. It is apposite to ask whether, in the present-day environment, they cause more problems than render useful services. The issue is topical, because federal and state government agencies are energetically pursuing programs to contain salinisation, protect native flora and fauna and deliver environmental flows (e.g. *The Living Murray Program*: http://www.mdbc. gov.au). No serious consideration has yet been given to the prospects for removal of some (if not all) weirs, or for structural modifications to ameliorate their effects, but changes of this nature are inevitable. In the meantime, manipulation of weir heights is one means to manage and enhance floods and partly restore flow variability (see, for example, Blanch et al., 1996; Siebentritt et al., 2005). Management protocols could target rates of rise and fall that encourage recruitment in populations of native flora and fauna and flush accumulated salt from the floodplain soil, using a succession of flood magnitudes and frequencies in appropriate seasons. In so doing, the decline of the Lower Murray may yet be arrested. Governments recently have made unprecedented commitments to environmental flow allocations and, if sufficient water is made available and delivered appropriately, and if action is taken soon enough, we may trust in the oft-repeated dictum to 'let nature do the work'. The river will never return to pristine condition, but actions now will determine its character, and its utility as a resource, at the end of the twenty-first century.

ACKNOWLEDGEMENTS

Much information here is drawn from an article in the proceedings of a conference, *The Way Forward on Weirs*, edited by Dr Stuart Blanch and published by the Australian Conservation Foundation (Inland Rivers Network, Sydney, 2001). This material is used with kind permission of the Foundation. The author also thanks Dr Richard Kingsford for his patience and editorial assistance.

REFERENCES

Akeroyd, M. D., Tyerman, S. D., Walker, G. R., Jolly, I. D. and Baker, P. D. (1998). Impact of flooding on the water use of semi-arid riparian eucalypts. *Journal of Hydrology*, **206**, 104–17.

Allison, G. B., Cook, P. G., Barnett, S. R. *et al.* (1991). Land clearance and river salinization in the western Murray Basin. *Journal of Hydrology*, **119**, 1–20.

Baker, P., Brookes, J. D., Burch, M. D., Maier, H. R. and Ganf, G. G. (2000). Advection, growth and nutrient status of phytoplankton populations in the Lower River Murray, South Australia. *Regulated Rivers: Research and Management*, **16**, 327–44.

Baldwin, D. (2003). Assessing environmental change and restoration potential of Loch Luna wetland: A palaeolimnological study. MEnvS (Geographical & Environmental Studies) thesis, University of Adelaide.

Blanch, S. J. and Walker, K. F. (1998). Littoral plant life history strategies and water regime gradients in the River Murray, South Australia. *Verhandlungen der Internationalen Vereinigung für Theoretische und Angewandte Limnologie*, **26**, 1814–20.

Blanch, S. J., Ganf, G. G. and Walker, K. F. (1998). Growth and recruitment in *Vallisneria americana* as related to the average irradiance over the water column. *Aquatic Botany*, **61**, 181–205.

Blanch, S. J., Ganf, G. G. and Walker, K. F. (1999a). Growth and resource allocation in responses to flooding in the emergent sedge *Bolboschoenus medianus*. *Aquatic Botany*, **63**, 145–60.

Blanch, S. J., Ganf, G. G. and Walker, K. F. (1999b). Tolerance of riverine plants to flooding and exposure indicated by water regime. *Regulated Rivers: Research and Management*, **14**, 43–62.

Blanch, S. J., Walker, K. F. and Ganf, G. G. (2000). Water regime preferences of plants in four weir pools of the River Murray, Australia. *Regulated Rivers: Research and Management*, **16**, 445–56.

Blanch, S. J., Burns, A., Vilizzi, L. and Walker, K. F. (1996). *Ecological effects of shallow winter-spring flooding in the Lower River Murray, 1995. Observations on organic matter, plant growth and macroinvertebrate abundance in the littoral zone, with recommendations for flow management.* Canberra: Report to Murray-Darling Basin Commission, with Department of Environment and Natural Resources and SA Water, Adelaide.

Blanch, S. J., Maheshwari, B. L., Walker, K. F. and Ganf, G. G. (1999c). An evaluation of the River Murray Hydraulic Model with regard for ecological research and environmental flow management. *Australian Journal of Water Resources*, **3**, 107–20.

Bormans, M. and Webster, I. T. (1999). Modelling the spatial and temporal variability of diatoms in the River Murray. *Journal of Plankton Research*, **21**, 581–98.

Bormans, M., Maier, H., Burch, M. and Baker, P. (1997). Temperature stratification in the lower River Murray, Australia: implications for cyanobacterial bloom development. *Marine and Freshwater Research*, **48**, 647–54.

Boulton, A. J. and Lloyd, L. N. (1991). Macroinvertebrate assemblages in floodplain habitats of the Lower River Murray, South Australia. *Regulated Rivers: Research and Management*, **6**, 183–201.

Boulton, A. J. and Lloyd, L. N. (1992). Flooding frequency and invertebrate emergence from dry floodplain sediments of the River Murray, Australia. *Regulated Rivers: Research and Management*, **7**, 137–51.

Bourman, R. P. and Barnett, E. J. (1995). Impacts of river regulation on the terminal lakes and mouth of the River Murray, South Australia. *Australian Geographical Studies*, **33**, 101–15.

Briggs, S. V. and Maher, M. T. (1985). Limnological studies of waterfowl in south-western New South Wales, II. Aquatic macrophyte productivity. *Australian Journal of Marine and Freshwater Research*, **36**, 707–15.

Bulpin, S., Gell, P. A., Wallbrink, P., Hancock, G. and Bickford, S. (2005). Tareena Billabong—a palaeolimnological history of an ever-changing wetland, Chowilla Floodplain, Lower Murray-Darling Basin, Australia. *Marine and Freshwater Research*, **56**, 441–56.

Burns, A. and Walker, K. F. (2000). The effects of weirs on algal biofilms in the River Murray, South Australia. *Regulated Rivers: Research and Management*, **16**, 433–44.

Chong, C., and Walker, K. F. (2005). Does lignum rely on a soil seed bank? Germination and reproductive phenology of *Muehlenbeckia florulenta* (Polygonaceae). *Australian Journal of Botany*, **53**, 407–15.

Close, A. F. (1990). The impact of man on the natural flow regime. In *The Murray*, ed. N. Mackay and D. Eastburn, pp. 61–74. Canberra: Murray-Darling Basin Commission.

Crabb, P. (1997). *Murray-Darling Basin Resources.* Canberra: Murray-Darling Basin Commission.

Craig, A. E., Walker, K. F. and Boulton, A. J. (1991). Effects of edaphic factors and flood frequency on the abundance of lignum (*Muehlenbeckia florulenta* Meissner) (Polygonaceae) on the River Murray floodplain, South Australia. *Australian Journal of Botany*, **39**, 431–43.

Cremer, K. W., Kraayenoord, C. V., Parker, N. and Streatfield, S. (1995). Willows spreading by seed: implications for Australian river management. *Australian Journal of Soil and Water Conservation*, **8**, 18–27.

Evans, W. R. & Kellett, J. R. (1989). The hydrogeology of the Murray Basin, south eastern Australia. *BMR Journal of Australian Geology and Geophysics*, **11**, 147–66.

EWS (Engineering and Water Supply) (1978). *The South Australian River Murray Salinity Control Programme*. Adelaide: Engineering and Water Supply Department.

Fluin, J. (2002). A diatom-based palaeolimnological investigation of the lower Murray River (south east Australia). Ph.D. (Geography & Environmental Science) thesis, Monash University.

Geddes, M. C., Musgrove, R. J. and Campbell, N. J. H. (1993). The feasibility of re-establishing the River Murray crayfish, *Euastacus armatus*, in the lower River Murray. *Freshwater Crayfish*, 9, 368–79.

Gehrke, P. C., Brown, P., Schiller, C. B., Moffatt, D. B. and Bruce, A. M. (1995). River regulation and fish communities in the Murray-Darling river system, Australia. *Regulated Rivers: Research and Management*, 11, 363–75.

Gell, P. A., Sluiter, I. R. and Fluin, J. (2002). Seasonal and interannual variations in diatom assemblages in Murray River connected wetlands in north-west Victoria, Australia. *Marine and Freshwater Research*, 53, 981–92.

Gell, P., Tibby, J., Fluin, J. *et al.* (2005). Accessing limnological change and variability using fossil diatom assemblages, south-east Australia. *River Research and Applications*. 21, 257–69.

Gill, E. D. (1973). Geology and geomorphology of the Murray River region between Mildura and Renmark, Australia. *Memoirs of the National Museum of Victoria*, 34, 1–98.

Grace, M. R., Hislop, T. M., Hart, B. T. and Beckett, R. (1997). Effect of saline groundwater on the aggregation and settling of suspended particles in a turbid Australian river. *Water – Engineering and Management*, 144, 13.

Greenwood, A. (2000). Inter-annual variability in climate and streamflow of the Lower Murray-Darling Basin: an environmental perspective. M. Appl. Sci. (Hydrology and Water Resources) thesis, University of Adelaide.

Harris, J. H. and Gehrke, P. C. (1994). Modelling the relationship between streamflow and population recruitment to manage freshwater fisheries. *Agricultural Systems and Information Technology*, 6, 28–30.

Harris, J. H. and Gehrke, P. C. (1997). *Fish and Rivers in Stress*. (The NSW Rivers Survey.) Cronulla: NSW Fisheries, with Cooperative Research Centre for Freshwater Ecology and NSW Resource and Assessment Council.

Hart, B. T., Bailey, P., Edwards, R. *et al.* (1990). Effects of salinity on river, stream and wetland ecosystems in Victoria, Australia. *Water Research*, 24, 1103–17.

Hötzel, G. and Croome, R. (1996). Population dynamics of *Aulacoseira granulata* (Ehr.) Simonson (Bacillariophyceae, Centrales), the dominant alga in the Murray River, Australia. *Archiv für Hydrobiologie*, 136, 191–215.

Humphries, P., King, A. J. and Koehn, J. D. (1999). Fish, flows and flood plains: links between freshwater fishes and their environment in the Murray-Darling River system, Australia (review). *Environmental Biology of Fishes*, 56, 129–51.

Jacobs, T. (1990). River regulation. In *The Murray*, ed. N. Mackay and D. Eastburn, pp. 39–60. Canberra: Murray-Darling Basin Commission.

Jensen, A. (1998). Rehabilitation of the River Murray, Australia: identifying the causes of degradation and options for bringing the environment into the management equation. In *River Rehabilitation: Principles and Implementation*, ed. L. C. de Waal, A. R. G. Large and P. M. Wade, pp. 215–36. Chichester: Wiley.

Jolly, I. D. (1996). The effects of river management on the hydrology and hydroecology of arid and semi-arid floodplains. In *Floodplain Processes*, ed. M. G. Anderson, D. E. Walling and P. D. Bates, pp. 577–609. Chichester: Wiley.

Jolly, I. D. and Walker, G. R. (1996). Is the field water use of *Eucalyptus largiflorens* F. Muell affected by short-term flooding? *Australian Journal of Ecology*, 21, 173–83.

Jolly, I. D., Walker, G. R., Hollingworth, I. D. *et al.* (1996a). The causes of decline in eucalypt communities and possible ameliorative approaches. In *Salt and Water Movement in the Chowilla Floodplain*, ed. G. R. Walker, I. D. Jolly and S. D. Jarwal, pp. 119–38. (Water Resources Series no.15.) Canberra: CSIRO Division of Water Resources.

Jolly, I. D., Walker, G. R. and Thorburn, P. J. (1996b). Salt accumulation in the floodplain soils, with implications for forest health. In *Salt and Water Movement in the Chowilla Floodplain*, ed. G. R. Walker, I. D. Jolly and S. D. Jarwal, pp. 90–104. (Water Resources Series no.15.) Canberra: CSIRO Division of Water Resources.

Kennedy, S. A., Ganf, G. G. and Walker, K. F. (2003). Does salinity influence the distribution of exotic willows (*Salix* spp.) along the Lower River Murray? *Marine and Freshwater Research*, 54, 1–7.

Kingsford, R. T. (2000). Protecting or pumping rivers in arid regions of the world? *Hydrobiologia*, 427, 1–11.

Kingsford, R. T., Jenkins, K. M. and Porter, J. L. (2004). Imposed hydrological stability imposed on lakes in arid Australia and effect on waterbirds. *Ecology*, 85, 2478–92.

Koehn, J., Brumley, A. and Gehrke, P. (2000). *Managing the Impacts of Carp*. Canberra: Bureau of Rural Sciences.

Lee, S. M., Mair, A. A. and Trebilcock, D. S. (1998). Numerical modelling of weir operations to restore instream habitat: the River Murray, South Australia. Fourth-Year Student Report, Department of Civil and Environmental Engineering, University of Adelaide.

Lloyd, L. N. and Walker, K. F. (1986). Distribution and conservation status of small freshwater fish in the River Murray, South Australia. *Transactions of the Royal Society of South Australia*, 106, 49–57.

Mackay, N. J., Hillman, T. J. and Rolls, J. (1988). *Water quality of the River Murray: Review of monitoring, 1978 to 1986*. (Water Quality Report no.1.) Canberra: Murray-Darling Basin Commission.

Maheshwari, B. L., Walker, K. F. and McMahon, T. A. (1993). *The Impact of Flow Regulation on the Hydrology of the River Murray and its Ecological Implications*. University of Melbourne: Centre for Environmental Applied Hydrology, and University of Adelaide: River Murray Laboratory, Department of Zoology.

Maheshwari, B. L., Walker, K. F. and McMahon, T. A. (1995). Effects of regulation on the flow regime of the River Murray, Australia. *Regulated Rivers: Research and Management*, 10, 15–38.

Maier, H. R., Burch, M. D. and Bormans, M. (2001). Flow management strategies to control blooms of the cyanobacterium *Anabaena circinalis* in the River Murray at Morgan, South Australia. *Regulated Rivers: Research and Management*, 17, 637–50.

Maier, H. R., Dandy, G. C. and Burch, M. D. (1998). Use of artificial neural networks for modelling Cyanobacteria (*Anabaena* spp.) in the River Murray, South Australia. *Ecological Modelling*, 105, 257–72.

Mallen-Cooper, M. (1996). *Fishways and freshwater fish migration in south-eastern Australia*. PhD thesis, Sydney University of Technology.

Margules and Partners, Smith, P., Smith, J. and Department of Conservation, Forests and Lands, Victoria (1990). *Riparian Vegetation of the River Murray*. Canberra: Murray-Darling Basin Commission.

McMahon, T. A. and Finlayson, B. L. (2003). Droughts and anti-droughts: the low flow hydrology of Australian rivers. *Freshwater Biology*, 48, 1147–60.

McMahon, T. A., Finlayson, B. L., Haines, A. T. and Srikanthan, R. (1992). *Global Runoff: Continental Comparisons of Annual Flows and Peak Discharges.* Cremlingen: Catena Verlag.

MDBC. (1997). *Salt trends – historic trend in salt concentration and salt load of stream flow in the Murray-Darling Drainage Division.* (Dryland Technical Report no.1.) Canberra: Murray-Darling Basin Commission.

MDBC, (Murray-Darling Basin Commission) (2002). *The Murray Mouth. Exploring the Implications of Closure and Restricted Flow.* Canberra: Murray-Darling Basin Commission.

MDBMC, (Murray-Darling Basin Ministerial Council) (1995). *An Audit of Water Use in the Murray-Darling Basin, June, 1995.* Canberra: Murray-Darling Basin Ministerial Council.

MDBMC. (Murray-Darling Basin Ministerial Council) (1999). *The Salinity Audit of the Murray-Darling Basin. A 100-year Perspective, 1999.* Canberra: Murray-Darling Basin Ministerial Council.

Munday, S., Sestokas, K. and Whyatt, R. (1998). An investigation into boat-wave induced bank erosion along the Lower River Murray. Fourth-Year Student Report, Department of Civil and Environmental Engineering, University of Adelaide.

NEC. (1988). *Chowilla Salinity Mitigation Scheme. Draft Environmental Impact Statement.* Adelaide: Report by National Environmental Consultancy to Engineering and Water Supply Department.

O'Malley, C. and Sheldon, F. (eds) (1990). *Chowilla Floodplain Biological Study.* Adelaide: Hyde Park Press for Nature Conservation Society of South Australia.

Pressey, R. L. (1986). *Wetlands of the River Murray below Lake Hume.* (Environmental Report **86/1**) Canberra: River Murray Commission.

Puckridge, J. T. and Walker, K. F. (1990). Reproductive biology and larval development of a gizzard shad, *Nematalosa erebi* (Günther) (Dorosomatinae: Teleostei), in the River Murray, South Australia. *Australian Journal of Marine and Freshwater Research,* **41**, 695–712.

Puckridge, J. T., Walker, K. F. and Costelloe, J. F. (2000). Hydrological persistence and the ecology of dryland rivers. *Regulated Rivers: Research and Management,* **16**, 385–402.

Puckridge, J. T., Sheldon, F., Walker, K. F. and Boulton, A. J. (1998). Flow variability and the ecology of large rivers. *Marine and Freshwater Research,* **49**, 55–72.

Reid, M., Fluin, J., Ogden, R., Tibby, J. and Kershaw, P. (2002). Long-term perspectives on human impacts on floodplain-river ecosystems, Murray-Darling Basin, Australia. *Verhandlungen der Internationalen Vereinigung für Theoretische und Angewandte Limnologie,* **28**, 710–16.

Robertson, A. I., Bunn, S. E., Boon, P. I. and Walker, K. F. (1999). Sources, sinks and transformations of organic carbon in Australian floodplain rivers. *Marine and Freshwater Research,* **50**, 813–24.

Schulze, D. J. and Walker, K. F. (1997). Riparian eucalypts and willows and their significance for aquatic invertebrates in the River Murray. *Regulated Rivers: Research and Management,* **13**, 557–77.

Sheldon, F. and Walker, K. F. (1989). Effects of hypoxia on oxygen consumption by two species of freshwater mussel (Unionacea: Hyriidae) from the River Murray. *Australian Journal of Marine and Freshwater Research,* **40**, 491–9.

Sheldon, F. and Walker, K. F. (1993). Pipelines as a refuge for freshwater snails. *Regulated Rivers: Research and Management,* **8**, 295–300.

Sheldon, F. and Walker, K. F. (1997). Changes in biofilms induced by flow regulation could explain extinctions of aquatic snails in the lower River Murray, Australia. *Hydrobiologia*, **347**, 97–108.

Shiel, R. J., Walker, K. F. and Williams, W. D. (1982). Plankton of the Lower River Murray, South Australia. *Australian Journal of Marine and Freshwater Research*, **33**, 301–27.

Siebentritt, M., Ganf, G. G. and Walker, K. F. (2005). Effects of an enhanced flood on riparian plants of the River Murray, South Australia. *River Research and Applications*, **20**, 765–74.

Simpson, H. J., Cane, M. A., Herczeg, A. L., Zebiak, S. E. and Simpson, J. H. (1993). Annual river discharges in south-eastern Australia related to El Niño Southern Oscillation forecasts of sea surface temperatures. *Water Resources Research*, **29**, 3671–80.

Skinner, R., Sheldon, F. and Walker, K. F. (2000). Animal propagules in dry wetland sediments as indicators of ecological health: effects of salinity. *Regulated Rivers: Research and Management*, **17**, 191–7.

Slavich, P. G., Walker, G. R., Jolly, I. D., Hatton, T. J. and Dawes, W. R. (1999). Dynamics of *Eucalyptus largiflorens* growth and water use in response to modified watertable and flooding regimes on a saline floodplain. *Agricultural Water Management*, **39**, 245–64.

Stephenson, A. E. (1986). Lake Bungunnia: a Plio-Pleistocene megalake in southern Australia. *Palaeogeography, Palaeoclimatology, Palaeoecology*, **57**, 137–56.

Suter, P. J., Goonan, P. M., Beer, J. A. and Thompson, T. B. (1993). *A biological and physico-chemical monitoring study of wetlands from the River Murray flood plain in South Australia.* (Report 7/93.) Adelaide: Australian Centre for Water Quality Research.

Thoms, M. C. and Walker, K. F. (1989). Preliminary observations of the environmental effects of flow regulation on the River Murray, South Australia. *South Australian Geographical Journal*, **89**, 1–14.

Thoms, M. C. and Walker, K. F. (1992a). Channel changes related to low-level weirs on the River Murray, South Australia. In *Lowland floodplain rivers: geomorphological perspectives*, ed. C. A. Carling and G. E. Petts, pp. 235–49. Chichester: Wiley.

Thoms, M. C. and Walker, K. F. (1992b). Sediment transport in a regulated semi-arid river – the River Murray, Australia. In *Aquatic Ecosystems in Semi-arid Regions: Implications for Resource Management*, ed. R. D. Robarts and M. L. Bothwell *(Environment Canada, NHRI Symposium Series* 7), pp. 239–50.

Thoms, M. C. and Walker, K. F. (1993). Channel changes associated with two adjacent weirs on a regulated lowland alluvial river. *Regulated Rivers: Research and Management*, **8**, 271–84.

Thorburn, P. J., Mensforth, L. J. and Walker, G. R. (1994). Reliance of creek-side river red gums on creek water. *Australian Journal of Marine and Freshwater Research*, **45**, 1439–43.

Thorncraft, G. and Harris, J. H. (2000). *Fish passage and fishways in New South Wales: a status report.* University of Canberra: Cooperative Research Centre for Freshwater Ecology. *(Technical Report* 1/2000.)

Tibby, J. and Reid, M. A. (2004). A model for inferring past conductivity in low salinity waters, derived from Murray River (Australia) diatom plankton. *Marine and Freshwater Research*, **55**, 587–607.

Twidale, C. R., Lindsay, J. M. and Bourne, J. A. (1978). Age and origin of the Murray River and gorge in South Australia. *Proceedings of the Royal Society of Victoria*, **90**, 27–42.

Vörösmarty, C. J., Meybeck, M., Fekete, B. et al. (2003). Anthropogenic sediment retention: Major global-scale impact from the population of registered impoundments. Global and Planetary Change, 39, 169–90.

Walker, G. R., Jolly, I. D. and Jarwal, S. D. (eds) (1996). Salt and water movement in the Chowilla floodplain. (Water Resources Series 15.) Canberra: CSIRO Division of Water Resources.

Walker, K. F. (1981). Ecology of freshwater mussels in the River Murray. Australian Water Resources Council Technical Paper 63.

Walker, K. F. (1986). The Murray-Darling river system. In The ecology of river systems, ed. B. R. Davies and K. F. Walker, pp. 631–59. Dordrecht: Dr W. Junk.

Walker, K. F. (1992). A semi-arid lowland river: the River Murray, Australia. In The Rivers Handbook, ed. P. A. Calow and G. E. Petts, vol. 1, pp. 472–92. Oxford: Blackwell Scientific.

Walker, K. F. (1996). The river snail Notopala hanleyi: an endangered pest. Xanthopus 14(1), 5–7.

Walker, K. F. and Thoms, M. C. (1993). Environmental effects of flow regulation on the River Murray, South Australia. Regulated Rivers: Research and Management, 8, 103–19.

Walker, K. F., Puckridge, J. T. and Blanch, S. J. (1997). Irrigation development on Cooper Creek, central Australia: prospects for a regulated economy in a boom-and-bust ecology. Aquatic Conservation: Freshwater and Marine Ecosystems, 7, 218–29.

Walker, K. F., Sheldon, F. and Puckridge, J. T. (1995). A perspective on dryland river ecosystems. Regulated Rivers: Research and Management, 11, 85–104.

Walker, K. F., Thoms, M. C. and Sheldon, F. (1992). Effect of weirs on the littoral environment of the River Murray, South Australia. In River Conservation and Management, ed. P. J. Boon,, P. A. Calow, and G. E. Petts, pp. 270–293. Chichester: Wiley.

Walker, K. F., Boulton, A. J., Thoms, M. C. and Sheldon, F. (1994). Effects of water-level changes induced by weirs on the distribution of littoral plants along the River Murray, South Australia. Australian Journal of Marine and Freshwater Research, 45, 1421–38.

Walker, K. F., Byrne, M., Hickey, C. W. and Roper, D. S. (2001). Freshwater mussels (Hyriidae) of Australasia. In Ecology and Evolutionary Biology of the Freshwater Mussels Unionoidea, ed. G. Bauer and K. Wächtel, pp. 5–31. Berlin: Springer.

Ward, J. V. and Stanford, J. A. (1995). Ecological connectivity in alluvial river ecosystems and its disruption by flow regulation. Regulated Rivers: Research and Management, 11, 105–19.

Webster, I. T., Maier, H. R., Baker, P. D. and Burch, M. D. (1997). Influence of wind on water levels and lagoon-river exchange in the River Murray, Australia. Marine and Freshwater Research, 48, 541–50.

Whiting, J., Lambert, M., Metcalfe, A. and Walker, K. F. (2003). Identification of persistence in River Murray streamflows. Twenty-Eighth International Hydrology and Water Resources Symposium (University of Wollongong, November 2003), pp. 157–64.

Whittington, J., Cottingham, P., Gawne, B., Hillman, T. J., Thoms, M. C. and Walker, K. F. (2000). Review of the operation of The Cap. Ecological sustainability of rivers of the Murray-Darling Basin. University of Canberra: Cooperative Research Centre for Freshwater Ecology, Technical Report.

Salinisation as an ecological perturbation to rivers, streams and wetlands of arid and semi-arid regions

P. C. E. BAILEY, P. I. BOON, D. W. BLINN AND W. D. WILLIAMS

INTRODUCTION

Almost no text on the pollution of fresh waters considers salt as a significant pollutant. Key scholarly monographs (e.g. Hynes, 1978; Abel, 1996; Laws, 2000; Mason, 2002) cover pollution, in terms of organic materials, heavy metals, pesticides, eutrophication processes, acidification and thermal pollution, but largely refer to 'salt' or 'salinity' in the context of estuarine systems. Connell's (1981) *Water Pollution* devotes one (small) chapter to salinity; it may be no accident that this Australian text reflects more accurately pollution issues important in parts of the world other than the well-watered, temperate northern hemisphere (see Williams, 1988, 1995). Even comprehensive treatises on the ecology of fresh waters commonly fail to consider impact and management of salinity. For example, the encyclopaedic *Rivers Handbook* (Calow & Petts, 1998) does not even list 'salt' or 'salinity' in the index.

What makes these omissions surprising is the fact that large areas of the globe are affected by secondary salinisation and salt can severely affect aquatic and fringing ecotonal communities. Salt has probably its greatest potential to be an aquatic pollutant in the arid and semi-arid

Ecology of Desert Rivers, ed. R. T. Kingsford. Published by Cambridge University Press.
© Cambridge University Press 2006.

zones covering 28% of the global terrestrial surface, with major areas in Africa, the Middle East and Australia (Chapter 1, this volume; Middleton & Thomas, 1997). Moreover, many of the world's major rivers not only flow through arid and semi-arid regions but their waters are used as water sources for substantial – and sometimes ancient – irrigation schemes (Williams, 1995; Bell, 1998; Chapter 7, this volume). About one third of such irrigated lands are affected by salt, which impairs plant growth (Peck, 1996). Although the salinity problem is currently severe in many places, projections for many arid and semi-arid areas (e.g. the Murray–Darling Basin of Australia) indicate that secondary salinisation could be a future catastrophe (Smith, 1998).

We contend that one reason for the lack of appreciation of salt as a significant aquatic pollutant is it is largely seen as an agricultural problem, affecting farm productivity with secondary impacts on potable water quality (see, for example, Heathcote, 1983). This bias is not surprising, given the obvious importance of water quality for irrigated food production and the supply of potable water in arid and semi-arid regions of the world (Chapter 11, this volume). The poor knowledge base on salinity impacts on aquatic systems in arid and semi-arid zones also contributes. The arid-zone literature deals overwhelmingly with terrestrial environments (see, for example, Ludwig *et al.*, 1997); the limnological literature deals overwhelmingly with temperate, well-watered environments, and the salinity literature deals overwhelmingly with agricultural issues (Fig. 10.1). To redress some of these imbalances, we provide a synthesis covering: sources of secondary salinisation; a brief overview of the global scale of secondary salinisation; and the impact that salinisation has on the structure and processes of aquatic systems in arid and semi-arid zones.

WHAT IS SALINISATION?

The term 'salinisation' refers to the *process* of increasing concentration of dissolved salts in a waterbody or landscape, and the ecological and economic *results* of this phenomenon (Williams, 1987). Salinisation is a natural process where endorheic drainage systems are common (i.e. where rivers discharge into terminal lakes) and the rainfall is low and erratic but the potential evaporation is high. Good examples include Lake Eyre in South Australia, Lake Chad in Africa and the Great Salt Lake in North America (Hammer, 1986; Williams, 1987; Davies *et al.*, 1993). Primary salinisation results from natural processes, whereas

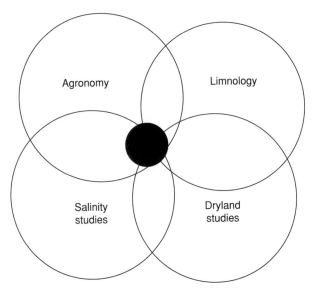

Figure 10.1. Stylised representation of the bias in the knowledge base of salinity impacts on aquatic systems in arid and semi-arid zones. The filled circle represents studies of salinity in dryland aquatic systems.

secondary salinisation is an anthropogenically induced or enhanced process.

Origins of primary salinisation

Salt-affected soils of primary origin result from the long-term influence of natural processes accumulating salts in a particular region. This may be from gradual concentration of the products of weathering, or a single geological event involving the ancient submergence of soils under seawater. In addition, salt in many regions is wind- and rain-borne, distributed by the prevailing pattern of westerly winds. For example, much of the marine-derived salt in underlying sediments in the semi-arid areas of southern Western Australia was rain-borne (Dimmock *et al.*, 1974). In contrast, much of the salt-containing sediments in the central and western regions of the Murray River Basin in eastern Australia were originally derived from at least three invasions of the sea during 30 million of the past 60 million years. The last inundation began about six million years ago and the seas only retreated between four and two million years ago (Evans *et al.*, 1990).

Origins of secondary salinisation

In contrast to primary salinisation, secondary (or anthropogenic) salinisation results from the mobilisation of salt stored in the soil profile and/or groundwater by human activities, principally irrigation and clearing of deep-rooted native vegetation (Burgman & Lindenmayer, 1998; Hatton & Nulsen, 1999; Williams, 2001; Cramer & Hobbs, 2002) (Fig. 10.2). These activities change the natural hydrological balance, raising water tables, increasing pressure of confined aquifers, and creating an upward leakage of water (and salt) to surface aquifers. When the water table approaches the soil surface, the water evaporates, leaving salts behind and causing secondary salinisation (Fig. 10.3). The mobilised salt or salty groundwater can also move laterally or vertically towards watercourses and increase the salinity of streams, rivers and wetlands from 'fresh water' (<0.3–0.5 g l^{-1}) to 'salty' (0.5–10 g l^{-1}). Hydrological disturbances can also *decrease* rather than *increase* the salinity of natural water bodies; this is especially significant for salt lakes and reverse estuaries (e.g. the Coorong in South Australia), where the biota may not be well adapted to a decrease in salinity (De·Deckker, 1983; Williams, 2001).

Irrigation and salinisation

Additional water is supplied to a crop via irrigation because rainfall is insufficient for a species to grow economically. This water may be taken up by the plant and eventually lost in transpiration, evaporate, drain off the land into adjacent streams or wetlands, or percolate through the soil and reach the groundwater (Fig. 10.2, Box 1). Unless the water percolating through the soil can move laterally in the aquifer, the level of the groundwater will rise, bringing stored salts up into the surface soils and the root-zone of the plants (Fig. 10.2, Box 2). Accordingly, soil waterlogging is often closely associated with salinisation (Fig. 10.4). Evaporation of these waterlogged surface soils or soils in deeper layers via capillary action can form scalds of salt, deposited on the soil surface (Fig. 10.2, Boxes 3 and 4). Ultimately, the more concentrated groundwater or mobilised salt in the soil profile can flow into nearby streams or wetlands by vertical or lateral movement, increasing their salinity (Peck, 1996; Smith, 1998).

Water for irrigation also causes salinisation in semi-arid and arid regions (Peck, 1996). Most irrigation water is more salty than rainwater and so its application will inevitably rapidly increase soil salinity unless

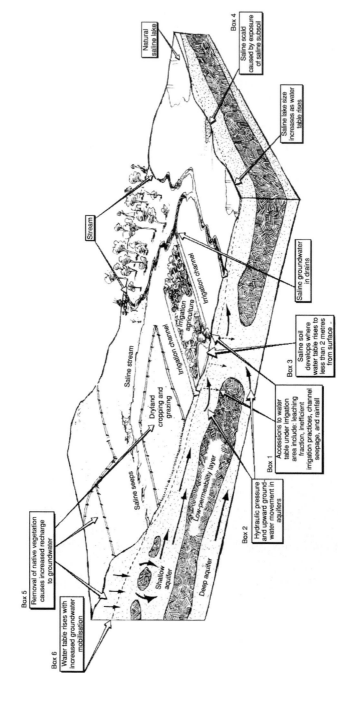

Figure 10.2. Stylised diagram of causes of secondary salinisation from irrigation and vegetation clearing disturbance (modified after Williams, 2001).

Box 5
Removal of native vegetation causes increased recharge to groundwater

Box 6
Water table rises with increased groundwater mobilisation

Box 1
Accessions to water table under irrigation area include: leaching fraction, inefficient irrigation practices, channel seepage, and rainfall

Box 2
Hydraulic pressure and upward ground-water movement in aquifers

Box 3
Saline soil develops where water table rises to less than 2 metres from surface

Box 4
Saline scald caused by exposure of saline subsoil

Natural saline lake

Stream

Saline stream

Saline seeps

Dryland cropping and grazing

Irrigation channel

Irrigation agriculture

Irrigation channel

Shallow aquifer

Deep aquifer

Low-permeability layer

Saline groundwater in drains

Saline lake size increases as water table rises

Figure 10.3. Dried salt on soil surface (arrowed) following evaporation of saline groundwater in Clydebank Morass, South Gippsland, Victoria, Australia (Photo P. Bailey).

Figure 10.4. Effect of prolonged waterlogging at Kow Swamp, north-central Victoria, Australia, showing the death of river red gum E. camaldulensis and black box E. largiflorens from an area that was previously a shallow wetland and is now used as a storage for irrigation water (Photo P. Boon).

the salts are leached out. In the Murray–Darling Basin of Australia, for example, the best irrigation waters have a salinity of about 0.1 g l^{-1}. If drainage through the soil is limited, the application of only *c*. 45 m of water renders the soil saline. Between 0.3 m (for pasture) and 1.2 m (for rice), of water is used annually to irrigate various crops (Smith, 1998). If the irrigation source water has a salinity of 1 g l^{-1}, only 4.5 m of water effects significant salinisation. For rice, this equates to only four or five years of cropping, compared with hundreds of years of rainfall (*c*. 0.01–0.05 g l^{-1} salinity) that might make the same soil saline (Peck, 1996; Smith, 1998). The only mechanism to prevent the accumulation of salts in the soil profile is to flush them out (i.e. leach) with more water. For irrigation water with salt content 0.5–1.0 g l^{-1} or 4–5 g l^{-1} (excluding Na$^+$), the soil has to be leached, respectively, every watering or one to two years (Heathcote, 1983).

Vegetation clearance and salinisation

Clearing of deep-rooted and perennial vegetation (grasses as well as trees) for dryland farming, or changes in vegetation for pastoral use in rangelands and semi-arid zones, results in secondary salinisation. This is because water use by plants, run-off, and recharge of groundwaters disturbs natural hydrological equilibria (Barr & Cary, 1992; Hatton & Nulsen, 1999; Cramer & Hobbs, 2002). This process is well understood in Australian systems (e.g. Williamson, 1990; Salma & Bartle, 1995; Ludwig *et al.*, 1997) (Fig. 10.2, Boxes 5 and 6). The native vegetation is generally well adapted for longevity and high water use. Its removal decreases transpiration and interception of rainfall and increases runoff and the rate of movement of rainfall past the root soil zone into the groundwater as recharge (Williamson *et al.*, 1997). Water tables can rise rapidly under these conditions: as much as 0.1–2.0 m per year in regions of southwestern Western Australia (Ruprecht & Schofield, 1991).

Increased groundwater mobilises salts in the soil profile which discharge to streams, wetlands, and seeps, increasing salinisation (Fig. 10.5). This produces often highly visible salinised tracts of land, leading to loss of terrestrial productivity, and saline water that flows into streams and wetlands. The subsurface expression of groundwater discharge to streams and wetlands is less visible, but more insidious, because it may take longer to appear and has severe long-term consequences of increasing riverine and wetland salinity (Williamson, 1990; Williams, 2001).

Figure 10.5. Effect of secondary salinisation at Kow Swamp, north-central Victoria, showing the death of black box *Eucalyptus largiflorens* trees and colonisation of the understorey by salt-tolerant plants (Photo P. Boon).

In Australia a common engineering practice to lower the water table involves groundwater pumping. Pumped saline wastewater is temporarily stored in ponds or wetlands for later release or discharged into irrigation drains or directly into creeks or rivers (Fig. 10.6). The environmental impacts of such releases on riverine or wetland habitats are poorly quantified (Hart *et al.*, 1991; Bailey & James, 1999; Marshall & Bailey, 2004).

GLOBAL EXTENT OF SALINISATION

The following section focuses on the arid and semi-arid regions within Australia and North and South America. It provides a brief overview of the extent of the problem in South Africa, Central Asia and China, and Egypt and Middle East.

Australia

In terms of area affected, dryland salinity is the major contributor to anthropogenic salinisation throughout Australia (NLWRA, 2001); all States and Territories are affected. About 5.7 million hectares (Mha) are currently within regions 'at risk' or 'affected' by dryland salinity

Figure 10.6. Groundwater pumping station in the MacAlister Irrigation District, Victoria, Australia. Water is pumped from saline groundwater into evaporation basins to ensure agricultural land does not become salinised by rising groundwater caused by irrigation (Photo P. Bailey).

(NLWRA, 2001) but this will increase to about 17 Mha by 2050 (Table 10.1). The largest areas currently affected are in south-western Western Australia, Victoria, and South Australia (Table 10.1). Irrigation-induced salinity is mainly concentrated in the Murray–Darling Basin (MDB) of southeastern Australia, which has about 75% of irrigated land in Australia (*c.* 2 Mha) (Crabb, 1997). Within the Murray–Darling Basin, the area predicted to be at risk will increase from approximately 152 000 ha to 1.3 Mha by 2050 (NLWRA, 2001).

In south-western Western Australia, 80% (by length) of the region's rivers and streams are 'seriously degraded' (Government of Western Australia, 1996), with the length of streams affected by salinity likely to nearly double over the next 50 years, posing substantial risks for riparian zones and in-stream water quality (NLWRA, 2001)

Table 10.1. *Estimates of areas (ha) and as percentage of national total (excluding Tasmania, Northern Territory and Australian Capital Territory) currently (in 2000) affected and 'at risk' (in 2050) by dryland salinity under current land use in mainland* Australia

State	2000	%	2050	%
Western Australia	4 363 000	77	8 800 000	52
Victoria	670 000	12	3 110 000	18
South Australia	390 000	7	600 000	3
New South Wales	181 000	3	1 300 000	8
Queensland	48 000[a]	—	3 100 000	18
Total	5 604 000	100	c. 17 000 000	100

[a] Does not include all land with groundwater levels within two metres of the surface. The value presented has not been included in the total.
Source: From NLWRA (2001).

Table 10.2. *Estimates of aquatic assets in arid and semi-arid zones of Western Australia and South Australia currently and potentially at risk (2020, 2050) due to salinisation*

Aquatic assets	2000	2020	2050
Western Australia			
Stream length (km)	1 520	1 700	2 850
Important wetlands (number)	21	21	21
Important wetlands (ha)	72 500	72 500	80 000
South Australia			
Ephemeral rivers (km)	160	190	210
Wetlands (ha)	45 000	52 000	57 000
Wetlands of national significance (number)	4	4	4

Source: From NLWRA (2001).

(Table 10.2). Salinity in the affected large rivers increases typically by about 0.5 g 1^{-1} each year. Salinisation in many small streams probably increases at much greater rates, but cannot be quantified because these streams are not gauged (Ghassemi *et al.*, 1995). Less than half (c. 48%) of the potentially divertible surface water resources of the region remain fresh, and the remainder have significantly increased salinities: more than a third (36%) are moderately saline or saline and no longer suitable for human consumption (Government of Western Australia, 1996). Forested catchments are generally not salinised (Ghassemi *et al.*, 1995;

Cramer & Hobbs, 2002). Virtually all wetlands in the wheat-belt of the State have become severely degraded or have disappeared, principally as a result of salinisation (Sanders, 1991; Froend & McComb, 1991); Lake Toolibin is the last remaining lake with any extensive surface area coverage (Boulton & Brock, 1999) (Table 10.2).

The southern regions of South Australia and Victoria are not arid or semi-arid regions but we have included them because of their 'Mediterranean' type climate and availability of sparse experimental work on salinisation. In South Australia, increasing stream salinisation is occurring on the Eyre Peninsula and the Middle River of Kangaroo Island. Biodiversity mapping has identified several areas at risk from rising watertables, including extensive *Melaleuca* and native grassland habitats in the Coorong District, and seasonal wetlands and watercourses in the Upper South-East (NLWRA, 2001). The River Murray floodplain in South Australia covers an area of *c.*100 000 ha, including the Ramsar-listed Chowilla and Lindsay Islands wetlands, with an estimated 26 000 ha (25%) currently influenced by salt (MDBMC, 1999). This is likely to increase to 35% by 2020 and 50% by 2100 (MDBMC, 1999). The Chowilla wetland is the single largest natural salt inflow zone along the River Murray; salt inflow is exacerbated by river regulation, reducing flood frequency, and the flushing frequency of salt from the floodplain (Walker *et al.*, 1996). The Chowilla wetlands represent the last part of the lower River Murray that has not been developed for irrigation and retain much of their natural beauty and environmental value, but more than 50% of vegetation in the Chowilla floodplain will be affected by salinity (Anon., 2001).

In Victoria, a three-fold increase in the length of stream or perimeter of reservoir, lake or wetland located in areas of shallow watertable and therefore affected by salinity, is predicted over the coming 50 years in north Central regions, Glenelg and Corangamite regions (Table 10.2). Shallow water tables are predicted to increase under more than 30 000 ha of land around Ramsar-listed wetlands of the Western District lakes during the next 20 years (NLWRA, 2001).

North America

Throughout the continental United States of America about 20 Mha (9%) of the pasture, croplands, and irrigated land is affected by salinisation (Ghassemi *et al.*, 1995). Much of the salinisation processes have occurred west of the 95°N meridian where semi-arid landscapes interface with extensive irrigation, dryland farming and ranching, and where soil profiles are not naturally flushed by rainfall.

The Colorado River system and its surrounding landscape have endured the ultimate decisions of politicians and management personnel (Blinn & Poff, 2005). Regulatory guidelines apportion the use of the river's water for agriculture, mining and municipal activities across semi-arid and arid landscapes from its headwaters in Wyoming and Colorado to its river delta in Mexico, some 2300 km downstream. There are 40 large flow-regulation structures along the Colorado River and six mainstream dams in the lower arid region below Glen Canyon Dam in Arizona (Patrick, 2000; Blinn & Poff, 2005). The high evaporation rates on desert reservoirs and irrigation-return flows have increased river salinities and adversely affected agriculture in the Lower Colorado River Basin. Almost 2.5×10^9 m^3 of annual evaporation occurs in reservoir basins on the Colorado River; by the time the river reaches the border between the USA and Mexico, its waters have specific conductance levels well over 3 g l^{-1} and annual mean streamflow is less than 100 m^3 s^{-1} (Blinn & Poff, 2005). The lower Red River in Texas and Oklahoma is another extreme case of salinisation: the water is saltier than seawater from its load of leached salts.

Two thirds of the USA has saline groundwater with salinity greater than 1 g l^{-1} at depths less than 150 m below the land surface, owing to natural saline groundwater, halite solution, seawater intrusion in the coastal states, oil and gas production, agricultural effluents, saline seeps and road salt (Richter & Kreitler, 1993). Some regions in North and South Dakota have aquifers that exceed 100 g l^{-1} at depths less than 150 m. In New Mexico, over 75% of the groundwater is estimated to be too saline for most uses (Ong, 1988; Richter & Kreitler, 1993). Road salt used for de-icing in cold sections of the USA also accounts for the salt contamination of surface and groundwater in at least 21 States (Richter & Kreitler, 1993).

One of the most productive and diverse agricultural areas in the USA lies in the San Joaquin Valley in the Central Valley of California (Tanji, 1990), but its productivity has come with considerable ecological cost (Chapter 7, this volume). Reduced runoff and salt accretion in natural drainages due to agriculture and irrigation for over a century have increased the salinity of the Upper San Joaquin River from 0.3 g l^{-1} to over 0.6 g l^{-1} total dissolved solids. Currently about 162 000 ha of land in the valley are affected by brackish water tables; with the present expansion rate, over 455 000 ha will become unproductive within the next half-century (www.gps.caltech.edu/~arid/salt/salt.html).

The Imperial Valley in the south-east corner of California near the Mexican border is also a rich agricultural region even though it lies in a hot desert climate. Since 1940, water has been imported from the lower Colorado River in the All-American Canal near Yuma, Arizona, to irrigate over 80 000 ha of crops (Reisner, 1993; Ghassemi *et al.*, 1995). Apart from the problems with salinity, selenium concentrations (up to 2 × 10^{-4} g l^{-1}) are more than twice the maximum permissible median concentration for drinking water (Richter & Kreitler, 1993; Ghassemi *et al.*, 1995).

Although ranching and wheat farming are prevalent, there are fewer signs of secondary salinisation in Canada with its colder climate, lower evaporation rates, and lower density of people (Hammer, 1986). Only 8% of the landmass of Canada is used for agriculture compared with over 45% in the United States (Gleick *et al.*, 2002). However, agricultural activities threaten the biodiversity of watercourses and wetlands through the addition of pesticides, fertilisers and sediment (Abell *et al.*, 2000).

South America

The largest semi-arid region in South America lies south of the Atacama Desert along a narrow 3400 km strip along the west coast in the interior of Chile and Argentina, between about 65 and 70°W longitude (Hammer, 1986). Two smaller semi-arid regions occur in north-eastern Brazil between approximately 38 and 45°W longitude, which extends to the sea, and there is a narrow 1400 km strip along the northern coastlines of Venezuela and Colombia between 62 and 76°W longitude (Chapter 1, this volume).

About 13% or 36 Mha of the landmass in Argentina is under cultivation, of which 1.7 Mha are irrigated (Ghassemi *et al.*, 1995; Gleick *et al.*, 2002). The Provinces of San Juan, Santiago del Estero, and Mendoza are most affected by salinisation, each with at least 55% of their landscapes influenced (Ghassemi *et al.*, 1995). Some rivers in Argentina are affected by natural salt loading and irrigation-return flows. The average salinity of water in the San Juan River is about 0.6 g l^{-1}. Much irrigation uses groundwater resources, owing to the lack of surface water (Ghassemi *et al.*, 1995). Groundwater aquifers are saline and range from 5 to 10 g l^{-1} in total dissolved solids. In addition to problems with salinity, the boron content of the groundwater in some regions is about 0.04 g l^{-1}, or 36 times the recommended upper limit for drinking water (Ghassemi *et al.*, 1995; Gleick *et al.* 2002).

South Africa

Southern and northern areas of South Africa are affected by salinisation. With eutrophication, salinisation is the major cause of the decline of water quality in South Africa (DWA, 1986; Allanson et al., 1990; Du Plessis & Van Veelen, 1991; Ghassemi et al., 1995; NSER, 1999). In industrial areas of the Witwatersrand and Pretoria, mining regions of eastern Transvaal and part of northern Natal, saline wastewater discharge is largely responsible for increases in salinity (Ghassemi et al., 1995). Salinity of discharged wastewater from heavy industry is often about 1 g l^{-1}; that from mining activities often has salinities of 10 g l^{-1} (Williams, 1987). In rural areas, increased salinisation results from natural runoff from marine sediments, exacerbated by irrigation runoff, particularly in semi-arid regions (Davies et al., 1993). For example, the Vaal–Hartz irrigation system has more than trebled the salt concentration of the Hartz River (Allanson et al., 1990). The Great Fish, Sundays, Berg and Breede Rivers, with naturally moderate to high salinity levels, are predisposed to secondary salinisation due to irrigation and river impoundment and show increasingly high salt concentration downstream (Ghassemi et al., 1995).

Central Asia and China

After the collapse of the Soviet Union, five Central Asian countries gained independence: Kazakhstan, Uzbekistan, Kyrgyzstan, Turkmenistan and Tajikistan. These five countries previously constituted the southern arc of the Soviet Union and have a total area of 3 882 000 km^2 and a population of over 53 million people (UNECE, 1998). Central Asia's natural environment is dominated by arid landscapes and all five countries have a significant environmental problem of secondary salinisation. Over 90% of the water from the two major rivers that discharge into the Aral Sea, the Amu Darya River and the Syr Darya River, is diverted for irrigation (Kiessling, 1998). As a consequence, salinity has increased from approximately 0.3 g l^{-1} in the upper reaches to 3.0 g l^{-1} downstream (Ghassemi et al., 1995), rendering the water useless for irrigation. The environmental impact for instream biota is not currently known. The salinity of the Aral Sea, formerly 10 g l^{-1}, has increased over the past 30 years and is currently c. 30 g l^{-1}. The ecological consequences of this diversion and increasing river salinity on the Aral Sea have resulted in an environmental catastrophe. (see, for example, Micklin, 1991; Aladin & Potts, 1992; Aladin *et al.*, 1996; Chapter 7, this volume).

The construction of the Kapchagay reservoir on the Ili River (Kazakhstan) has resulted in reduced water flow into Lake Balkhash and increased salinity. The lake level has fallen by over 2 m, decreasing the lake's surface area of 261 km^2 while increasing salinity from 2 g l^{-1} to 2.7 g l^{-1}. At the same time, aquatic productivity decreased (Peter, 1992). Future salinity may exceed 3 g l^{-1} in the fresher south-western basin and 6 g l^{-1} in the eastern basin (Peter, 1992).

The landmass of China can be divided into two parts along the 400 mm isohyet: east of the line is largely a humid area and to the west are the arid (31% of land mass) and semi-arid (22%) areas of central Asia (Ghassemi *et al.*, 1995). Many of the rivers in the northern and western regions exhibit endorheic drainage. Excessive groundwater extraction has resulted in seawater intrusion in a number of coastal areas (UNDP, 2002). Such intrusion is particularly acute in northern China, which is heavily dependent on groundwater.

Recent estimates put the area of salinised land in China between 80 and 100 million ha (UNDP, 2002) mostly concentrated in the groundwater-irrigated areas in the North China Plain and the irrigated areas in the northwest of Xinjiang (Chapter 7, this volume) and Inner Mongolia. There are no recent data available on the extent of aquatic resources affected by secondary salinisation for the whole of China.

Egypt and the Middle East

Following the construction of the Aswan High dam on the Nile River and significant development of irrigation schemes in the 1960s, water-logging and salinity increased at an alarming rate during the 1970s (Ghassemi *et al.*, 1995). As irrigation drainage water returns to the river, its salinity increases towards Lower Egypt and the delta, where it is about 2.3 g l^{-1}. Nevertheless, the water of much of the Nile is considered of good quality (El-Gabaly, 1977, cited in Ghassemi *et al.*, 1995). In contrast, other aquatic systems have been negatively affected (Chapter 7, this volume). Lake Qarun, southwest of Cairo, formerly a freshwater lake, has become saline from inflow of drainage water (Williams, 2001).

Approximately 42.5 % of the surface of the Middle East is affected by salinisation (Harahsheh & Tateishi, 2000 in UNEP, 2002). Some 20.5% and 33.6% of cultivated land in Saudi Arabia and Bahrain, respectively, are affected by salinisation (UNEP, 2002). Moreover, 8.5 Mha (64%) of the total arable land in Iraq is salinised, and 20%–30% of the irrigated land has been abandoned owing to salinisation. More than 50% of irrigated land of the Euphrates in Syria and Iraq has been significantly affected by salinisation (UNEP, 2002) with devastating ecological consequences (Chapter 7, this volume).

SALINITY AS AN ECOLOGICAL PERTURBATION OF AQUATIC ECOSYSTEMS

What is an ecological perturbation?

An ecological perturbation is a distinct and abnormal change in the properties of an ecological system created by a disturbance (Downes *et al.*, 2002). Perturbations consist of two elements: a *disturbance* and a *response*. The conceptual separation of disturbance and response allows comparisons of how different taxa, populations or ecosystems respond to a similar disturbance.

Disturbances take place over a wide range of spatial and temporal scales. *Pulse* disturbance is short-lived with a sharp peak in intensity, whereas *press* disturbance is a sustained, long-term event that generally builds at a constant rate. *Ramp* describes long-term disturbances that change in intensity with time (Downes *et al.*, 2002). Catastrophes are the most severe disturbance, involving major and widespread destruction of habitat with no potential for the biota to recover (Underwood, 1996). It is a mistake to believe that disturbances and responses occur on the same time frames. A disturbance may be acute (i.e. a pulse), but the ecological impact may be chronic. For example, the recovery of a Louisiana (USA) freshwater marsh from a single saline intrusion event varied markedly over a long period with salinity and water levels (Flynn *et al.*, 1995).

There are three components of biological responses: *inertia, resilience* and *stability* (Underwood, 1989, 1996). *Inertia* refers to the ability of a population to withstand disturbance; the abundance of a population with high inertia does not change with disturbance. *Resilience* is the ability of a population to recover from the disturbance; the abundance of a poorly resilient population will be permanently reduced by the disturbance. *Stability* is the rate at which the abundance of a population recovers from the disturbance; a population with high stability will rapidly return to pre-disturbance abundances. As with ecological disturbances, there is a strong effect of temporal scale on biotic responses. For example, repeated disturbances to a stable population will have little impact if the disturbances are spaced at intervals that allow the population to recover. In an unstable population, subsequent disturbances could occur before the population has recovered and the ecological impact could be much greater.

Responses can be examined at three levels: the *structure* of the ecosystem (usually loss of species), the *process* level and the *landscape* level. Salinity affects rates of key ecosystem-scale processes, such as primary production, nutrient cycling and decomposition. Spatial patterns and heterogeneity are the important guiding factors at the landscape scale (Fisher *et al.*, 2001).

Is NaCl the sole component of a salinity disturbance?

Salinisation is a complex process that often involves ions and synergistic effects on dissolved oxygen, pH, water density, nutrients and bioavailability of heavy metals. NaCl is one salt that contributes to the salinity of aquatic systems. Other ions make up source materials for a wide range of chemicals, fertilisers and other commodities (Watson, 1983), including evaporates based on sulphates (e.g. gypsum), halides (e.g. halite and carnallite), carbonates (e.g. natron), borax and nitrates (e.g. nitre). Nevertheless, rivers, streams, lakes and wetlands in arid and semi-arid zones are commonly dominated ionically by NaCl. For example, total dissolved salts (TDS) concentrations in the natural lentic surface waters of Western Australia ranged from 0.04 to over 350 g l^{-1} and Cl$^-$ was the dominant anion in 18 of the 22 waters, but for South Australian lentic waters Mg^{2+} or Na$^+$ were the most abundant cations and Cl$^-$ or CO$_3^{2-}$ the most abundant anions (Williams, 1967). In the semi-arid regions of North America, although Na$^+$ was the overwhelming dominant cation, lakes above latitude 47°N were typically dominated by SO$_4^{2-}$, and lakes below 47° were more commonly dominated by CO$_3^{2-}$ and Cl (Blinn, 1993). Interestingly, ionic dominance of a single lake can vary in time with freshwater inflows and within a lake (see Halse *et al.*, 1998).

Increases in salinity affect water-column physicochemistry, particularly that of dissolved oxygen, which is also strongly controlled by temperature (Table 10.3). Oxygen deficiencies in waters of arid and semi-arid regions are exacerbated if warm. Salty (260 g l^{-1}) and warm (30 °C) water

Table 10.3. *Predicted air-equilibrium oxygen concentrations (mg O_2 l^{-1}) at three temperatures and six NaCl concentrations*

	Temperature (°C)		
NaCl concentration (g l^{-1})	10	20	30
0	11.3	9.1	7.5
5	11.0	8.9	7.3
10	10.7	8.6	7.1
50	8.3	6.8	5.8
100	6.0	5.0	4.3
200	2.9	2.6	2.3

Source: Derived from Williams (1998a, p. 23).

produce an air-equilibrium oxygen concentration of only 1.6 mg l^{-1}, considerably below the tolerances of almost all invertebrates and fish.

Salinity is also linked to changes in water column pH. Although pH generally increases, as alkalinity increases (Williams, 1998a, b), we detected a significant fall in pH as riverine waters in southern Australia were subjected to increased salinity (in the form of largely NaCl additions). In these field experiments (1991–92, and 1995–96) (K. James, B. Hart, P. Bailey and N. Warwick, unpublished data), we examined the effects of increasing salinity on water plants, aquatic macroinvertebrates, and physicochemical variables in Raftery Swamp, an intermittent floodplain wetland on the Goulburn River in central Victoria (Australia). This site experiences hot, dry summers and extended dry periods (1997–2004) and so was a good model for semi-arid systems. In 1991, after flooding, (salinity of river water *c.* 0.14 g l^{-1}, pH 6.9), the salt concentration in six of nine mesocosms was increased: three replicate areas to 0.6 g l^{-1} (low salt treatment) and three to 1.8 g l^{-1} (high salt treatment). The remaining mesocosms were controls. As the water evaporated (Oct.–Feb. 1991–92), the concentration of salts increased more than three-fold and pH decreased. In the low salt treatment enclosures, pH decreased by 1.4 units to 5.5; in high salt treatment enclosures there was a reduction in pH to 3.8. In contrast, pH in controls and wetland decreased by 0.5 pH units. Similar results were observed during a second experiment conducted at the same location during 1995–96. Subsequent laboratory experiments using dry wetland sediment collected from the same site that was flooded with similar salt concentrations showed a similar decrease in pH. Similarly, pH decreased when dry sediment, collected from 20 different wetland sites from central and western Victoria, was flooded with salt water ($3–15\ g\ l^{-1}$) (P. Bailey, K. James and D. Blinn, unpublished data). Such decreases in pH with increasing salt concentration will synergistically affect wetland animal and plant communities. The disturbance caused by salt and pH may be a press or a ramp.

Salinity and water density are closely related and so increased salinity affects thermal characteristics of lakes, including stratification. Clear skies, intense solar radiation and low humidity produce thermal stratification in surface waters in arid and semi-arid zones. This is most severe in meromictic salt lakes, where the upper layer of less salty water (the mixolimnion) overlies the lower and more salty monimolimnion, producing an upper layer cooler than the deeper waters. This is offset in shallow saline lakes by wind, limiting the depth of the epilimnion (Williams, 1998a), with effects on dissolved oxygen regimes and nutrient fluxes in the lake.

Saline groundwater may also increase NaCl concentration and nitrate, but not nitrite or inorganic phosphorus (Kefford, 1998). Conversely, if waters come from irrigation drainage, the opposite pattern will occur. Given the nature of eutrophication, the disturbance caused by the interaction of salt and nutrients may be a press or a ramp. Sudden, episodic and acute inputs of saline water from irregular flooding or pumping saline groundwater would be repeated press disturbances.

Biotic responses to salinisation

Pre-adaptation of the extant biota

Early debate, at least in Australia, on the impacts of salinisation on the freshwater biota was concerned with the extent of the impact, not with how salinity would affect the biota (Close, 1990). Biota were assumed to be well adapted to the periodically highly saline environments that had existed over the past 6 million years in the Murray–Darling Basin. They were therefore considered 'pre-adapted' to high salinity resulting from secondary salinisation.

We know that freshwater organisms exhibit a range of tolerances to salt levels (Hart *et al.*, 1991; Bailey *et al.*, 2002; Kefford *et al.*, 2003; Nielson *et al.*, 2003). Extreme care is necessary for extrapolating such arguments to secondary salinisation. Take Australia for example. Recent hydrological changes have been considerable (Kingsford, 2000). Salinity does not affect all areas equally; some areas will experience significant increases in salinity (for example, the median salinity of the River Murray at Morgan is estimated to increase by 25% over the next 50 years). These result in the temporal and spatial exposure of freshwater ecosystems and organisms to salinity levels outside their preadaptive flexibility.

Synergistic effects of salinity (see above) affect biota. Changes in salinity and water density can affect the position of non-motile plankton in the water column and the associated energy required to move through a more viscous medium. Decreases in pH with increasing salt concentration will affect aquatic plants and animals. Low pH (pH 5 compared with pH 7.5) significantly affected sexual reproduction (reduced seed production, absence of inflorescences) and vegetative growth (fewer and smaller tubers) in two macrophyte species (*Najas flexilis* and *Vallisneria americana*) (Titus & Hoover, 1993). In general, biodiversity of all taxonomic groups declines as lakes (Schindler, 1994; Baker *et al.*, 1994) or streams (Rosemond *et al.*, 1992) acidify. Changes in behaviour also occur. The number of drifting *Baetis* mayfly nymphs increased as pH was decreased (pH 5 compared with pH 6.4) during a six hour pulse disturbance in streamside channels during a field experiment (Hopkins *et al.*, 1989). The interactions between salinisation and nutrient enrichment particularly affect aquatic plants. For example, nutrient enrichment enhanced growth of *Bolboschoenus medianus* and *Typha domingensis* but not *Baumea arthrophylla* under saline conditions (Morris, 1989). Competitive or fast-growing species with a high nutrient demand may benefit more from nutrient enrichment compared with slow-growing species, but benefits decline as salinities increase (Morris, 1989). Finally, NaCl has synergistic effects with other aquatic pollutants. Zinc and ammonia, for example, seem to be least toxic to fish at salinities of about 30–40% seawater (i.e. *c.* 10–14 g l^{-1}) but phenol and cadmium toxicity increases with salinity (Abel, 1996). Salinity also affects the biological availability of cadmium and mercury and bioaccumulation of selenium (McLusky *et al.*, 1986; Lewis *et al.*, 1999). Such perturbations could be pulse, press or ramp disturbances.

Impacts at the structural level

Our knowledge of salt sensitivity of freshwater organisms comes primarily from presence or absence biotic data and field conductivity data

(Hart *et al.*, 1991; Bailey & James, 1999), with limited experimental data (James & Hart, 1993; Nielsen *et al.*, 2003; Marshall & Bailey, 2004; Brock *et al.*, 2005). Relatively few data exist on the sensitivity of freshwater organisms to increases in salinity (Hart *et al.*, 1991; Bailey *et al.*, 2002; Kefford *et al.*, 2003; Nielsen *et al.*, 2003; Brock *et al.*, 2005). A web-based database contains information on over 1200 Australian taxa (microbes to vertebrates), with notes, references and a statistical summary of species groupings (Bailey *et al.*, 2002). We provide examples from four representative groups (molluscs, mayflies (Ephemeroptera), chirono-mids (Diptera), aquatic plants) from the database to illustrate structural responses of biota to increasing salinity.

There were 57 entries for molluscs from 13 reports, identifying 21 genera, mostly gastropods, with only eight entries for bivalves. Molluscs may be divided into two groups, one limited to salinities of less than 3 g l^{-1}, representing 48% of reported genera, and another with salinity maxima in excess of 8 g l^{-1}, representing 28% of the reported genera (Fig. 10.7a). *Pettancyclus* sp. and *Helicorbis australiansis* were the only taxa limited to salinities of less then 1 g l^{-1}. The most salt-tolerant genus was *Coxiella* sp., occurring at 124 g l^{-1}.

For mayflies, there were 33 entries from 11 reports identifying 11 genera. They were not reported at salinities greater than 6 g l^{-1}, and 45% of genera were only found at salinities less than 0.5 g l^{-1} (Figure 10.7b). *Cloeon* and *Tasmanocoenis* were the most salt-tolerant genera, occurring at *c.* 5 g l^{-1}. Kefford *et al.* (2003) reported a 72 h LC_{50} (lethal concentration to kill 50% of individuals) of over 9 g l^{-1} for *Tasmanocoenis* spp. collected from the Barwon River, Victoria (Australia).

There were 86 entries from ten reports identifying 35 genera of chironomids and other aquatic diptera. Fifty-one percent of the reported genera were limited to salinities less than 2 g l^{-1}; the majority of genera (68%) occurred at salinities of less than 4 g l^{-1} (Fig. 10.7c). Four genera had salinity maxima of 7–8 g l^{-1}, and six genera (17%) were found at salinities in excess of that of seawater (i.e. over 35 g l^{-1}). Chironomids and other aquatic Diptera could be divided into two groups, one restricted to low salinities and another small group tolerant of hypersaline conditions. Among the more salt-sensitive genera were *Microspectra*, *Pseudochironomus*, *Rheocricotopus*, *Tvenetia* and *Cordites*. The most salt-tolerant taxon was *Tanytarsus barbitarsis*, occurring at 100 g l^{-1}.

There were 49 entries for emergent, submerged and floating aquatic plants from 12 reports, identifying 26 genera, but 40% of the

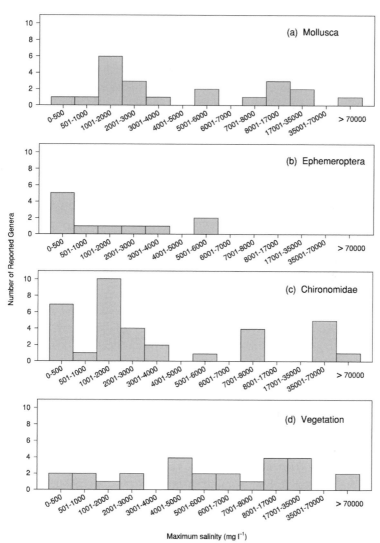

Figure 10.7. The maximum reported salinity for individual genera of three invertebrate groups and vegetation (emergent, submerged and floating). A genus can only occur in one salinity category (data from Bailey *et al.*, 2002).

entries were from two reports on saline wetlands. About 42% of the plant genera were restricted to salinities of less than 5 g l^{-1} (Fig. 10.7d) resulting in the conclusion that macrophytes with freshwater affinities occur at salinities less than 4–5 g l^{-1} (Brock & Lane, 1983). Forty percent of the reported plant genera were in brackish to hypersaline waters (more than 8 g l^{-1}). The most salt-tolerant species were *Ruppia tuberosa* and *Lepilaena preissii*, occurring at 230 g l^{-1} and 150 g l^{-1}, respectively, but able to tolerate large fluctuations in salinities (Brock & Lane, 1983). Other taxa present at high salinities were *Bolboschoenus caldwelli*, *Cyperus gymnocaulus*, *Pachycornia* (Tribe Salicornia) and *Suaeda* (25 g l^{-1}) and *Phragmites australis* (15 g l^{-1}).

At Raftery Swamp, in Victoria (Australia), the total combined biomass (m^{-2}) of all plant species in the wetland was not affected by salinities of up to 5.8 g l^{-1}, but there were differences among species in salinity tolerances. All the freshwater species present under non-saline conditions (*Amphibromus fluitans*, *Potamogeton tricarinatus*, *Triglochin procera* and *Myriophyllum crispatum*) persisted at salinities of 5.8 g l^{-1}. *Amphibromus fluitans* was unaffected by salt, whereas the biomass of *P. tricarinatus* and *M. crispatum* was significantly reduced (James *et al.*, in press). *Triglochin procera* was unaffected by increased salinity, but this may have been due to this species completing most of its growth before the application of salt. Epiphytic diatom communities from *Potamogeton* leaves were also negatively affected by salinity. Diatom densities (mean $\pm$ SE cells g^{-1} DW leaf material) were significantly reduced after 60 d in high-salt (5.8 g l^{-1}, 54 852 $\pm$ 11 850) but not in control (0.3 g l^{-1}; 327 912 $\pm$ 33 624) or low-salt (2 g l^{-1}; 289 620 $\pm$ 16 795) mesocosms. Moreover, the species diversity of diatoms was reduced from 32 and 25 in control and low-salt mesocosms, respectively, to 15 in high-salt mesocosms (James *et al.*, in press).

Sexual reproduction and allocation of biomass to underground vegetative storage organs, such as tubers or rhizomes, can also be reduced, compromising survival of macrophytes over periods of dormancy. At Raftery Swamp, flowering was prevented in *P. tricarinatus* by salt concentrations of 2–3 g l^{-1}; laboratory studies showed that the seed germination rate for *T. procera* was reduced by 50% at 6 g l^{-1}, and almost entirely eliminated at 12 g l^{-1}. Tuber number and biomass of *P. tricarinatus* were significantly reduced by salinity (up to 6 g l^{-1}), respectively by *c.* 70% and 50%. In contrast, *Bolboschoenus medianus* and *Baumea arthrophylla* shifted biomass allocation from culms to tubers with increasing salinity; this strategy may minimise reductions in underground storage tissues imposed by salinity (Morris & Ganf, 2001).

The invertebrate community at Raftery Swamp responded differentially to salt increase. Generally, high salinity reduced the abundances of all animal groups, particularly cladocerans, gastropod snails, some water beetles, dragonflies and some dipterans. Percentage reductions in abundance, compared with controls ranged from 10%–80% at 0.8 g l^{-1} to 32%–100% at 5 g l^{-1}. At the same time, the abundance of some salt-tolerant larvae such as mosquitoes (Culicidae) and some midgeflies (Chironomidae) increased substantially, some by as much as 400%. Such increases in abundances were noticeable at salt concentrations as low as 0.8–1.2 g l^{-1}. Increased salinity significantly reduced the abundance of individuals that emerged from the flooded substrate during recovery trials. Fewer individuals emerged from the high-salinity enclosures compared with low-salinity or control enclosures. This pattern was common across all major taxa including the water beetles, crustaceans and dipteran flies, which commonly rely on resting or resistant life stages to initiate colonisation following the next flood. No snails emerged from the substrate collected from the high-salt enclosures (P. Bailey, K. James and J. Radford, unpublished data).

Even though the salt concentration of flooding water may not cause immediate effect on biota, the concentrating of salt during drawdown imposes a press disturbance. Field experiments conducted in Hughes Creek, Victoria (Australia) (salinity 0.1–0.3 g l^{-1}) examined effects on macroinvertebrates of increased salt concentration (c. 1.0 g l^{-1} and 2.0 g l^{-1} TDS) and mode of salt water release (continuous press release of c. 1.5 g l^{-1} and four separate pulses of c. 3.4 g l^{-1} TDS (Marshall & Bailey, 2004). Results showed that the abundance of the gastropod *Ferrissia tasmanica*, the mayfly *Baetis* spp., and scraper and predator functional groups were significantly reduced at 1.5 g l^{-1}, with the effect exacerbated in pulse treatment channels. Further, the number of drifting animals in the pulse treatment channels significantly increased following saltwater release during daylight hours. In contrast, the abundance of 49 other macroinvertebrate taxa, collector–gatherer functional group and species diversity were unaffected by salt or release schedule. Marshall & Bailey (2004) concluded that much of the existing macroinvertebrate fauna appears to be halotolerant. However, delivering multiple, short pulses (in their case, 4 × c. 12 h duration over 6 days) is more detrimental than delivering the same salt load at a low concentration over a longer duration.

Impacts at the process level

Salinity probably has significant effects on ecosystem-scale biogeochemical processes, such as primary production, decomposition and nutrient cycling in arid and semi-arid regions, although studies are limited. After water, nutrients are possibly the most critical resource in arid and semi-arid zones (Louw & Seely, 1982). Nutrients in soils are usually poor, patchily distributed and often concentrated in the surface layers. There is intense demand for nutrients when plants, particularly annuals, grow rapidly after rainfall. Nutrient demands of medium and long-lived perennial shrubs and trees may also be affected by episodic flooding, responding quickly to increased water availability (Akeroyd *et al.*, 1998; Bornman *et al.*, 2002; Chapter 5, this volume).

Salinity affects osmotic pressure and ionic composition in plants (Taiz & Zeiger, 2002). Dissolved salts in the root zone lower (i.e. make more negative) the osmotic soil water potential, so plants have to maintain an even lower water potential for water to flow from the soil into the leaves. In salinised systems, there can be water physically present but only available at a low water potential. Plants adjust to this relative lack of water availability by increasing the concentration of osmotica (inorganic salts and organic 'compatible' solutes such as pro-line, glycine–betaine and sugars such as sorbitol and mannitol) in their tissues. Specific ion toxicities occur when injurious concentrations of ions, especially Na^+, Cl^- or SO_4^{2-}, occur in the root zone. This may manifest in abnormal $Na^+ : K^+$ ratios in the cytoplasm, inactivating protein synthesis, or progressive replacing of Ca^{2+} by Na^+ in the plasma membrane and changing membrane permeability (Taiz & Zeiger, 2002). As the primary cause of both osmotic effects and ion toxicity, salinisation could be evident as a press or a pulse disturbance, depending on the time frame of exposure.

Salinity probably increases the demand for nutrients by plants and, perhaps, microbes. With increased salinity, some plant families accumulate organic solutes, requiring substantial amounts of carbon for synthesis (up to 10% of plant mass (Taiz & Zeiger, 2002)). Proline and glycine–betaine require high rates of nitrogen acquisition from the sediments. This may alter prior 'balances' between nitrogen assimilation by plants and nitrogen metabolism by the soil microbial communities. For example, increased NH_4^+ uptake by plants may alter the availability of NH_4^+ to nitrifying bacteria (which oxidise NH_4^+ to NO_3^-) and consequently alter rates of denitrification in the soil (the bacterial reduction of NO_3^- to N_2 gas). Salinisation effects on nitrogen availability in aquatic environments in arid and semi-arid zones are probably highly system dependent. Plant growth could be nitrogen-limited only in well-drained sites. In water-logged areas, ammonium may accumulate in the soil with prevalence of reducing conditions (e.g. cessation of nitrification) and plant stresses related to low redox potential and/or flooding (Lewis *et al.*, 1999; Sanchez-Carrillo *et al.*, 2000).

The allocation of nitrogen to compatible osmotica may also reduce resources within the plant for partitioning to nutrient-rich materials such as leaf proteins and reproductive propagules. For example, salinity decreased protein content and decreased the grazing value of Egyptian desert plants (Migahid, 2003). This could affect terrestrial food webs, herbivory and detrital food, given that nitrogen (i.e. protein) is a critical limiting resource for animals, especially the young (Louw & Seely, 1982). Certainly, herbivory is a critical process in semi-arid regions (see, for example, Ludwig *et al.*, 1997).

Salinity probably also affects decomposition rates and pathways in arid and semi-arid zone aquatic systems. Decomposition proceeds by a biogeochemical sequence in aquatic systems, based on the relative availability of alternative electron acceptors. It begins with aerobic decay (O_2 as the oxidant or electron acceptor) and progresses through nitrification (NO_3^-, terminal electron acceptor with organic material as electron donor), metal reduction (Fe^{3-} to Fe^{2-} and Mn^{4-} to Mn^{2-}), sulphate reduction (SO_4^{2-} to H_2S) and, finally methanogenesis (production of CH_4 from acetate or the reduction of CO_2 by H_2). The various organic substrates used as electron donors are produced by fermentative bacteria, which operate at almost all levels of the sequence. Methanogenesis is the dominant anaerobic pathway in freshwater aquatic systems (Boon & Mitchell, 1995; Boon, 2000), but it is replaced by sulphate reduction as the salinity increases, owing to the abundance of SO_4^{2-} in seawater. Bartlett *et al.* (1987) showed a strong negative correlation between methane fluxes and long-term average salinity in a Virginia (USA) salt marsh; although we know of no comparable study for arid or semi-arid zone ecosystems, the pattern may be similar. It is not the NaCl that shifts decomposition pathways but the presence of SO_4^{2-}; methanogenic bacteria seem little affected directly by NaCl up to concentrations of about 0.5 mol l^{-1} (Liu & Boone, 1991). This has major implications for ecosystem-scale processes, primarily because of the extreme toxicity of H_2S to most aquatic organisms (see Ingold & Havill, 1984). H_2S also has biogeochemical significance, as the sulphide quickly reacts with reduced iron to produce ferrous sulphide (FeS) and pyrite (FeS_2) and precipitates cadmium, making such heavy metals unavailable to aquatic biota. Mercury may also be precipitated but, if the SO_4^{2-} concentration is low, it may be methylated by sulphate-reducing bacteria, greatly increasing its biological availability (Lewis *et al.*, 1999).

Increasing salinity also reduced leaf breakdown (*Triglochin procera*, *Phragmites australis*) and microbial activity (estimated from fluorescein diacetate hydrolysis) in shallow wetlands in southeastern Australia, at relatively low conductivities (up to 1.2 g l^{-1}) in field experiments (Roache, 2003). This was more pronounced in the laboratory at higher concentrations (up to 5.4 g l^{-1}) for *T. procera*. Moreover, NaCl inhibited mass loss and microbial activity relative to either sea salt or polyethylene glycol, indicating a possible toxic ionic effect. The additional ions in marine salt may promote alternative bacterial assemblages, altering microbial breakdown, or they may ameliorate the toxic effects of NaCl by catalysing microbial activity (Roache *et al.*, in press). Likewise, increases in salinity may also alter primary production in aquatic ecosystems. Photosynthesis of benthic algae was significantly higher in 50 g l^{-1} salinity-treated mesocosms, compared with those in 75, 100, 125 and 150 g l^{-1} (Herbst & Blinn, 1998).

Although impacts of salinity on trophic interactions and food web dynamics have been rarely studied, disturbance to natural landscapes and drainages, including salinisation, tends to diminish biodiversity in ecosystems (Biswas, 1997; Williams, 1999). The number of taxa within trophic levels of aquatic food webs is reduced dramatically as salinity increases (Hammer, 1986; Hart *et al.*, 1991; Blinn, 1993; Blinn & Bailey, 2001; Herbst, 2001). This may reduce the number of linkages in food webs and carbon transfer within these ecosystems (Martinez, 1994). Reduced biodiversity within primary and intermediate trophic levels reduces the resilience of function within food webs, because of a reduction in the number of alternative pathways available for carbon and energy flow. Community composition of invertebrates differed between three oligosaline (0.8–8.0 mS cm^{-1}) and three mesosaline (8.0–30.0 mS cm^{-1}) lakes in Wyoming High Plains (USA); large grazers / detritivores (gastropods and amphipods) were dominant in oligosaline lakes, whereas small planktivores and their insect predators were common in the mesosaline lakes. The direct physiological effects of increased salinity, as well as a shift in the form of primary production from macrophytes to phytoplankton, could explain changes in community composition; however, salinity effects on invertebrate communities were probably less important to top avian consumers than were the costs of osmoregulation (Wollheim & Louvern, 1995).

Increases in salt concentration from *c.* 0.14 g l^{-1} to 1.5 g l^{-1} significantly reduced the abundance of scraper and predator functional groups in Hughes Creek, Victoria (Australia) (Marshall & Bailey, 2004).

However, whereas the abundance of predators increased during the recovery period (5 days) the abundance of scrapers remained either low or absent in the press and pulse treatment channels, respectively. The abundance of scrapers in Hughes Creek (dominated by gastropods and baetid mayflies) is both spatially and temporally highly variable and can represent over 15% of all macroinvertebrates present (Vertessy, 1995). In this case, the gastropods represent a poorly resilient population whereas the baetid population has high stability. Of significance, however, is the spatial location of potential colonising populations. Biofilm consumption and primary productivity will be affected directly by the reduced population of scrapers resulting from increased salt.

Landscape-scale impacts

There is considerable overlap between ecosystem-scale and landscape-scale impacts of salinity, but we conceptually limit the latter to largely physical considerations and the former to biogeochemical processes. Salinisation could have massive impacts at the landscape level, affecting availability of refugia in arid and semi-arid zones. Opportunistic resources for biota are a key feature of the ecology of these regions after rainfall (see Kingsford & Norman (2002) for waterbirds). Critical refugia 'tide over' the biota during dry periods, (see Chapter 8, this volume), especially in arid and semi-arid regions. Vegetation clearing, cattle grazing, cats, foxes, rabbits, goats and camels, weed invasions, and tourism impacts degrade such refugia (Morton *et al.*, 1995) but, surprisingly, salinisation has rarely been listed as a management problem in arid or semi-arid regions of Australia. Perhaps salinity is not perceived as an issue, given that irrigation is not common.

Salinity affected diatom communities in 39 streams in Victoria, Australia, at a landscape scale. Diatom communities (245 taxa) were strongly correlated with land-use practices, i.e. secondary salinisation and clearing of vegetation. Analysis showed lower diatom species diversity and richness in lowland streams in areas with high secondary salinisation than in streams in catchments with moderate agricultural activity and salinisation (Blinn & Bailey, 2001).

CONCLUSIONS

Almost no text on the pollution of fresh waters considers salt as a significant pollutant. What makes these omissions surprising is the fact that very large areas of the globe are affected by secondary salinisation and that salt can have severe ecological impacts on aquatic and fringing ecotonal communities. We contend that one reason for the lack of appreciation of salt as a significant aquatic pollutant is that secondary salinisation is largely seen as an agricultural problem, with its major impacts being conceived in terms of farm productivity and secondary impacts on potable water quality, rather than as an ecological perturbation. To redress these imbalances, we have provided a synthesis covering (i) the sources of secondary salinisation in arid and semi-arid zone ecosystems; (ii) a brief overview of the global scale of secondary salinisation; and (iii) the impact that salinisation has on the structure and processes of aquatic systems in arid and semi-arid zones.

We contend that salinisation needs to be viewed as an ecological perturbation to freshwater ecosystems that is no different from other pollutants. Its impacts occur over a range of spatial scales, particularly at the landscape level, and are particularly insidious owing to their probable interactive effects with other components or processes within aquatic systems. Implications for biodiversity appear severe. We draw particular attention to likely significant effects on ecosystem processes and implications for trophic structure. Impacts range from changes in species composition to alterations to ecosystem-scale processes, such as primary production and decomposition. The future for arid and semi-arid zones is grim if predictions are correct regarding the extent of secondary salinisation over the next 50 years. In Western Australia, for example, 8 800 000 ha of land could develop dryland salinity by 2050, almost double the current estimate of affected land. Irrigation-derived salinisation is especially problematic as the source water is often saline and there is little ability for rainfall to leach salts out of the soil profile (Waisel, 2002).

The prognosis for irrigation-caused salinisation is so severe that there may be an inevitable succession of 'River diversion, irrigation, salinisation, desertification' (Keyser, 2002). Economic as well as environmental effects are significant. As river salinity increases, desalinisation processes will be required for agricultural and municipal uses because salinities greater than 1 g l^{-1} are generally useless for agriculture (Williams, 1999). This will have economic effects on food production and water consumption. For example, desalinisation of Colorado River water from 2.8 g l^{-1} to $c.\ 0.3 \text{ g l}^{-1}$ total dissolved solids by reverse osmosis cost well over US$250 per acre-foot two decades ago (Pillsbury, 1981). Current estimates for salt removal range from US$70 million to US$300 million dollars per desalinisation plant. Researchers at the Lawrence

Livermore National Laboratory in California are developing a more cost-effective method of desalinisation by using a carbon aerogel electrode, but this technique may not be commercially available for another decade (www.microirrigationforum.com/new/archives/desalt.html).

The challenge for us all is to address these impacts; after all, 47% of the earth's land surface is dryland (Middleton & Thomas, 1997) and secondary salinisation is a phenomenon with global implications.

ACKNOWLEDGEMENTS

Barry Hart and David Herbst are thanked for their insightful and constructive comments, which improved this manuscript.

REFERENCES

Abel, P. D. (1996). *Water Pollution Biology*, (2nd edn). London: Taylor and Francis.

Abell, R. A., Olson, D. M., Dinerstein, E. *et al.* (2000). *Freshwater Ecoregions of North America: a Conservation Assessment*. Washington, DC: Island Press.

Akeroyd, M. D., Tyerman, S. D., Walker, G. R. and Jolly, I. D. (1998). Impact of flooding on the water use of semi-arid riparian eucalypts. *Journal of Hydrology*, **206**, 104–7.

Aladin, N. V. and Potts, W. T. W. (1992). Changes in the Aral Sea ecosystem during the period 1962–1990. *Hydrobiologia*, **237**, 67–79.

Aladin, N. V., Plotnikov, I. S., Orlova, M. I. *et al.* (1996). Changes in the form and biota of the Aral Sea over time. In *The Aral Sea Basin*, ed. P. Micklin and W. D. Williams, pp. 33–55. Berlin: Springer, in cooperation with NATO Scientific Affairs Division.

Allanson, B. R., Hart, R. C., O'Keeffe, J. H. and Robarts, R. D. (1990). *Inland Water of Southern Africa; an Ecological Perspective*. Dordrecht: Kluwer Academic Publishers.

Anon, (2001). *Australia – a salt of the earth*. (Habitat Australia, series.) Melbourne: Australian Conservation Foundation.

Bailey, P. C. E. and James, K. A. (1999). Riverine and wetland salinity impacts – assessment of R & D needs. (Occasional Papers series, Paper No. 25/99.) Canberra: Land and Water Resources Research Development Corporation.

Bailey, P. C. E., Boon, P. I. and Morris, K. (2002). *Australian Biodiversity – Salt Sensitivity Database*. Canberra:Land and Water Australia. www.rivers.gov.au/research/contaminants/saltsen.htm.

Baker, J. P., Bohmer, J., Hartmann, A. *et al.* (1994). Physiological and ecological effects of acidification on aquatic biota. In *Acidification of Freshwater Ecosystems: Implications for the Future*, ed. C. E. W. Steinberg and R. F. Wright, pp. 275–312. New York: John Wiley and Sons.

Barr, N. and Cary, J. (1992). *Greening a Brown Land: the Australian Search for Sustainable Land Use*. South Melbourne: Macmillan.

Bartlett, K. B., Bartlett, D. S., Harriss, R. C. and Sebacher, D. I. (1987). Methane emissions along a salt marsh salinity gradient. *Biogeochemistry*, **4**, 183–202.

Bell, F. G. (1998). *Environmental Geology: Principles and Practice*. Oxford: Blackwell.

Biswas, A. K. (1997). *Water Resources: Environmental Planning, Management, and Development*. New York: McGraw-Hill.

Blinn, D. W. (1993). Diatom community structure along physiochemical gradients in saline lakes. *Ecology*, **74**, 1246–63.

Blinn, D. W. and Bailey, P. C. E. (2001). Land-use influence on stream water quality and diatom communities in Victoria, Australia: a response to secondary salinisation. *Hydrobiologia*, **466**, 231–44.

Blinn, D. W. and Poff, N. L. (2005). Colorado River Basin. In *Rivers of North America*, ed. A. C. Benke and C. E. Cushing, pp. 483–540. New York: Academic Press.

Boon, P. I. (2000). Carbon cycling in Australian wetlands: the importance of methane. *Verhandlungen der Vereinigung für Internationale Limnologie*, **27**, 1–14.

Boon, P. I. and Mitchell, A. (1995). Methanogenesis in the sediments of an Australian freshwater wetland: comparison with aerobic decay, and factors controlling methanogenesis. *FEMS Microbiology Ecology*, **18**, 175–90.

Bornman, T. G., Adams, J. B. and Bate, G. C. (2002). Freshwater requirements of a semi-arid supratidal and floodplain salt marsh. *Estuaries*, **25**, 1394–405.

Boulton, A. J. and Brock, M. A. (1999). *Australian Freshwater Ecology: Processes and Management*. Glen Osmond: Gleneagles Publishing.

Brock, M. A. and Lane, J. A. K. (1983). The aquatic macrophyte flora of saline wetlands in Western Australia in relation to salinity and permanence. *Hydrobiologia*, **105**, 63–76.

Brock, M. A., Nielson, D. L. and Crosslé, K. (2005). Changes in biotic communities developing from freshwater wetland sediments under experimental salinity and water regimes. *Freshwater Biology*, **50**, 1376–90.

Burgman, M. A. and Lindenmayer, D. B. (1998). *Conservation Biology for the Australian Environment*. Chipping Norton: Surrey Beatty and Sons.

Calow, P. and Petts, G. E. (1998). *The Rivers Handbook: Hydrological and Ecological Principles*. Oxford: Blackwell Science.

Close, A. (1990). River salinity. In *The Murray*, ed. N. Mackay and D. Eastburn, pp. 127–44. Canberra, ACT: Murray-Darling Basin Commission.

Connell, W. D (1981). *Water Pollution: Causes and Effects in Australia and New Zealand*. St Lucia, Queensland: University of Queensland Press.

Crabb, P. (1997). *Murray-Darling Basin Resources*. Canberra, ACT: Murray-Darling Basin Commission.

Cramer, V. A. and Hobbs, R. J. (2002). Ecological consequences of altered hydrological regimes in fragmented ecosystems of southern Australia – impacts and possible management responses. *Austral Ecology*, **27**, 546–64.

Davies, B. R., O'Keeffe J. H. and Snaddon C. D. (1993). Water chemistry. In *A Synthesis of the Ecological Functioning, Conservation and Management of South African River Ecosystems*, pp. 61–7. Pretoria: Water Research Commission.

De Deckker, P. (1983). Australian salt lakes: their history, chemistry, and biota – a review. *Hydrobiologia*, **105**, 231–44.

Dimmock, G. M., Bettenay, E. and Mulcahy, M. J. (1974). Salt content of lateritic profiles in the Darling Range, Western Australia. *Australian Journal of Soil Research*, **12**, 63–9.

Downes, B. J., Barmuta, L. A., Fairweather, P. G. *et al.* (2002). *Monitoring Ecological Impacts: Concepts and Practice in Flowing Waters*. Cambridge: Cambridge University Press.

Du Plessis, H. M. and Van Veelen, M. (1991). Water quality: salinisation and eutrophication time series and trends in South Africa. *South African Journal of Science*, **87**, 11–16.

DWA (Department of Water Affairs) (1986). *Management of the water resources of the Republic of South Africa*. Cape Town: CTP Book Printers.

Evans, R., Brown, C. and Kellett, J. (1990). Geology and groundwater. In *The Murray*, ed. N Mackay and D. Eastburn, pp. 76–93. Canberra, Australia: Murray-Darling Basin Commission.

Fisher, S. G., Welter, J., Schade, J. and Henry, J. (2001). Landscape challenges to ecosystem thinking: creative flood and drought in the American Southwest. *Scientia Marina*, **65**, 181–92.

Flynn, K. M., McKee, K. L. and Mendelssohn, I. A. (1995). Recovery of freshwater marsh vegetation after a saltwater intrusion event. *Oecologia*, **103**, 63–72.

Froend, R. H. and McComb, A. J. (1991). An account of the decline of Lake Towerrinning, a wheatbelt wetland. *Journal of the Royal Society of Western Australia*, **73**, 123–8.

Ghassemi, F., Jakeman, A. J. and Dix, H. A. (1995). *Salinisation of Land and Water Resources: Human Causes, Extent, Management and Case Studies*. Sydney: University of New South Wales Press.

Gleick, P., Burns, W. C. G., Chalecki, E. L., *et al.* (2002). *The World's Water: the Biennial Report on Freshwater Resources, 2002–2003*. Washington, DC: Island Press.

Government of Western Australia (1996). *Western Australian Salinity Action Plan*. Government of Western Australia, Perth.

Halse, S. A., Shiel, R. J. and Williams, W. D. (1998). Aquatic invertebrates of Lake Gregory, northwestern Australia, in relation to salinity and ionic composition. *Hydrobiologia*, **381**, 15–29.

Hammer, U. T. (1986). *Saline Lake Ecosystems of the World*. The Hague: W. Junk Publishers.

Harahsheh, H. and Tateishi, R. (2000). Environmental GIS database and desertification mapping of West Asia. Paper presented at the Workshop of the Asian Region Thematic Programme Network on Desertification Monitoring and Assessment, Tokyo, 28–30 June 2000. Tokyo: United Nations University.

Hart, B. T., Bailey, P., Edwards, R. *et al.* (1991). A review of the salt sensitivity of the Australian freshwater biota. *Hydrobiologia*, **210**, 105–44.

Hatton, T. J. and Nulsen, R. A. (1999). Towards achieving functional ecosystem mimicry with respect to water cycling in southern Australian agriculture. *Agroforestry Systems*, **45**, 203–14.

Heathcote, R. L. (1983). *The Arid Lands: the Use and Abuse*. London: Longman.

Herbst, D. B. (2001). Gradients of salinity stress, environmental stability and water chemistry as a template for defining habitat types and physiological strategies in inland salt waters. *Hydrobiologia*, **466**, 209–19.

Herbst, D. B. and Blinn, D. W. (1998). Experimental mesocosm studies of salinity on the benthic algal community of a saline lake. *Journal of Phycology*, **34**, 772–8.

Hopkins, P. S., Krants, K. W. and Cooper, S. D. (1989). Effects of an experimental acid pulse on invertebrates in a high altitude Sierra Nevada stream. *Hydrobiologia*, **171**, 45–58.

Hynes, H. B. N. (1978). *The Biology of Polluted Waters* (6th edn). Liverpool: Liverpool University Press.

Ingold, A. and Havill, D. C. (1984). The influence of sulfide on the distribution of higher plants in salt marshes. *Journal of Ecology*, **72**, 1043–54.

James, K. R. and Hart, B. T. (1993). Effect of salinity on four freshwater macrophytes. *Australian Journal of Marine and Freshwater Research*, **44**, 769–77.

James, K. R., Hart, B. T., Bailey, P. C. E. and Blinn, D. W. (2005). Impact of secondary salinisation on freshwater ecosystems: effect of experimentally increasing salinity on an intermittent floodplain wetland: water chemistry, aquatic plants and epiphytic diatoms. *Marine and Freshwater Research*, in press.

Kefford, B. J. (1998). Is salinity the only water quality parameter affected when saline water is disposed in rivers? *International Journal of Salt Lake Research*, **7**, 285–300.

Kefford, B. J., Papas, P. J. and Nugegoda, D. (2003). Relative salinity tolerance of macroinvertebrates from the Barwon River, Victoria, Australia. *Marine and Freshwater Research*, **54**, 755–65.

Keyser, D. (2002). River diversion, irrigation, salinisation, desertification – the inevitable succession. In *Sustainable Land Use in Deserts*, ed. S.-W. Breckle, M. Veste and W. Wucherer, pp. 14–23. Berlin: Springer-Verlag.

Kiessling, K. L. (1998). *Alleviating the Consequences of an Ecological Catastrophe*. (Conference on the Aral Sea – Women, Children, Health and Environment, Riksplan, Stockholm, 1998). Stockholm: Swedish UNIFEM Committee.

Kingsford, R. T. (2000). Ecological impacts of dams, water diversions and river management on floodplain wetlands in Australia. *Austral Ecology*, **25**, 109–27.

Kingsford, R. T. and Norman, F. I. (2002). Australian waterbirds – products of the continent's ecology. *Emu*, **102**, 47–69.

Laws, E. A. (2000). *Aquatic Pollution: an Introductory text* (3rd edn). New York: John Wiley and Sons.

Lewis, M. A., Mayer, F. L., Powell, R. L. *et al.* (1999). *Ecotoxicology and Risk Assessment for Wetlands*. Montana: SETAC Press.

Liu, Y. and Boone, D. R. (1991). Effects of salinity on methanogenic decomposition. *Bioresource Technology*, **35**, 271–3.

Louw, G. and Seely, M. (1982). *Ecology of Desert Organisms*. London: Longman.

Ludwig, J., Tongway, D., Freudenberger, D., Noble, J. and Hodgkison, K. (eds) (1997). *Landscape and Ecology: Function and Management. Principles from Australia's Rangelands*. Canberra: CSIRO.

Marshall, N. A. and Bailey, P. C. E. (2004). Impact of secondary salinisation on freshwater ecosystems: effects of contrasting, experimental short-term releases of saline wastewater on macroinvertebrates in a lowland stream. *Marine and Freshwater Research*, **55**, 509–23.

Martinez, N. D. (1994). Scale-dependent constraints on food-web structure. *American Naturalist*, **144**, 935–53.

Mason, C. F. (2002). *Biology of Freshwater Pollution* (4th edn). Harlow: Pearson Education.

McLusky, D. S., Bryant, V. and Campbell, R. (1986). The effects of temperature and salinity on the toxicity of heavy metals to marine and estuarine invertebrates. *Oceanography and Marine Biology Annual Review*, **24**, 481–520.

MDBMC (Murray-Darling Basin Commission) (1999). *The Salinity Audit of the Murray-Darling Basin*. Canberra, ACT: Murray-Darling Basin Commission.

Micklin, P. (1991) *The Water Management Crisis in Soviet Central Asia*. (The Carl Beck Papers in Russian and East European Studies, no. 905.) University of Pittsburgh Centre for Russian and East European Studies.

Middleton, N. J. and Thomas, D. S. G. (1997). *World Atlas of Desertification* (2nd edn). United Nations Environment Program. London: Edward Arnold.

Migahid, M. M. (2003). Effect of salinity shock on some desert species native to the northern part of Egypt. *Journal of Arid Environments*, **53**, 155–67.

Morris, K. (1989). Salinity and nutrients; growth and water use of aquatic macrophytes under controlled and natural conditions. Ph.D. thesis, University of Adelaide.

Morris, K. and Ganf, G. (2001). The response of an emergent sedge *Bolboschoenus medianus* to salinity and nutrients. *Aquatic Botany*, **70**, 311–28.

Morton, S. R., Short, J. and Barker, R. D. (1995). *Refugia for Biological Diversity in Arid and Semi-arid Australia*. Canberra: CSIRO Division of Wildlife and Ecology.

Nielsen, D. L., Brock, M. A., Crosslé, K., Harris, K., Healy, M. and Jarosinski, I. (2003). The effects of salinity on aquatic plant germination and zooplankton hatching from two wetland sediments. *Freshwater Biology*, **48**, 2214–23.

NLWRA (National Land and Water Resources Audit) (2001). *Australian Dryland Salinity Assessment 2000*. Canberra, ACT: National Land and Water Resources Audit.

NSER (1999). *Freshwater Systems and Resources - National State of the Environment Report*. Pretoria, South Africa: Water Research Commission.

Ong, K. (1988). New Mexico ground-water quality. In *National Water Summary 1986: Hydrological Events and Groundwater Quality*, eds. D. W. Moody, J. Carr, E. G. Chase, and R. W. Paulson, pp. 377–84. (United States Geological Supply Paper, 2325.) Washington, DC: US Geological Survey.

Patrick, R. (2000). *Rivers of the United States*, vol. 5, Part A, *The Colorado River*. New York: John Wiley and Sons.

Peck, A. J. (1996). Salinity. In *Land Degradation Processes in Australia*, ed. G. H. McTainsh and W. C. Broughton, pp. 234–70. Melbourne: Longman.

Peter, I. (1992). Lake Balkhash, Kazakhstan. *International Journal of Salt Lake Research*, **1**, 21–46.

Pillsbury, A. F. (1981). The salinity of rivers. *Scientific American*, **245**, 55–65.

Reisner, D. (1993). *Cadillac Desert: the American West and its Disappearing Water*. New York: Penguin Books.

Richter, B. C. and C. W. Kreitler. (1993). *Geochemical Techniques for Identifying Sources of Ground-water Salinisation*. Boca Raton, FL: CRC Press.

Roache, M. (2003). Investigation of the effects of increasing salt concentration on macrophyte litter breakdown in a freshwater wetland. B.Sc. Honours Thesis, Monash University.

Roache, M. C., Bailey, P. C. E. and Boon, P. I. (2006). Effects of salinity on the decay of the freshwater macrophyte, *Triglochin procerum*. *Aquatic Botany* (in press).

Rosemond, A. D., Reice, S. R., Elwood, J. W. and Mulholland, J. (1992). The effects of acidity on benthic invertebrate communities in the south-eastern United States. *Freshwater Biology*, **27**, 193–210.

Ruprecht, J. K. and Schofield, N. J. (1991). Effects of partial deforestation on hydrology and salinity in high storage landscapes. *Journal of Hydrology*, **129**, 19–55.

Salma, R. B. and Bartle, G. (1995). *Past, present and future groundwater level trends in the wheatbelt of Western Australia*. (Technical Memorandum No. 95.10.) Canberra: CSIRO, Division of Water Resources.

Sanchez-Carrillo, S., Alvarez-Cobelas, M., Cirujano, S. *et al.* (2000). Rainfall-driven changes in the biomass of a semi-arid wetland. *Verhandlungen der Vereinigung für Internationale Limnologie*, **27**, 1690–1694.

Sanders, A. (1991). *Oral histories documenting changes in wheatbelt wetlands*. (Occasional Paper no. 2/91). Perth, Australia: Department of Conservation and Land Management.

Schindler, D. W. (1994). Changes caused by acidification to the biodiversity, productivity and biogeochemical cycles of lakes. In *Acidification of Freshwater Ecosystems: Implications for the Future*, ed. C. E. W. Steinberg and R. F. Wright, pp. 275–312. New York: John Wiley and Sons.

Smith, D. I. (1998). *Water in Australia*. Oxford: Oxford University Press.

Taiz, L. and Zeiger, E. (2002). *Plant Physiology*, 3rd edn. Sunderland, MA: Sinauer Associates.

Tanji, K. K. (ed.) (1990). *Agriculture Salinity Assessment and Management*. (American Society of Civil Engineers Reports on Engineering Practice no. 71.) New York: American Society of Civil Engineers.

Titus, J. E. and Hoover, D. T. (1993). Reproduction in two submersed macrophytes declines progressively at low pH. *Freshwater Biology*, **30**, 63–72.

Underwood, A. J. (1989). The analysis of stress in natural populations. *Biological Journal of the Linnean Society*, **37**, 51–78.

Underwood, A. J. (1996). Spatial and temporal problems with monitoring. In *River Restoration*, ed. G. Petts and P. Calow, pp. 182–204. Oxford: Blackwell.

UNDP (United Nations Development Program) (2002). *China Human Development Report*. pp. 17–38. Stockholm: United Nations Development Program.

UNECE (United Nations Economic Commission for Europe) (1998). Central Asia: environmental assessment. Background document, Fourth Ministerial Conference, Arhus, 23–25 April, 1998. www.mem.dk/aarhus-conference/issues/NIS/assesasia.htm.

UNEP (United Nations Environment Program) (2002). *GEO-3 Total Environment Outlook 3: Past, Present and Future Prospects*. Stockholm: United Nations Environment Program.

Vertesy, D. (1995). Effects of increasing stream salinity on a typical lowland macroinvertebrate community, Hughes Creek, Victoria. M.Sc. thesis, Monash University.

Waisel, Y. (2002). Salinity: a major enemy of sustainable agriculture. In *Sustainable Land Use in Deserts*, ed. S.-W. Breckle, M. Veste and W. Wucherer, pp. 166–73. Berlin: Springer.

Walker, G. R., Jolly, I. D. and Jarwal, S. D. (eds) (1996). *Salt and Water Movement in the Chowilla Floodplain*. Canberra: CSIRO.

Watson, J. (1983). *Geology and Man*. London: George Allen and Unwin.

Williams, W. D. (1967). The chemical characteristics of some surface waters in Australia. In *Australian Inland Waters and their Fauna*, ed. A. H. Weatherley, pp. 18–77. Canberra: Australian National University Press.

Williams, W. D. (1987). Salinisation of rivers and streams: an important environmental hazard. *Ambio*, **16**, 180–15.

Williams, W. D. (1988). Limnological imbalances: an antipodean view point. *Freshwater Biology*, **20**, 407–20.

Williams, W. D. (1995). Dryland rivers: a brief global review. *Australian Biologist*, **8**, 175–80.

Williams, W. D. (1998a). *Guidelines of Lake Management*. Shiga: International Lake Environment Committee Foundation.

Williams, W. D. (1998b). Salinity as a determinant of the structure of biological communities in salt lakes. *Hydrobiologia*, **381**, 191–201.

Williams, W. D. (1999). Salinisation: a major threat to water resources in the arid and semi-arid regions of the world. *Lakes and Reservoirs: Research and Management*, **4**, 85–9.

Williams, W. D. (2001). Anthropogenic salinisation of inland waters. *Hydrobiologia*, **466**, 329–37.

Williamson, D. R. (1990). Salinity – an old environmental problem. In *Year Book Australia 1990*, pp. 202–11. Canberra, Australia: Government Publishing Service.

Williamson, D. R., Gates, G. W. B., Robinson, G. *et al.* (1997). *Salt trends. Historic trend in salt concentration and salt load of stream flow in the Murray-Darling Drainage Basin*. (Dryland Salinity Report no. 1). Canberra, ACT: Murray-Darling Basin Commission.

Wolheim, W. M. and Louvern, J. R. (1995). Salinity effects on macroinvertebrate assemblages and waterbird food webs in shallow lakes of the Wyoming High Plains. *Hydrobiologia*, **310**, 207–23.

Water scarcity: politics, populations and the ecology of desert rivers

M. J. WISHART

INTRODUCTION

In the world's desert regions, where rainfall is less than 500 mm yr^{-1} (Chapter 1, this volume), the spatial and temporal variability associated with water provides a complex template for ecosystem functioning (Davies *et al.*, 1995; Ward *et al.*, 1999; Sheldon *et al.*, 2002; Chapter 2, this volume). This has given rise to unique systems within which a diversity of specially adapted species live (Davies *et al.*, 1995; Chapters 4–7, this volume). Spatial and temporal variability in the availability of fresh water also results in a scarcity that has provided a focal foundation for the development and sustenance of human civilisation. The need to constrain this variability and secure water to sustain the human population has resulted in large-scale manipulation of the world's waterways (see McCully, 1996 for review). Increasingly, the historical legacy of human impoundment, storage and diversion of water is impinging upon the ecological integrity of these systems and placing human needs in direct competition with those required to sustain ecosystem functioning (see McCully, 1996; Postel, 1996; Dyson *et al.*, 2003; Chapter 8, this volume) (Fig. 11.1).

Lacking the specialised physiological adaptations to cope with the scarcity of water in desert environments, human populations developed

Ecology of Desert Rivers, ed. R. T. Kingsford. Published by Cambridge University Press.

Figure 11.1. People collect drinking water from deep within the dry bed of the Changara River in central Mozambique (Photo M. Wishart). Lack of water in the world's desert rivers, poverty and lack of access to potable water means many are forced to obtain water from within dry river beds.

sociopolitical, religious and cultural systems that were interwoven with the fabric and development of civilisation. In the absence of appropriate technologies, the distribution and scarcity of fresh water has historically limited the extent of human settlement and exploitation. Early desert communities were forced to lead a nomadic existence, searching for suitable supplies of potable water (Fig. 11.1). Subsequent development of more permanent agrarian societies was limited to areas with perennial sources of water and relatively stable climates (Postel, 1996). The permanence of these societies was determined largely by the predictability and availability of water (Fig. 11.2), with access and utilisation of water typically governed by strict rules and regulations. These have been central to the evolution of social, cultural and religious structures. The early Puebloan communities in north-western Mexico, for example, would avoid the constraints imposed by the scarcity of water by occupying their villages for one generation before abandoning them (DuMars et al., 1984).

The subsequent growth and development of more permanent agrarian societies was aided by developments in engineering that

Figure 11.2. Movements and use of water by nomadic societies were primarily governed by the natural variability of floods and droughts. During floods, water is plentiful in this wetland for this Botswana man (Photo M. Wishart).

enabled human societies to exert increasingly greater control over the distribution and variability of water (Figs. 11.3 and 11.4). In so doing they were able to ensure more secure supplies of water, which could also be used to dispose of waste materials. Such technological advances also required the development of more complex sociopolitical and cultural systems. Early Hohokam societies in North America developed elaborate canal systems along which villages were built on earthen platform-mounds roughly every five kilometres (Elson, 1998). Despite the relative technological sophistication of these systems, their complexity and orientation necessitated the development of intricate sociopolitical structures. These were centred on the recognition that water was a basic resource of critical importance. The theocratic or secular elite of these platform mound villages controlled the construction of these canals as well as the maintenance and allocation of water.

The demand for water has increased with the growth and development of human civilisation. Concomitant with this growth in demand has been the need to evolve new approaches to management that continue to recognise the importance of water and integrate with the sociopolitical and cultural fabric of society. With an increasing global population, the scarcity and variability of water associated with the

Figure 11.3. Increasing pressure is exerted on the development of desert rivers for irrigation. In arid Australia, large areas on the floodplains are developed for irrigating cotton, such as here on the Condamine–Balonne River (Photo R. T. Kingsford). Water is captured by pumps during floods and stored in large off-river storages for later use. This has considerable ecological effects on downstream ecosystems and also deprives floodplain graziers of the flooding necessary for rearing cattle.

natural flow regime of desert rivers has proved contrary to the security of supply required by human civilisations (Walker *et al.*, 1997). In response human civilisations have exerted increasing control over the world's rivers (Figs. 11.3 and 11.4). The resulting technologies have constrained the inherent spatial and temporal variability, permitting a level of control, storage and redistribution of water previously unseen (Boon *et al.*, 1992; McCully,1996; Davies *et al.*, 2000; Chapter 8, this volume).

The first evidence for such technological manipulations dates back to Jordan around 3000 BC, although it is likely that this history can be traced to the irrigation systems of the early agrarian societies of Mesopotamia, around 8000 years ago. Such developments reflected a significant shift in the way in which human civilisations manipulated water, with these initial simple rock and earth impoundments giving rise to a proliferation of more sophisticated dam developments. Today there are estimated to be more than 40 000 large dams (e.g. the Dartmouth Dam) (Fig. 11.4). About 87% of these have been built since the 1950s (ICOLD, 1989), reflecting the advent of modern supply-side

Figure 11.4. Dartmouth Dam in the headwaters of the River Murray in southeastern Australia (Photo R. T. Kingsford), with other dams, controls river flows in the desert regions of this river, allowing water to be diverted, primarily for irrigation.

paradigms that persist today. Such philosophies ensure that sufficient supplies of water are available to meet increasing human demands.

With today's large and rapidly increasing population, particularly in areas with low and variable rainfall, the challenges facing the management of water are greater than ever. Contemporary solutions to the problems of supply have been dominated by the manipulation and modification of the world's rivers, for consumptive purposes, transfers, augmentation, hydro-power, flood control or extraction for agriculture and industry (Chapter 8, this volume), at a scale never before experienced. However, the scarcity of water continues to affect, or be affected by, broad social, political, economic and, increasingly, environmental issues. In an era of increasing social awareness, the reality of our current inability to meet basic human demand is juxtaposed against increasing conflicts among urban, agricultural, industrial, indigenous and environmental requirements. The back drop for these competing uses of the world's limited water are increasingly intruding upon the human consciousness, manifest through declining water quality and availability of supply, famine, new and increasing outbreaks of water-related diseases, conflicts, and growing numbers of endangered and threatened species (Boon *et al.*, 2000; Gleick *et al.*, 2001; UN/WWAP,

Figure 11.5. Fishers process their catch of sharptooth catfish *Clarius gariepinus* in southern Zimbabwe. Fisheries provide an important source of protein in areas prone to cyclical droughts and crop failure (Photo M. Wishart).

2003). Upstream control of desert rivers affects communities that live along these rivers, rely predominantly on fish products (Fig. 11.5) and use the river for bathing and washing (Fig. 11.6).

Given the increasing competition between different water users, this chapter examines the challenges in managing rivers in desert environments. Issues pertaining to desert systems are typical of the issues facing rivers worldwide. Given the magnitude of the variability

Figure 11.6. River flows in desert rivers are essential for existence of communities. During low flows, people wash and collect drinking water from the receding waters of the Changara River in the central Tete Province, Mozambique (Photo M. Wishart).

associated with these systems and the scarcity of water in desert regions, manifestation of the problems and development of appropriate solutions will be, and are being, found among the world's desert river systems. Although agriculture remains the largest single user of water (Fig. 11.3), the focus of this chapter is on individual human requirements juxtaposed against the consequences facing the biodiversity of freshwater environments. It is argued that approaches to management need to evolve to again reflect the broad social and cultural values of society. The shifting political framework governing the management of water is examined to determine how much current practices reflect contemporary values and the realities of water scarcity.

SCARCITY

Traditionally, the concept of scarcity has been interpreted as the insufficient quantity of a particular resource to meet demand. However, scarcity depends on the context. Human scarcity is perceived differently from environmental scarcity, and these in turn can be variously defined to include temporal and spatial components that may be life-threatening, seasonal, temporary or cyclical. Such scarcity may be perceived in absolute terms, pertaining to quantity, or more relative terms, relating to quality.

Similarly, the point at which scarcity is realised will differ. Different needs at individual, community, provincial or national levels influence scarcity in the same way that species, habitats and ecosystems define scarcity. The influence of human populations over the hydrological cycle can also substantially influence the extent of scarcity. With almost all of the world's major rivers shared by two or more countries, scarcity within a catchment can be affected by national economic and development strategies. Scarcity often arises because socioeconomic trends typically have little to do with basic needs. At the other end of the continuum, people in many rural communities have to meet their daily needs on less than 20 l of water (Figs. 11.1, 11.4 and 11.5), often from sources that others would deem unfit for consumption. Access to potable water introduces another dimension; more than seven million people in Mozambique alone spend more than two hours a day collecting water (Fig. 11.1). In contrast, the average person in the United Kingdom uses 37 l of potable water a day with the flush of a conventional toilet.

The widespread distribution and visual plenitude of water, highlighted by the fact that water covers 70% of the earth's surface, belies the fact that only 2.53% is considered fresh and fit for human consumption (Groombridge & Jenkins, 1998). Two thirds is trapped within glaciers and permanent snow cover; the remainder is unevenly distributed across the world's continents. Only a minuscule 0.000 2% of the world's water flows through rivers and streams, representing 0.006% of the available freshwater (Shiklomanov, 2000). Of this, 15% is found within North and Central America, 26% in South America, 8% in Europe, 11% in Africa, 36% in Asia and 5% in the world's driest inhabited continent, Australia (UN/WWAP, 2003). Here the concept of scarcity needs to be juxtaposed with population levels and the degree of development. For example, Asia has more than a third of the world's available water but it is also home to nearly 60% of the world's population. Calculating thus, the relative availability of water per head of population in this region is lower than any other region. In contrast, Australia and the Oceanic region has only 5% of the world's total water but, with less than 1% of the world's population, has a high ratio of available water proportional to its population. Given the size of the European population and the available volume of water, Europe has a ratio of available water per head of population similar to that of Asia. Greater infrastructural development in Europe ensures that almost everyone has access to safe supplies of water. In contrast, Africa has a higher relative proportion of water available to the number of people on the continent, but poor infrastructure and low levels of development mean that large numbers of people are without access to safe potable supplies of water.

As a resource with increasing political value, scarcity is often perceived at political scales. There are 261 international rivers with catchments encompassing 145 different nations (Postel, 1996). In desert regions the nature of the drainage pattern is such that many of the rivers arise elsewhere, in more well-watered parts of the catchment. Different levels of development within these catchments accentuate the existing problems of a limited volume of available water. Occupying an area of 3 349 000 km^2, about one tenth of continental Africa, the Nile River is a classic example of many of the problems inherent in catchments that drain into these arid areas. Including parts of Tanzania, Burundi, Rwanda, Zaire, Kenya, Uganda, Ethiopia, the Sudan and Egypt, the river provides an important arterial link, transporting and delivering not only water but also important agricultural inputs. The rich sediments transported along the river ensure the annual renewal of some of Africa's most fertile plains. Developments within the catchment that deprive downstream nation states of these valuable resources are further accentuated in the more arid regions by evaporative losses. Efforts to reach a negotiated settlement on the distribution and access to water from within the basin are hampered by different degrees of development within the basin states and the limited available resources.

In a region where political tensions already run high, the Middle East exemplifies many of the perceptions of scarcity on a regional scale and highlights the constraints imposed by the uneven distribution of water. Whereas Turkey is often perceived by other countries in the Middle Eastern region as having abundant water, in reality it has about one fifth the water available in water-rich regions such as North America and Western Europe (10 000 m^3 per person per year) (Vallée & Margat, 2003). Rainfall in Turkey yields 501 billion cubic metres of water each year; however, just under two thirds of this is lost through evaporation. A total of 186 billion cubic metres remains as surface runoff, of which only 98 billion cubic metres is available for economic use. Much of this water supplies the once extensive Mesopotamian wetlands downstream (Chapter 8, this volume). Stored surface water is supplemented by a further 12 billion cubic metres of underground water. However, all of this is of little relevance as infrastructural constraints mean that only 40 billion cubic metres of water can actually be used. While the remaining 70 billion cubic metres are currently unused, this water represents a valuable resource for Turkey's economic development but is also a source of tension with downstream countries. Despite the absolute scarcity, Turkey is perceived as water-rich because of its position within the region and lack of infrastructural development.

The human context

The world's population has tripled in the last century, to number over six billion, while the use of water has grown six-fold (UN/WWAP, 2003). By the middle of this century, an additional seven billion people in sixty different countries may experience water scarcity of some kind (UN/WWAP, 2003). Even the most optimistic estimates place this number at around two billion people among forty-eight countries. About one in every four people will be living in a country affected by chronic or recurrent shortages of fresh water. This will probably worsen as populations gravitate towards the world's cities. These currently contain 48% of the world's population, but it is estimated that this will increase to around 60% by 2030. Concomitant with these changes, water use is predicted to increase by 50% over the next 30 years (UN/WWAP, 2003), following which half of the world's population will live in countries where water is scarce. These will be concentrated in Africa, the Middle East and Asia, where poor infrastructure and service delivery accentuate scarcity.

Combined with economic and development disparities, the scarcity and variability of fresh water has created a human tragedy. Despite efforts by the international community over the past three decades or more, 1.1 billion people still lack access to clean water and 2.4 billion people live without access to appropriate sanitation. Another four billion people do not have a safe and suitable system of wastewater disposal (WHO, 2000) and every year between 2 million and 12 million people die from preventable waterborne diseases (Hinrichsen *et al.*, 1997; WHO, 2000). Small children are the most vulnerable, with preventable diarrhoeal diseases (e.g. cholera) claiming the lives of three million children every year, or one child every ten seconds. In the time you take to read this chapter, nearly 300 children will have died as a result of preventable waterborne diseases.

The problems of insufficient quantities of water are exacerbated by pollution and its effects on water quality. Every day about two million tonnes of industrial, chemical, agricultural and human waste are dumped into rivers, with global production of waste water estimated at around 1 500 km^3 (UN/WWAP, 2003). Given the current systems and technologies for waste disposal, a single litre of waste water pollutes eight litres of fresh water. Based on these estimates, the present burden of pollution could be as great as 12 000 km^3 worldwide (UN/WWAP, 2003).

Water increasingly represents a valuable national commodity. As governments strive to meet development objectives and provide basic services to their rapidly expanding populations, available resources have to be shared among more and more users. Such are the pressures associated with the human dilemma facing some countries and so key is water to economic development, that there is a widely held belief that 'the next war will be over water' (Gleick, 1998; Davies & Wishart, 2000). Although the only recorded conflict over water took place some 4500 years ago, water undoubtedly causes tension, often straining relations between many countries (Gleick, 1998, 2003). Although conflicts are typically caused by other factors, water resources are increasingly targeted and used as weapons during war. There is a long history of conflicts and tensions surrounding access to and utilisation of water, which remains a security concern for many countries.

The environmental context

In addition to providing water for human consumption and development, the world's desert rivers and the water that flows through them maintain valuable aquatic ecosystems, supporting a rich and unique diversity of species (Chapters 4–7, this volume). This diversity in turn supports numerous 'services' with the hydrological cycle, forming important arterial links that quantitatively and qualitatively influence other ecosystems. They can directly affect the supply of water or, more subtly and indirectly, the local climate. The magnitude of human modifications of the world's waterways means that we now constitute the major component within the hydrological cycle.

The legacy of human manipulation and management of rivers, governed by the dominant supply-side paradigms, has reduced the quantity and quality of water. In the world's desert regions the variability of natural flow (Puckridge *et al.*, 1998; Chapter 2, this volume) highlights the contradictions of supply-side paradigms. Hydraulic structures now interrupt the streamflows of about 60% of the world's largest rivers (UNEP, 1999). Although data are not readily available for most of the world, three quarters of the 139 largest rivers in North America, Europe and the former Soviet Union are strongly or moderately affected by water regulation by dams, interbasin transfers or irrigation withdrawals (Dynesius & Nilsson, 1994). Less than 2% (< 100 000 km) of the 5.2 million kilometres of natural stream length in the United States of America can be regarded as being of sufficient quality to warrant federal protection. Only 42 high-quality free-flowing streams longer than 200 km still exist (Benke, 1990); in Japan, only two of 30 000 rivers are neither dammed nor modified (McAllister *et al.*, 1997). Rivers in the world's more arid areas have received less attention to date, but the pressure to secure water for supply against a backdrop of natural spatial and temporal variability will mean more dams, interbasin transfers and withdrawal greatly modifying natural flows.

It is difficult to differentiate the impact upon desert rivers in contrast to others, as rivers and wetlands the world over are suffering rates of extinction and imperilment far greater than those of any other ecosystem (Williams *et al.*, 1989, 1993; Kay, 1995; McAllister *et al.*, 1997; Groombridge & Jenkins, 1998; Masters *et al.*, 1998; Ricciardi & Rasmussen, 1999; Karr *et al.*, 2000; Lemly *et al.*, 2000; Chapter 8, this volume). The major threats to aquatic ecosystems include habitat loss and degradation, introduction and spread of exotic species, overexploitation, secondary extinctions, and chemical and organic pollution along with climate change (Allan & Flecker, 1993). As a result of these threats, 20% of all freshwater species have become or are in danger of becoming extinct (see Wishart *et al.*, 2003). An examination of global trends in freshwater biodiversity from a sample of more than 200 freshwater, wetland and water-margin vertebrate species, revealed that most species were in decline (Groombridge & Jenkins, 1998). Nearly one quarter of all mammal species and 12% of bird species associated with freshwaters are considered threatened, along with one third of the 10% of fish species studied in detail (UN/WWAP, 2003). Population data for 70 of these species documented declines of around 50% since 1970 (Groombridge & Jenkins, 1998). Of the species in the 2000 IUCN Red Data List, 20% of amphibians, 30% of fishes (mostly freshwater species), 27% of molluscs (mostly freshwater species) and 20% of crustaceans are classed as threatened (UNEP, 1999; Hilton-Taylor, 2000). When the status of major aquatic taxa such as fish, amphibians, crayfish and mussels are compared with that of terrestrial birds and mammals in North America, the rates of decline are typically three to eight times higher (Masters, 1990, in Angermeier, 1995). Similarly, in Africa, a preliminary assessment of amphibian populations suggests that 28% of the continent's 632 amphibian species are threatened at levels much greater than those observed for birds or mammals (UNEP, 1999).

Conflicting pressures on the world's freshwater systems, between the need to develop water resources for human needs and those required for maintaining ecosystem processes, have resulted in a degradation of ecological integrity and loss of species. Much of the current knowledge about functioning and diversity in rivers comes from research in the world's mesic areas (Williams, 1988). Lack of data and a poor understanding of the linkages between riverine biota and the processes that sustain them in desert regions, where water is scarce and spatially and temporally variable, are compounded by a lack of definitive taxonomy. Despite this, human demands require control and reduction in variability of river flows (Walker *et al.*, 1997). The challenge of meeting human demands without compromising ecosystem function and species protection is increasingly more difficult in desert regions than in mesic areas and this contrast will contribute to greater losses of biodiversity (Chapter 8, this volume).

The political context

Increasingly sophisticated engineering, enabling greater control over the spatial and temporal variability in desert river flows, has necessitated the development of increasingly complex and comprehensive frameworks for effective management. These reflect not only increasing technological development but also the shifting paradigms and societal values. Numerous measures aim to effectively govern the management of water resources. The international community recognises the scarcity and importance of water as a resource for development and human survival. This began more than 30 years ago with the first International Conference on Water and culminated with proclamation of the International Year of Freshwater in 2003. The first International Conference initiated a series of global activities in water, including proclamation of the International Drinking Water and Sanitation Decade (1981–90), extending basic water and sanitation services to the poor. Although the success of this initiative remains questionable, it did focus the world's attention on water for more than a decade and was the foundation for further initiatives during the 1990s.

These initiatives were largely successful and widely accepted, reflecting a shift in thinking that better recognised and included environmental concerns. The International Conference on Water and the Environment (ICWE, 1992), a precursor for the UN Conference on the Environment and Development (UNCED, 1992) or Rio Earth Summit, gave rise to four key guiding principles: the Dublin Principles.

- Fresh water is a finite and vulnerable resource, essential to sustain life, development and the environment.
- Water development and management should be based on a participatory approach, involving users, planners and policy-makers at all levels.
- Women play a central part in the provision, management and safeguarding of water.
- Water has an economic value in all its competing uses and should be recognised as an economic good.

Following the Dublin meeting, the Rio Earth Summit released Agenda 21, a blueprint for environment and sustainable development into the twenty-first Century, with seven areas for action in freshwater. These helped herald a new era in water management and were the platform for subsequent initiatives, such as the World Water Forum, held in Marrakech (1997), The Hague (2000) and Japan (2003). The International Conference on Freshwater in Bonn in 2001 built on this process and culminated in the World Summit on Sustainable Development in Johannesburg (2002).

The most significant recent global initiative is the United Nations Millennium Assembly's announcement in 2000 of the Millennium Development Goals (MDGs) for 2015. These detail eight overarching, time-bound and measurable goals with eighteen specific targets for combating poverty, hunger, disease, illiteracy, environmental degradation and discrimination against women (United Nations, 2000). Each goal has specific targets, with indicators to measure performance and impact.

Targets 9 and 10 of MDG Goal Seven (Table 11.1) are most relevant to the current crisis facing the world's rivers. In a statement similar to proclamations made during the International Drinking Water and Sanitation Decade, these targets call for commitment by the world's governments to halve the proportion of people without access to clean

Table 11.1. *Goal seven (Ensure Environmental Sustainability) of the Millennium Development Goals for 2015*

Goals and Targets	Indicators
Target 9. Integrate the principles of sustainable development into country policies and programmes and reverse the loss of environmental resources	**25.** Proportion of land area covered by forest **26.** Land area protected to maintain biological diversity **27.** GDP per unit of energy use (as proxy for energy efficiency) **28.** Carbon dioxide emissions (per capita)
Target 10. Halve, by 2015, the proportion of people without sustainable access to safe drinking water	**29.** Proportion of population with sustainable access to an improved water source

Source: United Nations (2000).

water, reduce child mortality and ensure sustainable management of the underlying resource that reverses loss of environmental resources.

For the MDGs, sustainable access is at least 20 l per person per day (basic human requirement) from a source of water within one kilometre of a user's dwelling. About 1.1 billion people do not currently have access to a suitable supply of water (WHO, 2000) and 11×10^6 m^3 of water every day or 40.15×10^6 m^3 per year will be needed to meet Target 10 by 2015. This is just over 25 times the total annual surface runoff for the world's 12 most water poor countries. It would require an annual increase in spending of US$180 billion (Terry & Calaguas, 2003), more than double current spending levels.

Although the targets are aimed at integrating sustainable development into country policies and programmes, they will result in an increase in the use of water and yet the indicators are largely terrestrial (Table 11.1). There is an identified need for more water but there are no provisions to monitor the effects of increased utilisation on freshwater ecosystems. Similarly, there is no reference among the MDG proclamations of how increased water access will be managed sustainably. This is against a background of overwhelming evidence that the decline in biodiversity of freshwaters is far greater than in other systems, and where diversions result in predictable and devastating ecological consequences (see Chapter 8, this volume).

Many international agreements and frameworks recognise the need for sustainable development and integration of environmental requirements, but few countries have translated this international goodwill into provision of water for the environment (Dyson et al., 2003). In Australia and South Africa, progressive provisions are appearing in legislation. South Africa's 1998 National Water Act reflects the social and environmental needs of society, by recognising water resources as a public good, under state control. Further, a 'reserve' may be established with quality and quantity of water. This reserve includes a basic human reserve, providing for drinking, food and personal hygiene (Figs. 11.1, 11.2 and 11.6), and an ecological reserve with sufficient water to protect aquatic ecosystems. Once allocated, water in the reserve remains protected from competition with other water uses. Similarly, in Federal Australia, most jurisdictions responsible for river management have strong environmental provisions in relatively new legislation.

THE CONSERVATION AND MANAGEMENT CONFLICT

The protection of aquatic ecosystems and dependent biota requires different approaches to traditional paradigms of environmental protection (see Boon et al., 1992; Wishart & Davies, 2003). The nature of the catchment and the dendritic connectivity of rivers means that aquatic ecosystems do not conform or respond to traditional conservation approaches based on principles of exclusion and preservation (Wishart & Davies, 2003). Integrated solutions that meet and reconcile often conflicting demands of human population and ecosystem processes are required. Measures to protect river systems must determine the fine line between utilisation and conservation (Dyson et al., 2003). The implicit assumption underlying such measures is that aquatic systems have surplus water that can be 'harvested' within sustainable limits.

Irrespective of the governance and management, the world's 43 750 km^3 of available fresh water is finite (UN/WWAP, 2003). It has to be shared among the competing demands of human consumption, agriculture and industry and meet the requirements for maintaining functional ecosystems. As human populations grow and become increasingly affluent, our demands will increasingly impinge on water for the environment. Whereas the present challenge lies in allocating sufficient water for different user groups, the future challenge will be to reduce the growth in demand and balance competing needs. Future management and development of water resources needs to reflect contemporary social values and the realities of water scarcity.

The traditional response to increasing demand for water is to develop supply-side solutions. Such thinking contends that sufficient global or regional supplies exist where socioeconomic and technological improvements can overcome distribution and security of supply problems (UN/WWAP, 2003). This invariably translates into regulation and storage of water by building dams, increasing supply infrastructure and interbasin transfers. Although suitable sites for new dams are diminishing (McCully, 1996) there are still about 1 700 dams considered or planned (WCD, 2000).

Supply-side infrastructure has provided the foundations for the development of modern human society. However, the management of water resources has always been dynamic, reflecting societal and cultural values. Humans usually need the services that water can provide more than they need the water itself (Wolff & Gleick, 2003); these needs can be met in various ways, often with radically different implications for water. Increasing recognition of the environmental, economic and social costs of supply-side water management have stimulated new integrated approaches that aim to reduce demand.

The new paradigm for water has to focus on efficiency and greater equity, reflecting water's true value and environmental costs. Measures that reduce the pressure on this finite resource allow more for the environment. For example, simple dual and low-flush toilets use 6 l instead of the 16 l of conventional toilets. Providing 80% of American bathrooms with more efficient showerheads would reduce water consumption by nearly 3000 million litres per day (Jones & Dyer, 1993). Recycling of water and reduction in pollution also reduce pressure on scarce water supplies.

Improved use of existing resources and infrastructure, realistic pricing and regulated markets can also improve efficiency and availability of water. Vast amounts of water are lost through distribution networks. For example, more than 30% of Europe's water supply is lost through leaking pipes and over 20% of Cape Town's (South Africa) water is unaccounted for because of leaks, theft and unmetered consumption (Davies & Day, 1998). In Manila (Philippines), the estimates are about 60%. Such losses may be technically and economically reduced to 10% or 20% (World Bank, 1995). The recovery of Cape Town's losses would provide more than 12 million cubic metres; recovery of losses in

England would delay the building of new storage dams by 45 years (Davies & Day, 1998).

There is also an increasing need to protect freshwater biodiversity. Increasing conflict between growing human populations and environmental needs requires quantitative answers about water for aquatic ecosystems and their services. New and innovative approaches to conservation and management of rivers are emerging, particularly in the world's desert regions, where the inescapable pressures on water are greatest for humans and aquatic ecosystems.

Integrated catchment management is the start but we need to identify priority areas for protection, based on biogeography. Catchments within biogeographic areas need to be differentiated according to development pressures, conservation value and diversity (see Wishart *et al.*, 2003). Some catchments need to be protected as reserves with little or no development whereas other catchments should be identified as nodes for integrated development. For each catchment, conservation status needs to be defined so developments take place within predetermined and acceptable thresholds of change.

CONCLUSIONS

Sustainable use and management of water has always reflected the social, cultural and political fabric of society. New political initiatives are laying the foundations for action and education among policy makers, but gaps remain. The UN Millennium Development Goals provide important quantitative targets for development but fail to measure their efficacy or impacts on aquatic ecosystems. Furthermore, rather than focusing on large-scale irrigation or hydro-electric developments, priority should be given to meeting unmet human needs for water. Political rhetoric does not provide water to the world's poor, nor does it protect important ecosystem processes.

The dominance of the supply-side paradigm has been intertwined with the successful development of human civilisation. However, one of the most pressing challenges facing the twenty-first century is to satisfy the water demands of an ever-expanding human population and protect aquatic ecosystems and their ecological services. The scarcity of water in desert regions and increasing competition between users requires new approaches. The foundations for these new management paradigms are being forged in the world's desert regions. These will be the foundations for future management in the world's more well-watered regions.

REFERENCES

Allan, J. D. and Flecker, A. S. (1993). Biodiversity conservation in running waters: Identifying the major factors that threaten destruction of riverine species and ecosystems. *BioScience*, **43**, 32–43.

Angermeier, P. L. (1995). Ecological attributes of extinction-prone species: loss of freshwater fishes of Virginia. *Conservation Biology*, **9**, 143–58.

Benke, A. C. (1990). A perspective on America's vanishing streams. *Journal of the North American Benthological Society*, **91**, 77–88.

Boon, P. J., Calow, P. and Petts, G. E. (eds) (1992). *River Conservation and Management*. Chichester: John Wiley and Sons.

Boon, P., Davies, B. R. and Petts, G. E. (2000). *Global Perspectives in River Conservation: Science, Policy and Management*. Chichester: John Wiley and Sons.

Davies, B. R. and Day, J. A. (1998). *Vanishing Waters*. Cape Town: University of Cape Town Press, South Africa.

Davies, B. R. and Wishart, M. J. (2000). River conservation in the countries of the Southern African Development Community (SADC). In *Global Perspectives in River Conservation: Science, Policy and Management*, ed. P. Boon, B. R. Davies and G. E. Petts, pp. 179–204. Chichester: John Wiley and Sons.

Davies, B. R., Snaddon, C. D., Wishart, M. J., Thoms, M. and Meador, M. (2000). A biogeographical approach to inter-basin water transfers: implications for river conservation. In *Global Perspectives in River Conservation: Science, Policy and Management*, ed. P. Boon and B. R. Davies, pp. 431–44. Chichester: John Wiley and Sons.

Davies, B. R., Thoms, M. C., Walker, K. F., O'Keeffe, J. H. and Gore, J. A. (1995). Dryland rivers: their ecology, conservation and management. In *The Rivers Handbook: Hydrological and Ecological Principles*, Vol. 2, ed. P. Calow and G. E. Petts, pp. 484–511. Chichester: John Wiley and Sons.

DuMars, C. T., O'Leary, M. and Utton, A. E. (1984). *Pueblo Indian Water Rights: Struggle for a Precious Resource*. Arizona: The University of Arizona Press.

Dynesius, M. and Nilsson, C. (1994). Fragmentation and flow regulation of river systems in the northern third of the world. *Science*, **266**, 753–62.

Dyson, M., Bergkamp, G. and Scanlon, J. (eds) (2003). *The Essentials of Environmental Flows*. Cambridge: World Conservation Union (IUCN).

Elson, M. D. (1998). Expanding the view of Hohokam platform mounds: An ethnographic perspective. *The University of Arizona Anthropological Papers*, **63**, 1–160.

Gleick, P. H. (1998). Conflict and cooperation over freshwater. In *The World's Water 1998–1999: The Biennial Report on Freshwater Resources*, ed. P. H. Gleick, pp. 107–35. Washington: Island Press.

Gleick, P. H. (2003). *Water Conflict Chronology*: Environment and Security Water Conflict Chronology: www.worldwater.org/.

Gleick, P. H., Singh, A. and Shi, H. (2001). *Threats to the World's Freshwater Resources*. Oakland, California: Pacific Institute for Studies in Development, Environment, and Security, in cooperation with the United Nations Environment Programme.

Groombridge, B. and Jenkins, M. (1998). *Freshwater Biodiversity: a Preliminary Global Assessment*. Cambridge: World Conservation Monitoring Centre.

Hilton-Taylor, C. (compiler) (2000). *IUCN Red List of Threatened Species*. Gland and Cambridge: World Conservation Union (IUCN).

Hinrichsen, D., Robey, B. and Upadhyay, U. D. (1997). Solutions for a water-short world. *Population Reports*, Series M, no. 14. Baltimore: Johns Hopkins School of Public Health, Population Information Program.

ICOLD (International Commission on Large Dams) (1989). *World Register of Dams, 1985 update. (Régistre Mondial des Barrages, mise à jour 1988.)* Paris: International Commission on Large Dams.

ICWE (International Conference on Water and the Environment) (1992). *The Dublin Principles.* International Conference on Water and the Environment, Dublin, 26–31 January 1992. Available in full at www.wmo.ch/web/homs/icwedece.html.

Jones, A. and Dyer, J. (1993). The water efficiency revolution. *River Voices*, Spring, 1993.

Karr, J. R., Allan, J. D. and Benke, A. C. (2000). River conservation in the United States and Canada. In *Global Perspectives in River Conservation: Science, Policy and Management*, ed. P. Boon and B. R. Davies, pp. 3–39. Chichester: John Wiley and Sons.

Kay E. A. (1995). Status report on molluscan diversity. In *The Conservation Biology of Molluscs*, ed. E. A. Kay. Proceedings of a Symposium held at the 9th International Malacological Congress, Edinburgh, Scotland, 1986. SSC Occasional paper 9. Cambridge: IUCN Publication Services.

Lemly, A. D., Kingsford, R. T. and Thompson, J. R. (2000). Irrigated agriculture and wildlife conservation: conflict on a global scale. *Environmental Management*, **25**, 485–512.

Masters, L. L. (1990). The imperiled status of North American aquatic animals. *Biodiversity Network News*, **3**, 7–8.

Masters, L. L., Flack, S. R. and Stein, B. A. (1998). *Rivers of Life: Critical Watersheds for Protecting Freshwater Biodiversity.* Arlington, Virginia: The Nature Conservancy.

McAllister, D. E., Hamilton, A. L. and Harvey, B. (1997). Global freshwater biodiversity: striving for the integrity of freshwater ecosystems. *Sea Wind*, **11**, 1–140.

McCully, P. (1996) *Silenced Rivers: the Ecology and Politics of Large Dams.* London: Zed Books.

Postel, S. (1996). *Dividing the waters: food security, ecosystem health, and the new politics of scarcity.* (Worldwatch Paper no. 132.) Washington: Worldwatch Institute.

Puckridge, J. T., Sheldon, F., Walker, K. F. and Boulton, A. J. (1998). Flow variability and the ecology of arid zone rivers. *Marine and Freshwater Research*, **49**, 55–72.

Ricciardi, A. and Rasmussen, J. B. (1999). Extinction rates in North American freshwater fauna. *Conservation Biology*, **13**, 1220–2.

Sheldon, F., Boulton, A. J. and Puckridge, J. T. (2002). Conservation value of variable connectivity: aquatic invertebrate assemblages of channel and floodplain habitats of a central Australian arid-zone river, Cooper Creek. *Biological Conservation*, **103**, 13–31.

Shiklomanov, I. (2000). World water resources and water use: present assessment and outlook for 2005. In *World Water Scenarios, Analyses*, ed. F. R. Rijbersman, pp. 160–203. London: Earthscan Publications.

Terry, G. and Calaguas, B. (2003). *Financing the Millennium Development Goals for Domestic Water Supply and Sanitation.* United Kingdom: Water AID.

UN/WWAP (United Nations/World Water Assessment Programme) (2003). *UN World Water Development Report: Water for People, Water for Life.* Paris, New York and Oxford: United Nations Educational, Scientific and Cultural Organisation and Berghahn Books.

UNCED (1992). *The United Nations Conference on Environment and Development* (UNCED), Rio de Janeiro, 3–14 June 1992.

UNEP (United Nations Environment Program) (1999). *Global Environment Outlook 2000.* New York: Oxford University Press.

United Nations (2000). *The Millennium Goals.* United Nations Millennium Declaration A/RES/55/2 8 Sept. 2000.

Vallée, D. and Margat, J. (2003). *Review of world water resources by country.* (FAO Water Reports 23). Rome: Food and Agricultural Organisation.

Walker, K. F., Puckridge, J. T. and Blanch, S. J. (1997). Irrigation development on Cooper Creek, central Australia: prospects for a regulated economy in a boom-and-bust ecology. *Aquatic Conservation: Marine and Freshwater Ecosystems,* **7**, 63–73.

Ward, J. V., Tockner, K. and Schiemer, F. (1999). Biodiversity of floodplain river ecosystems: ecotones and connectivity. *Regulated Rivers: Research and Management,* **15**, 125–39.

WCD (World Commission on Dams) (2000). *Dams and Development: A New Framework for Decision-Making.* Report of the World Commission on Dams. London: Earthscan Publications.

WHO (World Health Organisation) (2000). *Global Water Supply and Sanitation Assessment 2000 Report.* Available in full at www.who.int/water_sanitation_health/Globassessment/GlobalTOC.htm.

Williams, J. D., Warren, M. L., Cummings, K. S., Harris, J. L. and Neves, R. J. (1993). Conservation status of freshwater mussels of the United States and Canada. *Fisheries,* **18**, 6–22.

Williams, J. E., Johnson, D. A., Hendrickson, S., *et al.* (1989). Fishes of North America endangered, threatened or of special concern: 1989. *Fisheries,* **14**, 2–20.

Williams, W. D. (1988). Limnological imbalances: an antipodean viewpoint. *Freshwater Biology,* **20**, 407–20.

Wishart, M. J. and Davies, B. R. (2003). Beyond catchment considerations in the conservation of lotic biodiversity. *Aquatic Conservation: Marine and Freshwater Ecosystems,* **13**, 429–37.

Wishart, M. J., Davies, B. R., Stewart, B. A. and Hughes, J. M. (2003). *Examining catchments as functional units for the conservation of riverine biota and maintenance of biodiversity.* (Water Research Commission Report No 975/1/02.) Pretoria: Water Research Commission.

Wolff, G. and Gleick, P. H. (2003). The soft path for water. In *The World's Water 2002-2003,* ed. P. H. Gleick, W. C. G. Burns, E. L. Chalecki *et al.,* pp. 1–32. Washington, DC: Island Press.

World Bank (1995). *Managing Urban Water Supply and Sanitation.* Washington, DC: Operations and Evaluation Department.

12

Changing desert rivers

R. T. KINGSFORD

Desert rivers fascinate us, sustain many of us and support a complex ecology whose variation reflects the dynamics of these rivers. Although dryland regions with many desert rivers occupy almost half the world's land surface, we know relatively little about the ecology of desert rivers, compared with rivers in mesic regions (Kingsford, 1995; Chapter 3, this volume). This book attempts to redress this imbalance because many of these rivers are being changed forever to meet our needs (Chapter 11, this volume).

Analysis at different temporal scales reveals that desert rivers have gone through major fluctuations within the past 100 years or more (Labat *et al.*, 2004). Climatic changes from glacial cycles impose even larger-scale changes in variability over thousands of years on rivers and wetlands (e.g. Australia (Kershaw *et al.*, 2003)). Current effects of climate change from carbon dioxide are predicted to increase aridity in dryland regions of the world (Labat *et al.*, 2004; Manabe *et al.*, 2004), possibly increasing the severity of dry periods (Nicholls, 2004). We know that desert rivers, their ecosystems and biota are extraordinarily resilient to the disturbances of droughts and floods (Chapters 2 and 4–7, this volume). But our species is rapidly changing the nature of desert rivers around the world. Pressure to take water from desert rivers is

Ecology of Desert Rivers, ed. R. T. Kingsford. Published by Cambridge University Press.
© Cambridge University Press 2006.

increasing inexorably (Chapters 8 and 11, this volume). River development is altering rivers in ways that may be much more rapid and far-reaching than those predicted by climate change (McMahon & Finlayson, 2003). For many nations, the building of dams and diversion of water from desert rivers is still seen as essential for development, mainly through irrigation (Lemly *et al.*, 2000). The costs of such developments are accumulating.

What have we learnt about desert rivers and their ecology and our interactions so that we can plan for their long-term future? Such knowledge may be crucial not only to the dependent ecosystems but also for indigenous human societies that depend on the service delivered by desert rivers. Perhaps the natural and human disturbances to desert rivers can improve our understanding of ecology of more mesic rivers and allow us to predict consequences of human changes that may ultimately affect mesic rivers, as our need for water and its products increases.

DESERT RIVER ECOLOGY

There is an incredible diversity in types of desert rivers around the world, with one overriding commonality: highly variable hydrology (Puckridge *et al.*, 1998; Chapters 2 and 3, this volume). Some river flows are more predictable, governed by seasonal factors such as snow (Chapters 1 and 2, this volume). Highly variable climate, flows and geomorphology shape the ecology of desert rivers (Chapters 2 and 3, this volume). We are only beginning to understand some of this complexity (Nanson *et al.*, 2002); many desert rivers have unique hydrological signatures of variability (Puckridge *et al.*, 1998; Chapter 2, this volume). This flow variability is linked to geomorphological processes with transfers of mass and energy ultimately determining the extent and quality of habitat available in a desert river, much more so than in more mesic rivers (Chapter 3, this volume). High sediment loads in dryland rivers provide habitat heterogeneity and inorganic material for downstream habitats. So variability in the distribution of riparian tree species on the Sabie River in South Africa reflects sedimentation processes mediated by hydrogeomorphic variability (Chapter 3, this volume). Highly variable geomorphology with braided channels, macro-channels and floodplains may increase in complexity downstream and often change between flood events (Nanson *et al.*, 2002; Chapter 3, this volume). Such geomorphological and hydrological complexities that change in space and time produce environmental gradients for different biota. It is this spatial dimension that contributes much to the highly productive and sometimes diverse biota sustained by desert rivers. Desert rivers expand and contract considerably in response to floods and subsequent dry periods.

Flood frequency and distribution structures plant and animal communities. There is a wide range of different types of plant, equipped with different structures to respond to floods; some are perennial and others are temporary, growing from seeds that germinate with the floods (Chapter 5, this volume). Plant species most tolerant of frequent flooding live closest to the river channel and those least tolerant occur on the floodplains (Chapters 3 and 5, this volume). The way in which floods interact with turbidity, salinity and depth also strongly influences plant distribution on desert rivers (Chapter 5, this volume). Flood periods are a time of dispersal, germination, growth and recruitment in most plant species, sometimes depending on the season (Chapter 5, this volume). These plants provide habitat and food, directly or indirectly, for many other aquatic organisms or terrestrial species that feed and rest along dryland rivers (Chapter 7, this volume).

High aquatic productivity underpins the 'boom' periods that occur with flooding. Algal processes are particularly important for dryland rivers, where terrestrial sources may be more limited than in mesic rivers (Chapter 4, this volume). This productivity following the arrival of floods, particularly benthic processes, may begin slowly on the floodplain (Chapter 4, this volume) but rapidly build up. River channels also possess their own small floodplain benches where organic and inorganic sediment can collect and become available for ecological processes during floods (Chapter 3, this volume). Such primary productivity stimulates high secondary invertebrate productivity (Jenkins & Boulton, 2003; Chapter 6, this volume). Invertebrate densities in some desert rivers can be positively associated with aquatic macrophytes (Kingsford & Porter, 1994).

Densities of invertebrates can be extremely high in desert rivers, providing food for other aquatic organisms and even terrestrial invertebrates such as spiders (Chapter 6, this volume). Waterbird numbers on a saline lake with abundant invertebrates and aquatic macrophytes were much higher than on a freshwater lake of similar size nearby (Kingsford & Porter, 1994). Invertebrate communities seem to be as variable in composition and abundance, as the hydrology of rivers on which they depend (Chapter 6, this volume). They are highly resilient to flash flooding in upper parts of rivers, recolonising rapidly (Chapter 6, this volume). Similarly, floodplain invertebrates respond within days from diapause resting stages, even if they have remained dry for up to two decades (Chapter 6, this volume). Some invertebrate species (e.g. hemipteran bugs, coleopteran beetles, chironomids) are highly mobile and can colonise habitats newly created by floods (Chapter 6, this volume).

Floods create considerable feeding and breeding areas for dependent aquatic vertebrates (e.g. reptiles, fish, waterbirds) as desert rivers expand (Chapters 4 and 7, this volume). These species colonise newly flooded habitats where there are abundant invertebrates and aquatic plants that provide food (Chapter 7, this volume). With abundant food, pulses of breeding for many vertebrates occur (e.g. fish: Puckridge *et al.*, 2000) and this is often highly correlated with flood size (e.g. for waterbirds: Kingsford & Johnson, 1998). Floods in desert regions can also be a magnet for terrestrial species in search of abundant food (Chapter 7, this volume). There are aquatic subsidies for many terrestrial invertebrates and vertebrates (Chapter 4, this volume). Many subsistence human communities depend on the highly productive flood processes to sustain their livelihoods (Chapters 4, 5, 7 and 8, this volume).

Inevitable dry periods follow floods, with biota responding accordingly to the restricted distribution of water. Aquatic plants rely on physiological characteristics (e.g. water storage organs), dispersal, different growth forms or persistent seed banks produced during the flood that remain dormant until subsequent flooding (Chapter 5, this volume). Similarly, many invertebrate species have diapause resting stages that remain viable for decades, and others are able to reduce their water requirements considerably (Chapter 6, this volume), as do some frog species (Chapter 7, this volume). Many adult invertebrates die without water, relying on their offspring to recolonise future aquatic habitats. Some escape to nearby aquatic refuges (Chapter 6, this volume), a response that is also prevalent among many vertebrates (Chapter 7, this volume). These aquatic refuges are often permanent, sustained by high algal productivity (Chapter 4, this volume) and harbouring vertebrates (e.g. fish, turtles) and invertebrates (e.g. snails) that can colonise the floodplain during flooding (Chapters 6 and 7, this volume). During the drying phase, invertebrate and vertebrate species, concentrate in refugia where predation may be high (Chapters 6 and 7, this volume). Such interactions occur across a range of temporal and spatial scales, depending on the size and sequence of the flood and the interaction of a particular organism's life history with the physical and biological parts of the river.

THE HUMAN FOOTPRINT

Humans live in most places in the world and inhospitable desert regions are no exception, where desert rivers provide the means for existence (Chapter 11, this volume). For many thousands of years, indigenous communities relied on desert rivers for their livelihoods. These rivers were the trade routes for Aboriginal communities in Australia (Veth *et al.*, 1990) and remain a fundamental part of biological and spiritual existence of Australian indigenous communities. Demand to access and control desert rivers, primarily to produce food and fibre through irrigation and supply growing urban centres, has fundamentally changed the structure and function of many desert rivers. Biota can withstand periodic and unpredictable flooding and drying regimes and evolve in response to geological climate change but they are probably unable to cope with damming, diversion and levees. Regional estimates for biodiversity loss in dryland regions are rising (Chapter 11, this volume).

The most devastating effects have occurred on rivers that have had most of their flows diverted for irrigation, upstream of some of the major floodplains (Chapter 8, this volume). The irrigation areas take the water that once sustained complex floodplains (Kingsford, 2000). There is a growing list of ecological disasters that have occurred on desert rivers (Chapter 8, this volume). Complex ecological communities no longer exist or are a fragment of their original state. Aquatic ecosystems become terrestrial environments without water. Perennial plant species 'migrate' towards water with species less tolerant of high frequency floods invading areas vacated by plants needing frequent flooding that have 'moved' to the channels (Chapters 5 and 9, this volume). Similarly, some mollusc and crustacean species have changed their distributions. Floodplain species are encroaching on the range of river species (Chapter 9, this volume). Short-lived plant and invertebrate species that depend on water will inevitably become locally extinct on the floodplain, whereas some aquatic plants and invertebrates that can persist for decades as seed and egg banks will remain in the soil but eventually degrade without water. It is not simply hydrological processes that are affected.

Hydrological and geomorphological processes that sustain and make the floodplains of desert rivers among the most fertile habitats in the world are being denied sediment by large dams in the catchments

(Vörösmarty *et al.*, 2003). River channels may experience less variability, and the release of dissolved organic carbon from floodplains can be severely diminished (Chapter 3, this volume). This inevitably affects aquatic and terrestrial food webs that rely on geomorphological and hydrological processes. Our needs for reliable water supplies have imposed a regime that has significantly reduced the natural variability of many desert rivers, making them more similar to mesic rivers with constant flows (McMahon & Finlayson, 2003). Such systems can often favour exotic vertebrate and plant species (Chapters 5 and 7, this volume).

Even the building of weirs and control structures in the lowland parts of rivers considerably changes river flows and dependant ecological communities and biota. For example, subtle hydrological changes effected by the weirs and their locks along the lower River Murray have significantly changed the distribution and abundance of plants and invertebrates (Chapter 9, this volume). Freshwater snail species once abundant in the lower part of this river are now reduced to a third in number. Their demise is linked to alienation of the floodplain, increased predation by introduced carp *Cyprinus carpio* and obliteration of the biofilms on which they depend by changes in water levels caused by weir operation (Chapter 9, this volume).

Along with diminishing river flows, water quality is declining in many of the world's rivers. Tonnes of industrial, chemical, agricultural and human waste are dumped into rivers (Foote *et al.*, 1996; Chapter 11, this volume). Increasing human populations live on the world's rivers and rely on the rivers as waste disposal system. This inevitably contributes to serious waterborne diseases (Chapter 11, this volume). The most insidious problem of all is rising salinity, a natural factor made even more prominent by diminishing river flows and irrigation practices but still not well researched (Chapter 10, this volume). Lower parts of desert rivers are particularly affected by increasing salinity, as floods no longer occur as frequently to flush soils (Chapters 9 and 10, this volume). Increasing salinity reduces diversity and abundance of most aquatic species, with relatively few species able to tolerate the highest salinity levels (Chapter 10, this volume). In the Lower Murray River, the hydraulic head on weirs forces saline groundwater downwards but it eventually resurfaces to reach the river (Chapter 9, this volume). Anthropogenically derived salinity (secondary salinity) affects much of the dryland region, degrading structure and processes of aquatic systems and agricultural practices, limiting their long-term longevity (Chapter 10, this volume). All of the major ecological disasters on desert rivers are also accompanied by increasing salinity (Chapter 8, this volume). Without flooding, aquatic and floodplain vegetation disappears, allowing saline soils to be blown to more fertile areas of the landscape, where they detrimentally affect the ecology and agriculture.

Rapidly increasing human populations create almost insurmountable challenges for management of desert rivers: – a world whose population has tripled in the last century but whose water use has grown six-fold (Chapter 11, this volume). Water use is predicted to double in the next 30 years (Chapter 11, this volume) and most will inevitably come from the world's rivers. The problem of sovereignty introduces another layer of complexity. Many of the world's desert rivers flow between different nations (Postel, 1996; Chapter 11, this volume), each with different political priorities for water. Even in desert regions within countries, jurisdictional governments can quarrel over their rivers (Kingsford et al., 1998). Until rivers are managed within whole catchments with recognition of their connectivity, political imperatives will always overcome ecological constraints and costs, particularly when these do not occur in your country.

The scale of ecological disasters on desert rivers should be a wake-up call for many of us that live in richer and wetter parts of the world. We are losing some of the most diverse and productive ecosystems in the world. Often, dislocation of indigenous peoples accompanies the ecological loss (Chapter 8, this volume). There are two implications. Much of the irrigation production (e.g. cotton) that occurs on desert rivers is probably destined for markets of rich nations. Those living in wet and affluent parts of the world are probably partly responsible for the demise of these ecosystems. Second, the process and scale of destruction should be relevant to all rivers around the world, as human populations increase and there is more demand for water resources (Chapter 11, this volume).

CONCLUSIONS

The message about ecological interactions is fascinating, variable and complex and that of human interactions with desert rivers is sobering. Much still needs to be learnt about the behaviour and dependencies to be found on desert rivers. There are unique challenges for work on desert rivers. Sometimes the environments are inhospitable but arguably the most difficult challenge is trying to use short-term projects to understand rivers whose variable behaviour sometimes demands years of data more than is needed on a mesic river (Chapter 2, this volume). The investment needs to go on; modelling tools that can be applied to large temporal and spatial scales are likely to be particularly useful. For many of these rivers, our opportunities to understand their ecology are rapidly diminishing.

Whereas many indigenous communities lived with the variability of desert rivers, modern human society is reducing much of this variability and changing much of the nature of desert rivers. There is some hope that increasing understanding will lead to new ways of managing rivers that better recognise the human and ecological dimension. For example, decisions have been made in Australia to protect whole desert river systems from major water resource development (Cooper Creek, Paroo River). New methods of managing rivers adaptively are also gaining prominence around the world (Poff et al., 2003). Informative scientific processes that integrate with management are essential. Adaptive management has become the mantra for managing rivers and many other ecosystems but its effective implementation demands rigour, time and resources. Natural processes may occur aseasonally with little predictability, at time frames that extend beyond funding cycles. Organisms are poorly known with dynamic life cycles, operating at unknown spatial and temporal scales. There is a need to manage rivers with a multidisciplinary approach (Chapter 3, this volume) to realise the benefits of adaptive management.

International political processes have at least begun to recognise the importance of water management and its competing uses (Chapter 11, this volume). It remains to be seen whether this recognition will extend to the ecological functions of desert rivers and their long-term importance in sustaining indigenous communities and other uses. It is encouraging that some of the legislation in two desert countries, Australia and South Africa, probably leads the world in its recognition of the environmental services of desert rivers (Chapter 11, this volume). We need to recognise that increasing demand for water will be met either by the demise of ecological systems of desert rivers or increased water use efficiency, unless our populations decrease. We should not be shy about using the 'ecological disasters' of desert rivers to illustrate the path of development for those rivers not yet heavily exploited. It is my hope that my descendants will also enjoy these magnificent desert rivers.

REFERENCES

Foote, A. L., Pandey, S. and Krogman, N. T. (1996). Processes of wetland loss in India. *Environmental Conservation*, **23**, 45–54.

Jenkins, K. M. and Boulton, A. J. (2003). Connectivity in a dryland river: short-term aquatic microinvertebrate recruitment following floodplain inundation. *Ecology*, **84**, 2708–23.

Kershaw, P., Moss, P. and van der Kaars, S. (2003). Causes and consequences of long-term climatic variability on the Australian continent. *Freshwater Biology*, **48**, 1274–83.

Kingsford, R. T. (1995). Occurrence of high concentrations of waterbirds in arid Australia. *Journal of Arid Environments*, **29**, 421–5.

(2000). Ecological impacts of dams, water diversions and river management on floodplain wetlands in Australia. *Australian Ecology*, **25**, 109–27.

Kingsford, R. T. and Johnson, W. J. (1998). The impact of water diversions on colonially nesting waterbirds in the Macquarie Marshes in arid Australia. *Colonial Waterbirds*, **21**, 159–70.

Kingsford, R. T. and Porter, J. L. (1994). Waterbirds on an adjacent freshwater lake and salt lake in arid Australia. *Biological Conservation*, **69**, 219–28.

Kingsford, R. T., Boulton, A. J. and Puckridge, J. M. (1998). Challenges in managing dryland rivers crossing political boundaries: lessons from Cooper Creek and the Paroo River, central Australia. *Aquatic Conservation: Marine and Freshwater Ecosystems*, **8**, 361–78.

Labat, D., Goddéris, Y., Probst, J. L. and Guyot, J. L. (2004). Evidence for global runoff increase related to climate warming. *Advances in Water Resources*, **27**, 631–42.

Lemly, A. D., Kingsford, R. T. and Thompson, J. R. (2000). Irrigated agriculture and wildlife conservation: conflict on a global scale. *Environmental Management*, **25**, 485–512.

Manabe, S., Wetherald, R. T., Milly, P. C. D., Delworth, T. L. and Stouffer, R. J. (2004). Century-scale change in water availability: CO_2-quadrupling experiment. *Climate Change*, **64**, 59–76.

McMahon, T. A. and Finlayson, B. L. (2003). Droughts and anti-droughts: the low hydrology of Australian rivers. *Freshwater Biology*, **48**, 1147–60.

Nanson, G. C., Tooth, S. and Knighton, A. D. (2002). A global perspective on dryland rivers: perceptions, misconceptions and distinctions. In *Dryland Rivers: Hydrology and Geomorphology of Semi-arid Channels*, ed. L. J. Bull and M. J. Kirkby, pp. 17–54. Chichester: John Wiley and Sons.

Nicholls, N. (2004). The changing nature of Australian droughts. *Climate Change*, **63**, 323–36.

Poff, N. L., Allan, J. D., Palmer, M. A. *et al.* (2003). River flows and water wars: emerging science for environmental decision making. *Frontiers in Ecology and Environment*, **1**, 298–306.

Postel, S. (1996). *Dividing the waters: food security, ecosystem health, and the new politics of scarcity.* (Worldwatch Paper no. 132). Washington DC: Worldwatch Institute.

Puckridge, J. T., Sheldon, F., Walker, K. F. and Boulton, A. J. (1998). Flow variability and the ecology of arid zone rivers. *Marine and Freshwater Research,* **49,** 55–72.

Puckridge, J. T., Walker, K. F. and Costelloe, J. F. (2000). Hydrological persistence and the ecology of dryland rivers. *Regulated Rivers: Research and Management,* **16,** 385–402.

Veth, P., Hamm, G. and Lampert, R. J. (1990). The archaeological significance of the lower Cooper Creek. *Record of the South Australian Museum,* **24,** 43–66.

Vörösmarty, C. J., Meybeck, M., Fekete, B. *et al.* (2003). Anthropogenic sediment retention: major global-scale impact from the population of registered impoundments. *Global and Planetary Change,* **39,** 169–90.

Index